Soil Degradation-Consolidation Theory and Its Applications

Yunmin Chen

Soil Degradation-Consolidation Theory and Its Applications

Springer

Yunmin Chen
Institute of Geotechnical Engineering
Zhejiang University
Hangzhou, Zhejiang, China

ISBN 978-981-97-7984-0 ISBN 978-981-97-7985-7 (eBook)
https://doi.org/10.1007/978-981-97-7985-7

Jointly published with Zhejiang University Press
The print edition is not for sale in China (Mainland). Customers from China (Mainland) please order the print book from: Zhejiang University Press.

This Springer imprint is published by the registered company Springer Nature Singapore Pte Ltd.
The registered company address is: 152 Beach Road, #21-01/04 Gateway East, Singapore 189721, Singapore

Preface

In classical soil mechanics, the soil skeleton is considered inert, and soils are categorised into types such as gravel, sand, silt, and clay according to particle size. However, in addition to these inert components, active components are often present in the soil skeleton. Soils containing such active components are referred to as skeleton-degradable soils. In such soils, the soil skeleton experiences mass loss and particle breakage owing to external loads and biochemical reactions, leading to changes in particle gradation and pore structure. Consequently, these alterations affect soil properties such as shear strength, compressibility, and permeability.

Skeleton-degradable soils are frequently encountered in various fields of geotechnical engineering, presenting diverse challenges. The imbalance in the complex coupled processes of biochemical reactions, physical changes, and mechanical movement in skeleton-degradable soils, accompanied by the failure and instability of engineering structures, can result in serious environmental disasters. However, a unified fundamental theory addressing the coupled effects of biochemical reactions, physical changes, and mechanical movement in skeleton-degradable soils is currently lacking, as are effective solutions for preventing and controlling the resulting geoenvironmental disasters.

To analyse the complex multifield interactions in skeleton-degradable soils, this book, 'Soil Degradation–Consolidation Theory and Its Applications', first summarises three typical degradation processes of soil skeletons: soil particle breakage, chemical degradation, and biochemical degradation. Corresponding degradation and constitutive models are proposed for each process, and a degradation–consolidation theory encompassing biochemical degradation, skeleton deformation, liquid–gas flow, solute migration, and heat conduction is established. The concepts of degradation degree and consolidation degree are introduced to describe the stabilisation process of skeleton-degradable soils under load and the engineering significance of the process. The degradation–consolidation theory overcomes the limitations of classical soil mechanics, such as the assumption of a constant solid-phase mass, and addresses the neglect of the effect of consolidation on degradation in environmental engineering. This theory addresses scientific challenges such as the constitutive models for skeleton-degradable soils, and the coupled analysis of solid–liquid–gas-phase transformations, skeleton deformation, liquid–gas flow, and solute migration.

This book explores the intricacies of soil degradation–consolidation theory, providing detailed insights into soil skeleton degradation, constitutive modelling, liquid–gas flow, and the production and drainage of leachate and landfill gas. By combining theoretical foundations with practical applications, the book demonstrates how this theory can support efficient approaches to municipal solid waste management, landfill stabilisation, and environmental sustainability. In addition, soil degradation–consolidation theory provides key theoretical foundations and new analytical perspectives for understanding phenomena such as the degradation–consolidation processes in the decomposition and exploitation of natural gas hydrates and the behavior evolution of geomaterials under the influence of particle breakage.

This book offers detailed theoretical insights, rich experimental data, and advanced numerical and physical simulation methods compiled from 30 years of research and practical experience. In particular, the book describes the application of centrifugal hypergravity and interdisciplinary experiments in preventing and controlling environmental disasters in municipal solid waste landfills. This book can be a valuable resource for researchers, engineers, and practitioners involved in soil constitutive relations, multifield coupling problems, and related fields, paving the way for innovative solutions to pressing environmental challenges. Moreover, the book can serve as a catalyst, inspiring more geotechnical engineers to explore skeleton-degradable soils, thereby collectively advancing the development of a new generation of soil mechanics.

Finally, I acknowledge the support from the NSFC Basic Science Center Program – Multiphase Evolution in Hypergravity (No. 51988101) and the NSFC Funds for International Cooperation and Exchange (No. 51010008). I extend my gratitude to Prof. Hywel Rhys Thomas from Cardiff University and Prof. William Powrie from the University of Southampton for their valuable academic collaboration and support. I would also like to thank Prof. Charles W. W. Ng from the Hong Kong University of Science and Technology for reviewing the book and providing constructive suggestions. Additionally, I am grateful to the following individuals for their invaluable contributions to this book: Prof. Liangtong Zhan, Prof. Daosheng Ling, Prof. Bin Zhu, Prof. Xuecheng Bian, Prof. Han Ke, Prof. Yuchao Li, Prof. Wenjie Xu, Assoc. Prof. Lujun Wang, Dr. Wu Gao, Dr. Jie Hu, Dr. Pengcheng Ma, and my former Ph.D. students.

Hangzhou, China Yunmin Chen

Contents

1 Introduction 1
- 1.1 Background 1
- 1.2 Book Purpose 4
- 1.3 Book Organization 4
- References 6

2 Degradation of the Soil Skeleton 7
- 2.1 General Introduction 7
- 2.2 Breakage of Soil Particle 9
  - 2.2.1 Static Loading 9
  - 2.2.2 Dynamic Loading 12
  - 2.2.3 Model for Particle Breakage 16
- 2.3 Chemical Degradation 19
  - 2.3.1 Decomposition Equation of Hydrates 19
  - 2.3.2 Solid Mass Loss of Hydrates 21
- 2.4 Biochemical Degradation 24
  - 2.4.1 Unified Equations 24
  - 2.4.2 Anaerobic Degradation of Municipal Solid Waste 27
  - 2.4.3 Aerobic Degradation of Municipal Solid Waste 37
- References 43

3 Constitutive Models for Skeleton-Degradable Soils 45
- 3.1 General Introduction 45
- 3.2 Modified Cam-Clay Model 46
  - 3.2.1 Compressive Hardening 47
  - 3.2.2 Friction 47
  - 3.2.3 Dilatancy 48
  - 3.2.4 Yield Function 49
- 3.3 Unified Hardening Model 50
  - 3.3.1 Mechanical Properties of Over-Consolidated Soils 50
  - 3.3.2 Yield Surface 51
  - 3.3.3 Hardening Rule 52
- 3.4 Constitutive Model for Breakable Soils 53
  - 3.4.1 Influence of Particle Size Distribution on Stress–Strain Relationship of Soil 53
  - 3.4.2 Yield Surface and Loading/Unloading Criteria 54

3.4.3 Elastoplastic Constitutive Relationship 57
3.4.4 Model Evaluation 58
3.5 Constitutive Model for Natural Gas Hydrates 60
3.5.1 Volumetric Strain and Modified Elastic Model 64
3.5.2 Model Validation and Parameter Analysis 67
3.6 Constitutive Model for Municipal Solid Waste 71
3.6.1 Influences of Biodegradation on the Compressibility of Municipal Solid Waste 71
3.6.2 Compression Line Equations of Degraded Municipal Solid Waste 72
3.6.3 Yield Surface and Loading/Unloading Criteria 74
3.6.4 Elastoplastic Constitutive Relationship 76
3.6.5 Model Evaluation 76
References 79

4 Liquid–Gas Seepage Constitutive Models for Skeleton-Degradable Soils 83
4.1 General Introduction 83
4.2 Intrinsic Permeability 84
4.2.1 Effects of Compression and Degradation 84
4.2.2 Anisotropy 86
4.3 Water Retention Curve 87
4.3.1 Test Setup 87
4.3.2 Effects of Compression and Degradation 89
4.4 Relative Permeability Coefficients of Liquid and Gas 91
4.5 Dual-Porosity Flow Model 93
4.5.1 Dual-Porosity Water Retention Curve 93
4.5.2 Dual-Porosity Well Flow Model 94
4.6 Governing Equations for Two-Phase Flow 96
4.7 Simulation of Liquid–Gas Pressure Formation in Landfills 97
References 100

5 Soil Degradation–Consolidation Models 103
5.1 General Introduction 103
5.2 Governing Equations of Degradation–Consolidation Models 104
5.2.1 Degradation–Consolidation Models 104
5.2.2 Finite Element Numerical Solution Framework 109
5.3 Degradation–Consolidation Behavior of Landfilled Municipal Solid Waste 111
5.3.1 Stabilization Indicators 111
5.3.2 Anaerobic Stabilization Behavior 113
5.3.3 Aerobic Stabilization Behavior 116
5.4 Large-Scale Bioreactor Experiment and Field Survey Verification 120
5.4.1 Experimental Setup 121
5.4.2 Experimental Procedure and Results 123
5.4.3 Numerical Verification and Analysis 129

5.4.4 Field Survey and Evaluation of the Degree of Stabilization in Landfills 133
5.5 Analysis of Two-Dimensional Landfill Slopes 134
5.5.1 Bio-Hydro-Mechanical Coupling Behaviors in the MSW Slope. 135
5.5.2 Effective Stress Paths of MSW 137
5.5.3 Global Deformation Mode of the MSW Slope 138
5.6 Decomposition and Consolidation Behavior of Natural Gas Hydrates 139
5.6.1 Model Verification. 139
5.6.2 Decomposition and Consolidation Behaviors of Hydrate Reservoirs 146
References. 151

6 Production and Drainage of Leachate and Landfill Gas. 153
6.1 General Introduction 153
6.2 Leachate Production 154
6.3 Landfill Gas Production 156
6.3.1 Calculation Method for Landfill Gas Production 156
6.3.2 Landfill Gas Production Behavior 159
6.4 Three-Dimensional Drainage Technology for Leachate and Landfill Gas 161
6.4.1 Horizontal Well Dewatering Test 162
6.4.2 Vertical Well for Vacuum Pressure Pumping Tests 166
6.5 Efficient Collection of Landfill Gas. 169
6.5.1 Relationship Between Landfill Gas Collection Rate and Leachate Level 171
6.5.2 Application in Shenzhen Xiaping Landfill. 173
References. 175

7 Landfill Settlement and Instability 177
7.1 General Introduction 177
7.2 Landfill Settlement and Capacity 178
7.2.1 Compression Model for Municipal Solid Waste . . . 178
7.2.2 Landfill Settlement Calculation 180
7.2.3 Landfill Capacity Calculation 181
7.3 Landfill Stability 183
7.3.1 Shear Strength of Municipal Solid Waste 183
7.3.2 Instability Modes. 184
7.3.3 Critical Leachate Level 185
7.3.4 Critical Landfill Gas Pressure 186
7.4 Development of High Gas Pressure in Deep Layer and Its Effect on Slope Stability 188
7.4.1 High Gas Pressure Development in Deep Layer During Multi-stage Landfilling 188
7.4.2 Variation of Safety of Factor During Multi-Stage Landfilling. 190

7.5 Centrifugal Hypergravity Modeling of Landfill Failures Induced by Rising Water Levels 191
7.5.1 Test Setup 192
7.5.2 Material Selection and Synthesis 195
7.5.3 Rising of Water Level and Critical Water Level 200
7.5.4 Typical Slope Failure Modes 203
7.6 Centrifugal Hypergravity Modeling of Landfill Failures Induced by Rising Landfill Gas Pressures 206
7.6.1 Test Setup 206
7.6.2 Model Consolidation and Liquid–gas Injection Process 208
7.6.3 Development of Pore Pressure and Deformation 208
7.6.4 Stability Analysis and Critical Landfill Gas Pressure 212
7.7 Expansion of Landfill Capacity Based on Stability Control 215
References 218

8 Sustainable Landfilling of Municipal Solid Waste 221
8.1 General Introduction 221
8.2 Approach to Sustainable Landfilling 222
8.2.1 Large-Scale Intensive Landfilling 222
8.2.2 Recycling of Landfill Capacity 223
8.3 Mass Transformation and Carbon Emissions of MSW 225
8.3.1 Mass Transformation Pathways of Landfilled MSW Throughout Its Lifespan 225
8.3.2 Assessment Method for Carbon Emissions from Landfills 225
8.4 Assessing and Improving Mining Potential of Landfills 228
8.4.1 Effects of Stabilization on Fine Particles and Pollutant Content 229
8.4.2 Assessment Method for Mining Potential 230
8.4.3 Aerobic Accelerated Stabilization and In-Situ Dehydration Treatment 231
8.5 Mining and Screening of Landfilled MSW 236
References 243

Introduction

1

Abstract

Classical soil mechanics often encounters challenges in precisely capturing the complexities of skeleton-degradable soils, characterized by intricate interactions among biochemical reactions, physical transformations, and mechanical transport mechanisms. This chapter delineates the background, purpose, and structure of the book. For the first time, this book presents a comprehensive soil degradation–consolidation theory, surpassing the constraints of common classical soil mechanics assumptions, such as those considering a constant solid-phase mass and disregarding consolidation effects on degradation, within the realm of environmental engineering. The ultimate goal of this book is to tackle scientific challenges in the integrated analysis of solid–liquid–gas phase transformations, skeleton deformations, liquid–gas flows, and solute migration within skeleton-degradable soils.

Keywords

Skeleton-degradable soils · Degradation–consolidation theory · Soil mechanics · Municipal solid waste · Landfill · Multiphysics coupling

1.1 Background

Soil, a porous medium composed of fragmented particles, comprises two primary components: the solid skeleton and voids. The solid skeleton represents the solid-phase of soil, while the voids can hold liquid and gas phases, rendering soil a multiphase medium. In Terzaghi (1923), introduced the principle of effective stress and the theory of one-dimensional consolidation, thus establishing soil mechanics as an independent discipline. The realm of soil mechanics is fundamentally characterized by the effective stress principle, deformation analysis based on elastic mechanics, and failure analysis based on rigid–plastic models. Collectively, these principles constitute the core framework of classical soil mechanics.

The classical soil mechanics rely on the assumption of a time-invariant solid-phase, particularly focusing on the analysis of stress, pore structure, and saturation to comprehend soil characteristics. Notably, particles constituting the soil skeleton comprise various chemical components, classified into inert and active ones. In classical soil mechanics, the complete soil skeleton is considered inert. Furthermore, soils are categorized based on their particle size into types such as gravel, sand, silt, and clay. However, in addition to inert components, the soil skeleton also features active components. Soils with such active components are labeled as

Y. Chen, *Soil Degradation-Consolidation Theory and Its Applications*,
https://doi.org/10.1007/978-981-97-7985-7_1

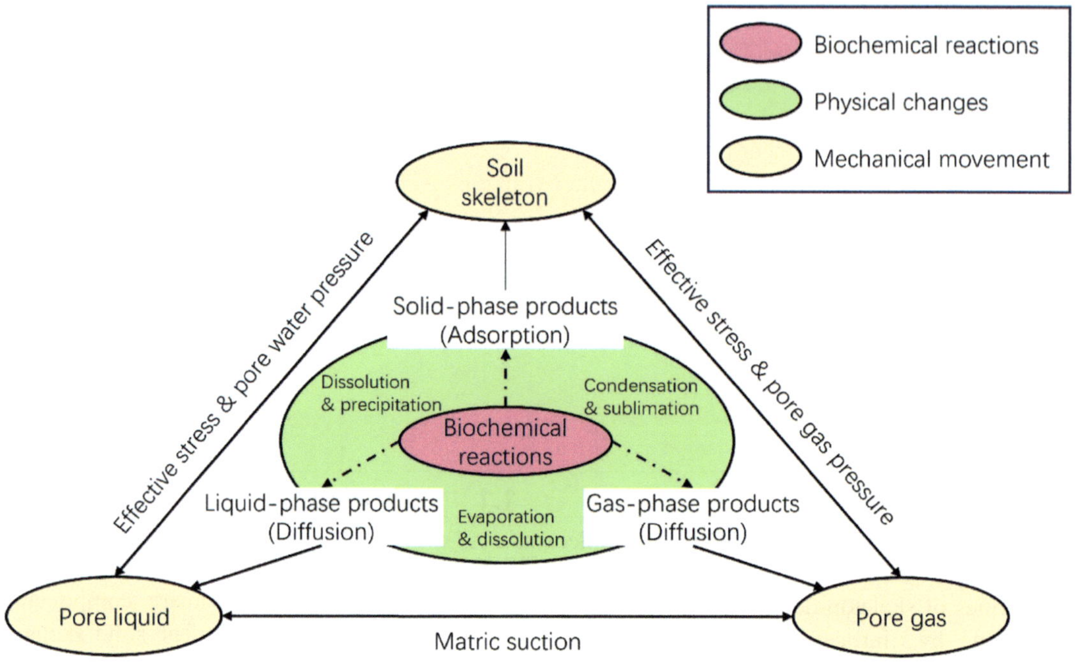

Fig. 1.1 Processes of mass change and substance transport within skeleton-degradable soils (reprinted from Chen 2014, with permission from the Chinese Journal of Geotechnical Engineering)

skeleton-degradable soils (or degradable soils). The soil skeleton of such soils undergoes mass loss and particle breakage owing to external loads and biochemical reactions, leading to particle gradation and pore structure variations. These alterations, in turn, influence soil properties such as shear strength, compressibility, and permeability.

Skeleton-degradable soils are commonly found in various geotechnical engineering fields, posing diverse engineering challenges. For instance, in energy geotechnics, processes such as solid-to-liquid and solid-to-gas transformations typically occur when extracting natural gas hydrates (NGHs). Meanwhile, in environmental geotechnics, municipal solid waste (MSW) degradation results in transformations from solid to liquid, gas, and solute phases. Furthermore, in transportation geotechnics, external loads cause subgrade material particles to break, thus altering their gradation characteristics. Similarly, in embankment dams, aggregate particles in high dams may undergo changes in gradation variations owing to high confining pressures. Unfortunately, classical soil mechanics encounter difficulties in accurately describing the deformation, strength, and seepage characteristics of skeleton-degradable soils.

Research on skeleton-degradable soils not only focuses on their solid, liquid, and gas phases but also considers their constituent active substances. Furthermore, it centers on understanding the degradation and breakage mechanisms of these substances, as well as the resulting changes in soil engineering properties and effects. As illustrated in Fig. 1.1, the processes of mass change and active substance transport within skeleton-degradable soils can be categorized into three levels (Chen 2014): (1) Biochemical reactions: At this level, one or more substances transform into different products through biochemical reactions. This process involves the consumption of original substances and the generation of new ones. (2) Physical changes: At this level, substances (i.e., solid, liquid, and gas) transform through processes such as the adsorption of liquid-phase solutes by solid-phase substances and the diffusion of liquid-phase solutes and gas phase substances into pores. (3) Mechanical movements: At this level, soil substances move relative to a reference time frame and configuration under the influence of various forces. This movement encompasses the macroscopic mechanical movement of the three primary components of soil (i.e., the solid skeleton, pore liquid, and pore gas) through several processes such as skeleton deformation, water–gas flow, and solute migration driven by diverse forces (such as skeleton stress, pore water pressure, and pore gas pressure). The coupling effects between biochemical reactions, physical changes, and mechanical movement in the soil represent the key scientific issue concerning the mass change and transport of degradable soil skeletons.

The coupling effects of biochemical reactions, physical changes, and mechanical movements within skeleton-degradable soils predominantly manifest in the following six ways: (1) Biochemical reactions and skeleton deformation: Biochemical reactions alter the properties of solid-phase substances constituting the soil skeleton, affecting its resistance to deformation and the stress conditions of other elements owing to variations in skeleton mass. (2) Biochemical reactions and water–gas flows: Pore water and gas participate in biochemical reactions as either reactants or products, influencing the source or sink of water–gas flow. Furthermore, biochemical-reaction-induced changes in the skeleton structure modify the structural properties of the porous medium (such as its porosity and pore radius), thereby impacting the water–gas flow. Meanwhile, the water–gas flow modifies the material or environmental conditions of biochemical reactions (such as moisture content), thereby influencing their progression. (3) Biochemical reactions and solute migration: Liquid-phase solutes participate in biochemical reactions as either reactants or products, thus acting as sources or sinks for solute migration. Conversely, solute migration alters the material or environmental conditions of biochemical reactions (e.g., reactant and product concentrations). (4) Skeleton deformation and water–gas flows: Skeleton deformation changes the structural properties (e.g., porosity) of the pore space, impacting water–gas flow through the porous medium. This water–gas flow, in turn, influences water and gas pressures, thereby affecting the mechanical equilibrium of the skeleton and inducing deformation. (5) Skeleton deformation and solute migration: Skeleton deformation alters the characteristics of the porous medium (such as its porosity, tortuosity, and saturation), thereby influencing the migration of liquid-phase solutes. (6) Water–gas flow and solute migration: The water–gas flow modifies the gradients of pore water and pore gas pressure, impacting convective transport of solutes. Furthermore, the water–gas flow alters soil saturation, further impacting solute migration.

Notably, the evolution of skeleton-degradable soils involves the multifield coupling of biochemical, mechanical, hydrological, and thermal processes. Acquiring a theoretical understanding of this complex interplay is crucial for accurately predicting the evolution of multiphase substances and enhancing the effectiveness of engineering disaster prevention and control measures. Professor Yunmin Chen's team at Zhejiang University has made significant contributions in this area by developing a degradation–consolidation theory that integrates various processes such as biochemical degradation, skeleton deformation, liquid–gas flow, solute migration, and heat conduction coupling (Chen et al. 2016, 2020; Li et al. 2021; Xiao et al. 2022; Hu et al. 2024; Wang et al. 2024). Within this theory, governing equations for biochemical degradation, liquid–gas flow, and solute migration are established according to the principle of mass conservation. Furthermore, governing equations for skeleton deformation are derived based on the principle of momentum balance, while those for heat conduction are developed according to the principle of energy balance.

Based on the constitutive relationships of skeleton-degradable soils (Ma 2022; Ma et al. 2024; Hu et al. 2020; Ke et al. 2022; Cai et al. 2024), numerical methods such as the finite element analysis approach can be employed to solve the corresponding governing equations. This approach facilitates tracking of the evolution patterns of various variables during the degradation–consolidation process of soil. These variables include the masses of all substances involved in the biochemical degradation process, liquid–gas production, pore fluid pressure, displacement, solute concentration, gas concentration, and temperature. Remarkably, this theoretical framework surpasses the constraints of classical soil mechanics assumptions such as those considering a constant solid-phase mass or disregarding consolidation effects on degradation in environmental engineering. The theory resolves scientific challenges associated with the coupled analysis of processes such as solid–liquid–gas phase transformations, skeleton deformation, liquid–gas flow, and solute migration.

1.2 Book Purpose

The objective of this book is to outline three primary processes observed in soil skeletons: particle breakage, chemical degradation, and biochemical degradation. To this end, constitutive models for these processes are introduced, and fundamental equations governing the coupling interactions among biochemical degradation, skeleton deformation, liquid–gas flow, solute migration, and heat conduction within skeleton-degradable soils are established. Furthermore, indicators characterizing the degradation and consolidation states of skeleton-degradable soils are outlined, forming the basis of the degradation–consolidation theory. This theory lays the groundwork for understanding the degradation and stabilization behaviors of MSW within landfills, decomposition and consolidation behaviors of NGHs, and stability and safety evolution behaviors of high earth-rock dams.

For instance, in the context of MSW landfill engineering, this theory posits that substantial leachate and landfill gas (LFG) production occurs during the initial stages of waste degradation within landfills containing high food waste content. Furthermore, the theory identifies mutual hindrance in liquid–gas flow as the primary factor leading to elevated leachate levels within the landfill body. Analytical methods rooted within the degradation–consolidation theory are proposed to address challenges such as landfill body instability or sliding, inadequate LFG collection efficiency, and potential breaches of containment barriers by pollutants under elevated leachate levels. This theoretical framework enables calculations of leachate and LFG production, thus guiding the practical designs of leachate treatment facilities, LFG collection systems, containment barriers, and cover layers. Moreover, it facilitates a quantitative analysis of degradation and stabilization processes within landfills, offering theoretical support for regulatory compliance, closure planning, and optimal resource management throughout landfill operations.

1.3 Book Organization

Subsequent chapters of this book cover the following topics.

Chapter 2: Degradation of the soil skeleton. This chapter focuses on skeleton-degradable soils, which undergo mass loss and particle breakage due to external loads and biochemical reactions. It categorizes the degradation processes into three primary processes: particle breakage (both static and dynamic), chemical degradation, and biochemical degradation. Models are introduced to capture these processes, particularly the anaerobic and aerobic degradation of municipal solid waste (MSW). The chapter highlights the significant impacts of mass loss and particle breakage owing to external loads and biochemical reactions, leading to particle gradation and pore structure variations. These alterations, in turn, influence soil properties such as shear strength, compressibility, and permeability.

Chapter 3: Constitutive models for skeleton-degradable soils. Beginning with an overview of the Modified Cam-clay and unified hardening models, this chapter introduces new state variables to facilitate the development of stress–strain constitutive models for skeleton-degradable soils, including soils subjected to particle breakage, chemically degraded soil (such as NGHs), and biochemically degraded soil (such as MSW). In particular, the constitutive model for soils subjected to particle breakage emphasizes the influence of particle size distribution variations on the stress–strain relationship. The constitutive model for NGHs underscores the significant influence of depressurization and hydrate decomposition rates on volumetric strain behavior. Additionally, the constitutive model for MSW considers the influence of degradation on waste hardening behavior.

Chapter 4: Liquid–gas seepage constitutive models for skeleton-degradable soils. This chapter explores the properties of liquid–gas flows and the saturated and unsaturated permeability characteristics of skeleton-degradable soils. Using MSW as an example, it examines how soil skeleton degradation affects porosity, pore size, tortuosity, intrinsic permeability, water retention curves, and relative permeability. The chapter also addresses the impact of compression, degradation, and anisotropy on MSW flow characteristics. A dual-porosity flow model is introduced for developing new water retention curve and to analyze two-phase flow. The evolution of pore water and gas pressures in high-food-waste-content (HFWC) and low-food-waste-content (LFWC) MSW landfills is analyzed, showing faster saturation and higher gas pressures in HFWC MSW landfills.

Chapter 5: Soil degradation–consolidation models. This chapter presents degradation–consolidation models relevant to MSW and NGHs, deriving governing equations and establishing a numerical solution framework. The chapter focuses on the degradation and stabilization behaviors of HFWC and LFWC landfilled MSWs, validated through large-scale experiments and field surveys. Indicators characterizing the degradation and consolidation states of MSW landfills are introduced. It also examines two-dimensional landfill slopes, analyzing the coupling effects of landfill gas (LFG) production and deformation due to the degradation of MSW on the landfill slope stability. Additionally, NGH degradation and consolidation behaviors are explored by the numerical model and validated by centrifugal hypergravity modeling. The model applies to multifield coupling processes in porous media, demonstrating its relevance to various skeleton-degradable soils.

Chapter 6: Production and drainage of leachate and landfill gas. Based on the degradation–consolidation model, this chapter introduces modified formulas for leachate and LFG production. The chapter compares LFG production in HFWC and LFWC MSW landfills, discusses a three-dimensional collection and drainage technique for leachate and LFG, and examines optimization strategies for LFG collection, accompanied by detailed discussions regarding horizontal well dewatering and vacuum pressure vertical well pumping tests. A case study on the Xiaping Landfill in Shenzhen highlights the effectiveness of these methods, showing an increase in LFG collection efficiency from 23 to 90%.

Chapter 7: Landfill settlement and instability. Beginning with an examination of long-term degradation compression tests, this chapter subsequently introduces compression models for MSW, along with methods for landfill settlement and capacity calculations. The discussion then focuses on landfill stability analysis, emphasizing factors such as the shear strength of MSW, instability modes, critical water levels, and critical LFG pressures. Furthermore, centrifugal hypergravity modeling of MSW landfill failures triggered by rising water levels and LFG pressures is presented, proposing critical water level and critical gas pressure values for landfill stability assessments. Strategies for expanding landfill capacity based on stability control are also discussed, including techniques such as dewatering and gas venting for landfill slopes and geogrid reinforcements, along with landfill capacity expansion rate calculations.

Chapter 8: Sustainable landfilling of municipal solid waste. This chapter first introduces approach to sustainable landfilling, leveraging insights from degradation–consolidation models. It outlines strategies like large-scale intensive landfilling and landfill capacity recycling to minimize environmental impact and maximize capacity. The chapter covers mass transformations, carbon emissions, and assessment methods throughout the MSW lifecycle. It also examines the mining potential of MSW landfills, addressing fine particle and pollutant effects, and presents methodologies for aerobic accelerated stabilization and in-situ dehydration to improve the mining potential of landfills. Lastly, it discusses the mining and screening of landfill MSW based on the project of the Woqishan Landfill in Wenzhou, China.

References

Cai ZZ, Ke H, Zhang PZ, Lan JW, Ma PC, Hu J, Xiao B, Chen YM (2024) Strength behaviour of stacked phosphogypsum incorporating dissolution–recrystallisation equilibrium. Géotechnique:1–13

Chen YM (2014) A fundamental theory of environmental geotechnics and its application. Chin J Geotechn Eng 36(1):1–46 (in Chinese)

Chen YM, Guo RY, Li YC, Liu HL, Zhan LT (2016) A degradation model for high kitchen waste content municipal solid waste. Waste Manage 58:376–385

Chen YM, Xu WJ, Ling DS, Zhan LT, Gao W (2020) A degradation–consolidation model for the stabilization behavior of landfilled municipal solid waste. Comput Geotech 118:103341

Hu J, Chen YM, Xu WJ, Ke H (2024) A bio-hydro-chemical model for gas pressure development in municipal solid waste landfills. Géotechnique 74 (7):647–660

Hu J, Ke H, Lan JW, Chen YM, Meng M (2020) A dual-porosity model for coupled leachate and gas flow to vertical wells in municipal solid waste landfills. Géotechnique 70(5):406–420

Ke H, Ma PC, Chen YM, Dong D, Zhao SY, Yang YQ, Nhema CC (2022) Instability of municipal solid waste along the constant deviatoric stress path and its engineering significance. Géotechnique 72(11):1025–1034

Li K, Chen YM, Xu WJ, Zhan LT, Ling DS, Ke H, Hu J, Li JL (2021) A thermo-hydro-mechanical-biochemical coupled model for landfilled municipal solid waste. Comput Geotech 134:104090

Ma PC (2022) Study on constitutive models for soils with degradable solid skeleton and their applications. Ph.D. thesis, Zhejiang University, Hangzhou, China (in Chinese)

Ma PC, Chen YM, Xu WJ, Hu J (2024) A novel constitutive model for degradable soil: municipal solid waste as a case study. Computers and Geotechnics (under review)

Terzaghi K (1923) Die Berechnung der Durchlässigkeitsziffer des Tones aus dem Verlauf der Hydrodynamischen Spannungs. ercheinungen. Sitzungsber. Akad. Wiss. Math. Naturwiss. Kl. Abt. 2A, 132, pp 105–124 (in German)

Wang LJ, Wang P, Zhu B, Kong DQ, Wang XB, Chen YM (2024) Physical modelling of hydrate dissociation in sandy sediment by depressurization under hyper-gravity and normal-gravity conditions. J Geotechn Geoenviron Eng 150(10):04024096

Xiao DK, Chen YM, Xu WJ, Zhan LT, Ke H, Li K (2022) Biochemical-thermal-hydro-mechanical coupling model for aerobic degradation of landfilled municipal solid waste. Waste Manage 144:144–152

Degradation of the Soil Skeleton 2

Abstract

This chapter discusses skeleton-degradable soils, which are soils whose skeletons can experience mass loss and particle breakage owing to external loads and biochemical reactions. These processes ultimately alter the particle size distributions and pore structures of the soils, thereby influencing soil properties such as strength, deformation, and permeability. First, the soil skeleton degradation process is categorized into the following based on distinct mechanisms: particle breakage, chemical degradation, and biochemical degradation. Among these, soil particle breakage is examined under both static and dynamic loading scenarios, followed by the introduction of a model capturing the evolution of the particle breakage process. Subsequently, the chemical degradation of soils is delineated, particularly the decomposition processes of hydrates and patterns of the associated solid-phase mass loss. Thereafter, the biochemical degradation of MSW is elucidated, emphasizing the development of unified equations. These equations form the basis for constructing models to capture the anaerobic and aerobic degradation behaviors of MSW, thus facilitating the investigation of degradation patterns under both conditions. By comprehensively exploring these degradation pathways, this chapter offers valuable insights into the profound impacts of soil skeleton degradation on soil properties, distinguishing skeleton-degradable soils from conventional ones.

Keywords

Skeleton-degradable soils · Particle breakage · Chemical degradation · Biochemical degradation · Ballast · Hydrates · Municipal solid waste

2.1 General Introduction

Soil, a porous medium comprising dispersed particles, consists of the soil skeleton and pore space. The soil skeleton constitutes the solid framework of soil, whereas the pore space contains both liquid and gas phases, rendering soil a multi-phase medium. When analyzing soil properties, classical soil mechanics assumes that the soil skeleton remains constant over time and solely focuses on the effects of stress, pore structure, and saturation. Notably, particles constituting the soil skeleton feature various chemical components, categorized into inert and active ones. However, classical soil mechanics considers the entire soil skeleton to exclusively comprise inert components, further classifying soils into gravel, sand, silt, and clay based on particle size.

In reality, the soil skeleton often comprises active components, in addition to inert components. Soils with such active components are

Y. Chen, *Soil Degradation-Consolidation Theory and Its Applications*,
https://doi.org/10.1007/978-981-97-7985-7_2

Table 2.1 Active components and reaction processes of typical solid waste

Solid waste	Reaction processes	Influencing factors	Time required for stabilization
MSW	Hydrolysis fermentation, hydrogen and acetate production, and anaerobic methane production	Temperature, water content, pH, salinity, and microorganisms	Decades
Construction waste	Degradation of paint residue and anaerobic fermentation of waste gypsum		Decades
Industrial waste	Oxidation of chromium, degradation of polycyclic aromatic hydrocarbons, and decomposition of phosphogypsum (Cai et al. 2024)		Several years to several hundred years
Radioactive waste	Nuclear fission and neutron transport		Over 10,000 years

termed as skeleton-degradable soils. These soils are particularly susceptible to mass losses and particle breakage owing to external factors such as loads and biochemical reactions. This leads to changes in soils' particle size distributions (PSDs) and pore structures, thereby altering soils' properties, such as strength, deformation, and permeability (Chen et al. 2020). Typical examples of skeleton-degradable soils include MSW, construction waste, industrial waste, radioactive waste, NGH-bearing sediments, calcareous soils, and seasonally frozen soils. Table 2.1 lists several types of active components of solid waste and reaction processes that occur in solid waste, in which biochemical degradation of active components produces liquid, gas, and pollutants. The latter three types of products are the root causes of three major types of geoenvironmental disasters that can occur in landfills, namely instability and landslides, leachate contamination, and landfill gas leakage.

Based on varying mechanisms, the degradation of the soil skeleton can be classified into particle breakage, chemical degradation, and biochemical degradation. Among these, particle breakage refers to the gradual disintegration of large particles into smaller ones under external forces, resulting in a macroscopic increase in the proportion of fine particles within the solid phase. This process alters the particle size distribution of the solid phase but does not cause it to lose mass. As depicted in Fig. 2.1, typical particle breakage processes encompass the breakage of pile-end soil, aggregates from earth-rock dams, railway ballast, and calcareous sand under high stress and cyclic loading conditions (Xiao et al. 2020). Notably, particle properties (including hardness, shape, weathering degree, surface roughness, and specific gravity) and external factors (such as stress levels and paths, loading duration, temperature, number of dynamic loading cycles, and frequency) primarily influence their susceptibility to breakage. Ballast breakage can lead to changes in subgrade strength and modulus, performance degradation, and acceleration of track-bed settlement, which poses significant challenges to railway maintenance and safe train operation. Meanwhile, in agricultural contexts, particle breakage can reduce soil permeability and aeration and is thus identified as one of the eight major risks threating soil quality by the European Thematic Strategy for Soil Protection (Blum 2011).

During both chemical and biochemical degradation, active chemical components from the soil skeleton undergo phase transformations through reactions, transitioning from the solid phase to a liquid phase containing solutes and a gas-phase containing multiple components. These transitions lead to mass losses from the solid phase and alter the particle size distribution of the soil skeleton. Furthermore, the involved chemical reactions often release or absorb energy (as depicted in Fig. 2.2).

Biochemical degradation differs from chemical degradation in that the former involves microorganisms, which directly affect the

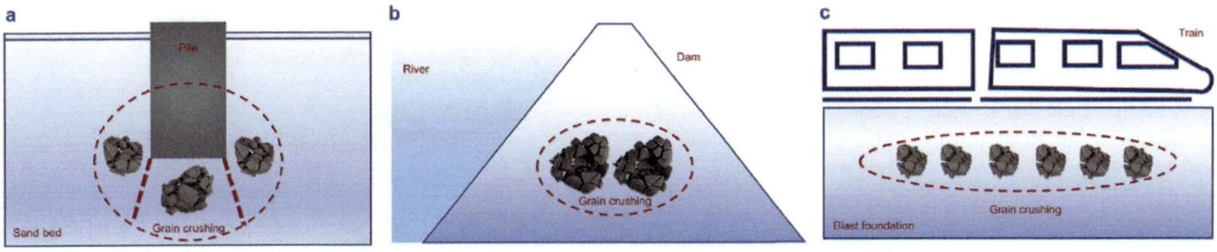

Fig. 2.1 Schematic of particle breakage in several engineering applications: **a** driven pile engineering, **b** rockfill dam engineering, and **c** railway engineering (Xiao et al. 2020)

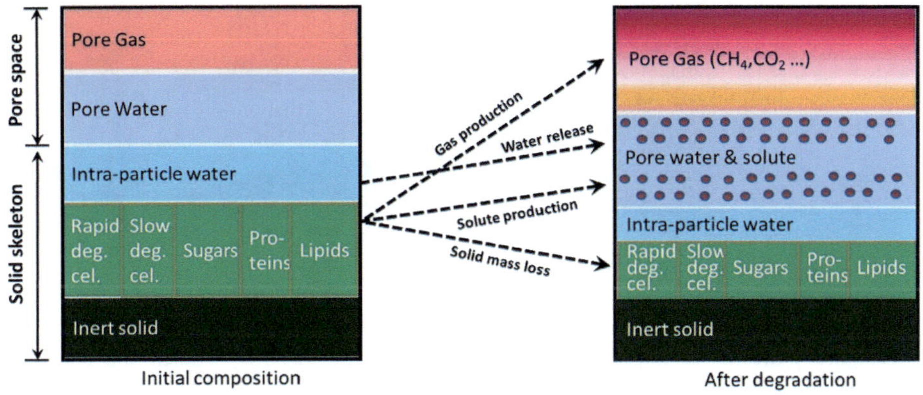

Fig. 2.2 Transformation of substances during biochemical degradation (Chen et al. 2020)

kinetics of reactions. Quantitative characterizations of both chemical and biochemical degradation rely on three key factors: the content of active chemical components within the soil skeleton, substance transformation relationships (reaction equations), and degradation rates. Typical chemical degradation processes involve the chemical decomposition of NGHs into CH_4 gas and water molecules. Under standard conditions, approximately 0.8 volumes of water and 164 volumes of gas are produced per volume of decomposed hydrates (Dai et al. 2012). Meanwhile, typical biochemical degradation processes include the conversion of solid phase MSW into leachate and LFG. For instance, Guo (2017) experimentally simulated the degradation of MSW with a high content of food waste. They showed that over 1000 days, the mass of solid phase per unit volume decreased from the initial value of 200–110 kg/m^3, equating to a mass loss rate of 45%. Furthermore, the rates of CH_4 and leachate production per unit mass of MSW were recorded as 50 m^3/t and 320 kg/t, respectively. Notably, biochemical degradation was also observed to reduce the particle size of MSW. Specifically, over a degradation age range of 0.7–10.9 years, the content of particles larger than 10 mm decreased from 65% to 18%, while that of particles smaller than 1 mm increased from 16% to 50% (Wei 2007).

2.2 Breakage of Soil Particle

2.2.1 Static Loading

Daouadji et al. (2001) noted that particle breakage in soils mainly occurs via one of three modes: (a) fracture, in which a particle breaks into smaller grains of similar sizes; (b) attrition, in which a particle breaks into one grain of a slightly smaller size and several other grains of much smaller sizes; and (c) abrasion, in which a particle breaks into fine particles but its

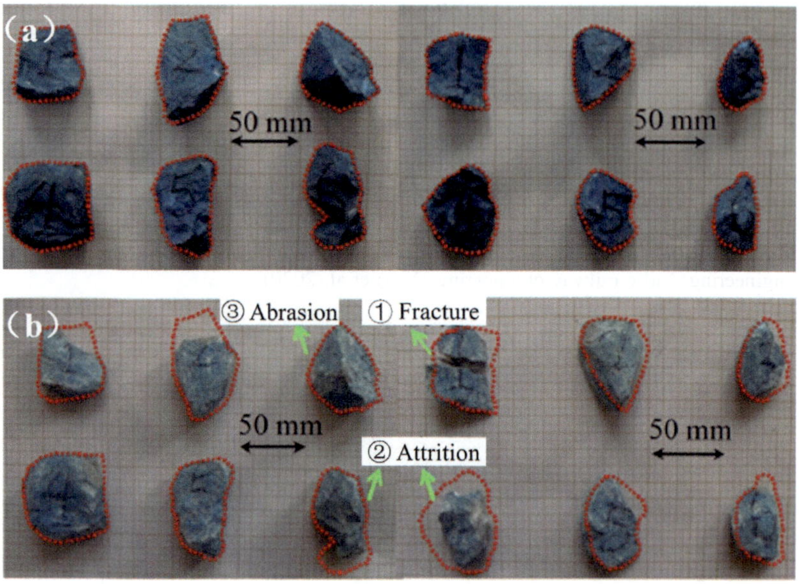

Fig. 2.3 Comparison of particle appearance **a** before test and **b** after test (reprinted from Kong et al. 2014, with permission from Rock and Soil Mechanics)

granulometry remains almost constant. Kong et al. (2014) conducted triaxial tests to study the particle breakage patterns of Zipingpu rockfill material under monotonic and cyclic loading conditions. For this experiment, 12 particles with diameters ranging from 20 mm to 60 mm were dyed, as depicted in Fig. 2.3, to distinctly identify the above three modes of particle breakage.

Particle breakage alters soil gradation. Turcotte (1986) proposed that as particle breakage progresses, soil particles with any initial distribution eventually tend toward a self-similar fractal distribution. This distribution is represented by the following PSD model, which is based on the relationship between the cumulative mass of soil particles and particle size:

$$F(d) = \left(\frac{d}{d_M}\right)^{3-D} \tag{2.1}$$

where $F(d)$ denotes the cumulative mass of soil particles smaller than d, d_M represents the maximum particle size, and D indicates the fractal dimension.

Figure 2.4 depicts soil particle size distribution curves corresponding to different fractal

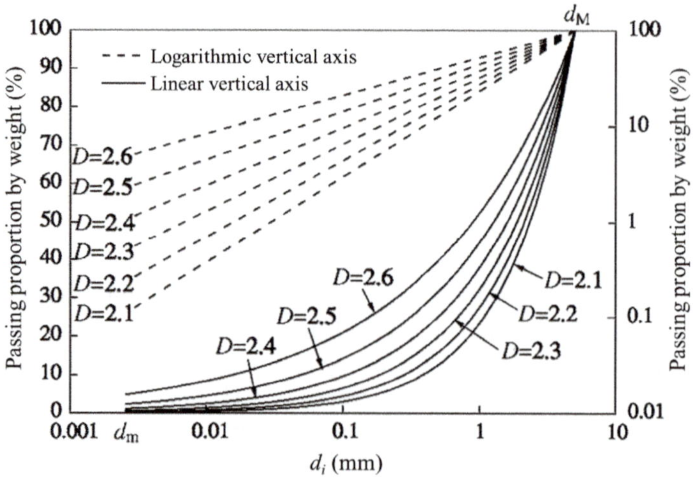

Fig. 2.4 Particle size distribution curves derived from Eq. (2.1) (reprinted from Zhang et al. 2015, with permission from the Chinese Journal of Geotechnical Engineering)

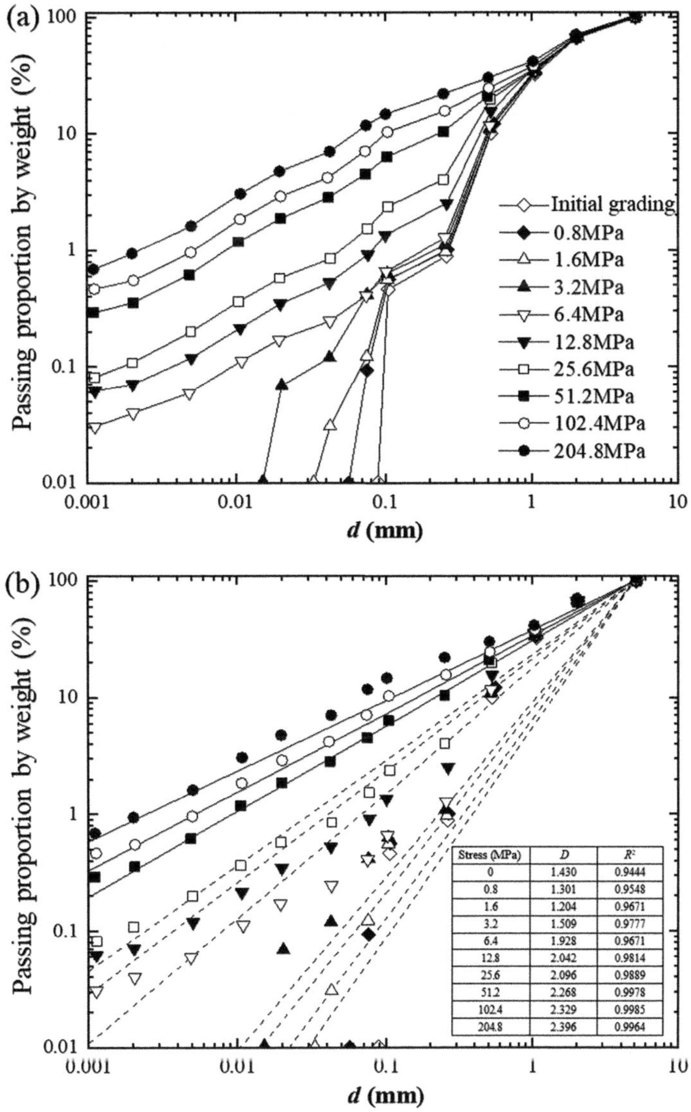

Stress (MPa)	D	R^2
0	1.430	0.9444
0.8	1.301	0.9548
1.6	1.204	0.9671
3.2	1.509	0.9777
6.4	1.928	0.9671
12.8	2.042	0.9814
25.6	2.096	0.9889
51.2	2.268	0.9978
102.4	2.329	0.9985
204.8	2.396	0.9964

Fig. 2.5 Evolution of the particle size distribution curves of quartz sand: **a** particle size distribution curves under various loads and **b** fitted by a fractal model (reprinted from Zhang et al. 2015, with permission from the Chinese Journal of Geotechnical and Engineering)

dimensions, as described by Eq. (2.1). In semi-logarithmic coordinate systems, curves conforming to the fractal relationship exhibit concave profiles, while in double-logarithmic coordinate systems, they appear as straight lines. Furthermore, curves with higher fractal dimensions lie above those with lower fractal dimensions.

Zhang et al. (2015) performed high stress confined compression tests on quartz sand, using nine vertical stress levels ranging from 0.8 MPa to 204.8 MPa. The load was maintained constant during each loading stage, and deformation measurements were recorded until the end of deformation before proceeding to the next loading stage. Figure 2.5a depicts the particle size distribution curve after each loading stage. It can be seen that particle breakage occurred as the load increased, resulting in the content of fine particles in the quartz sand gradually increasing and the particle size distribution curve gradually shifting towards the upper left. Fitting the particle size distribution curves at different stress levels using the fractal model (Eq. 2.1), as shown in Fig. 2.5b reveals that increasing stress levels and particle breakage lead to a gradual increase in the fractal dimension, *D*. Furthermore, Eq. (2.1) accurately models the particle size distribution curves of soils subjected to stresses of 51.2 MPa, 102.4 MPa, and 204.8 MPa ($D > 2.2$), indicating that the particle size distribution of quartz sand tends to exhibit a self-similar fractal distribution at a high degree of breakage. However, the gradation patterns of soils experiencing lower degrees of particle breakage deviate significantly from the fractal distribution. Furthermore, as the degree of particle breakage increases, variations in the grading curve become less pronounced (i.e., the rate of increase in *D* diminishes), suggesting the existence of an ultimate particle size distribution resulting from particle breakage.

Generally, the extent of particle breakage is defined by variations in the soil gradation curve. Hardin (1985) defined the particle breakage potential (B_p) as the ratio of the area enclosed between the current and initial gradation curves to the area enclosed between the initial gradation curve and the 0.074-mm-particle-size line (as depicted in Fig. 2.6a). According to this definition, B_p equals zero in the absence of particle breakage. Moreover, the greater the degree of particle breakage, the larger the value of B_p. A few years later, Einav (2007) revised the definition of B_p by replacing the 0.074-mm-particle-size line with the ultimate particle size distribution curve, thus defining the ratio of the areas as a new particle breakage index (B_r):

$$B_r = \frac{\int_{d_m}^{d_M}[F(d) - F_0(d)]\mathrm{d}d}{\int_{d_m}^{d_M}[F_u(d) - F_0(d)]\mathrm{d}d} \quad (2.2)$$

where d_m denotes the minimum particle size, $F_0(d)$ represents the initial particle size distribution curve, $F_u(d)$ indicates the ultimate particle size distribution curve, and $F(d)$ signifies the current particle size distribution curve.

As particle breakage progresses, the value of B_r increases from zero in the initial state to one in the ultimate state. Compared to the particle breakage potential (B_p) defined by Hardin (1985), the calculation of B_r relies on different limiting particle size distribution curves. As illustrated in Fig. 2.4, particles with size d_M are prevalent across numerous soil samples subjected to varying degrees of fragmentation. Therefore, the ultimate particle size distribution curve described by the fractal model is more appropriate than the 0.074-mm-particle-size line adopted in the definition of B_p.

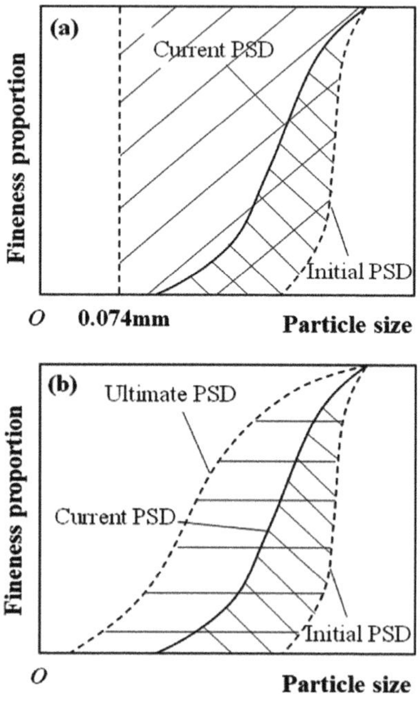

Fig. 2.6 Parameters used to describe the breakage degree of soil particles: **a** B_p and **b** B_r (Ma 2022)

2.2.2 Dynamic Loading

Railway ballast experiences severe breakage under long-term traffic dynamic loading, leading to changes in particle size and morphological characteristics. Various indicators to describe these changes in the particle size and morphology of railway ballast have been proposed. Among these, the particle size distribution captures particle size variations, while indicators such as the flat and elongated ratio (FER), sphericity, convexity, and angularity index characterize the changes in particle morphology. Precise definitions of these indicators and their

Fig. 2.7 Morphological analysis of dyed top ballast under a sleeper (Gu 2022)

computer-vision-based testing methods can be found in Shi (2020).

Bian et al. (2021) and Gu et al. (2022) conducted full-scale physical model tests on the ballasted track substructures of high-speed railways (as illustrated in Fig. 2.7). Subsequently, they utilized this railway ballast particle morphology analysis system to test and analyze the particle size distribution and morphology characteristics of the ballast during service. The device measures 15 m × 5 m × 6 m in length, width, and height, respectively, and uses a high-frequency vibrator array composed of eight servo-hydraulic actuators to reproduce the dynamic loading exerted by trains at fastening points. Moreover, by precisely controlling the phase difference between vibrators, axle movements and loading patterns on the track bed caused by train wheel passes are accurately simulated. In an indoor environment, train axle loads of up to 25 t and train speeds of up to 360 km/h can be achieved, meaning that the device can replicate all dynamic loads that occur on high-speed railways in China. The track structure used in these experiments is identical to actual high-speed railway lines. This similarity enables the simulation of realistic conditions where each test can replicate millions of train wheel passes and loading cycles.

Figure 2.8 depicts images comparing fresh and degraded ballast particles after being subjected to 500,000 train carriages (where one train carriage equals to four wheel passes or four axle loads). As indicated, initially, the diameters of all ballast particles from the newly constructed railway line exceeded 25 mm. However, prolonged exposure to train loads led to the subsequent generation of numerous fine particles with diameters smaller than 25 mm, indicating significant ballast particle breakage beneath the sleepers. This particle breakage involved severe abrasion and attrition and was demonstrated by the fading of the red dye on the surfaces of the particles.

Figure 2.9 depicts the particle size distribution curves of the ballast at various loading stages during the experiment. As illustrated, consistently increasing axle loads led to continuous ballast particle breakage, resulting in a gradual leftward shift of the particle size distribution curve. Specifically, the proportion of large particles with diameters ranging from 45 mm to 63 mm decreased by 16.97%, while the proportion of small particles with diameters below

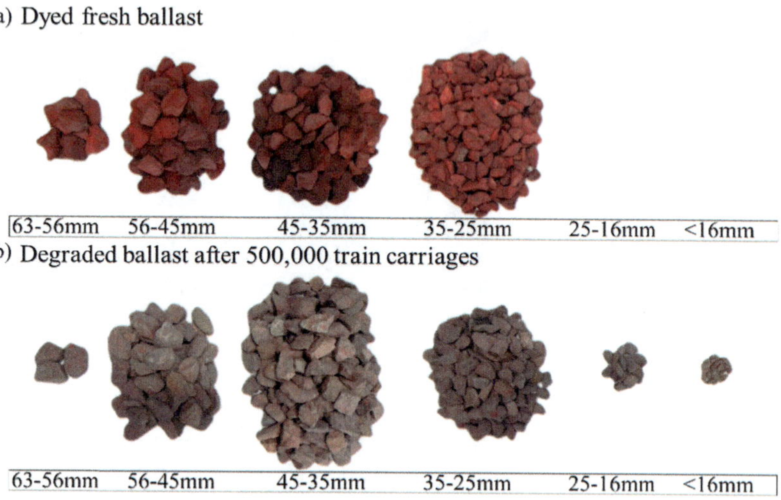

Fig. 2.8 Comparison of **a** fresh ballast, and **b** degraded ballast (i.e., after a cumulative loading of 500,000 train carriages) (Gu 2022)

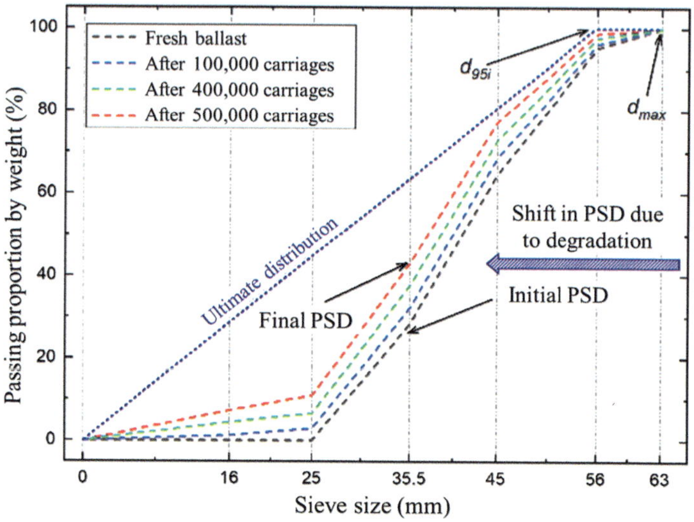

Fig. 2.9 Evolution of the PSD of the dyed ballast under the sleeper (Gu 2022)

25 mm increased by 10.84%. This indicates that under prolonged exposure to train loading, ballast particles beneath train sleepers disintegrate into smaller particles.

Figure 2.10 depicts the variation in B_r with the cumulative number of loading cycles. Notably, in the first stage, under an axle load of 17 t and a train speed of 100 km/h, the value of B_r increased to 0.1 after experiencing 100,000 carriages. In the second stage, with an increased train speed of 360 km/h, B_r continued to increase to 0.25 after a cumulative loading of 1.6 million axle loads (400,000 carriages), albeit at a slower rate. Finally, in the third stage, under an increased axle load of 25 t, the value of B_r sharply increased to 0.4 after a cumulative loading of 500,000 carriages. This trend is attributed to the fact that higher axle loads produce greater impact energies and higher internal dynamic stresses within the track bed, leading to more severe ballast breakage.

Figure 2.11 presents the variation trends of several ballast particle morphology parameters with different particle sizes following exposure to train loading. Among these parameters, the FER is defined as the ratio of the longest to the shortest dimension of a particle from an assembly of flat, elongated, and cubical particles.

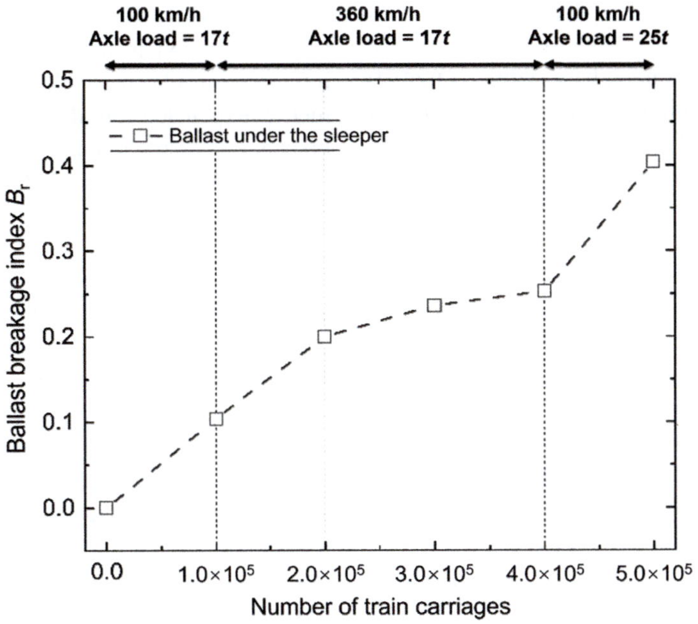

Fig. 2.10 Evolution of the B_r of the dyed ballast under the sleeper (Gu 2022)

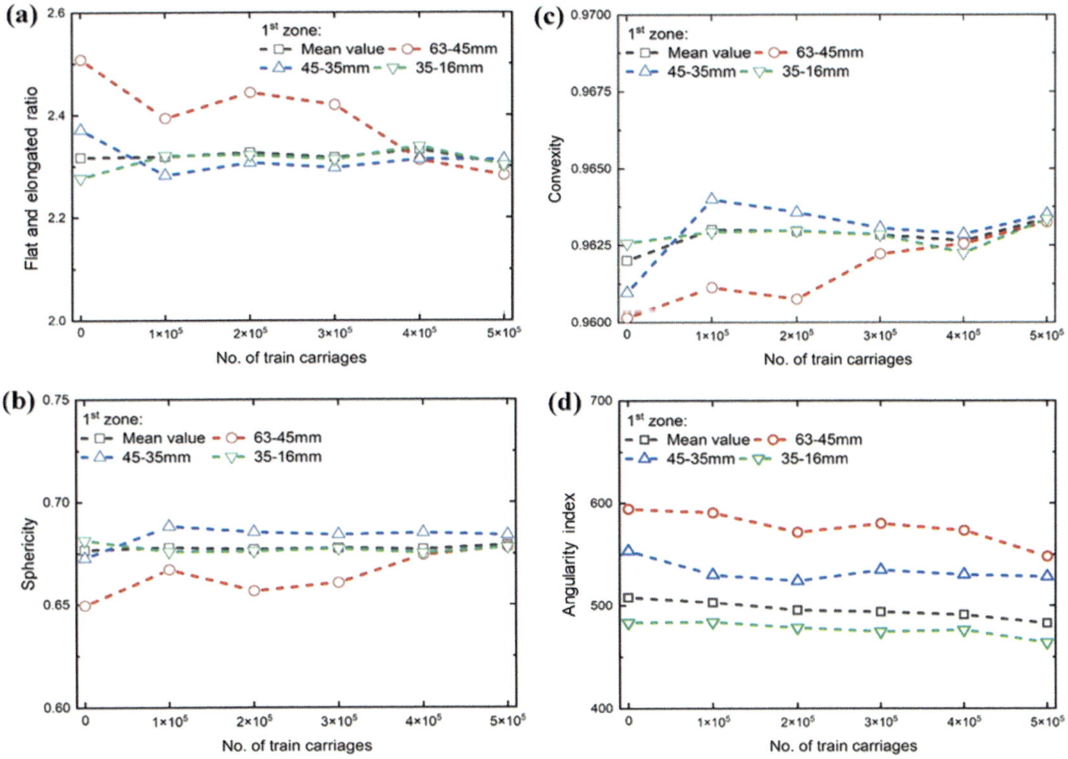

Fig. 2.11 Variations in the morphological parameters of dyed ballast subjected to long-term loading: **a** FER, **b** sphericity, **c** convexity, and **d** angularity index (reprinted from Gu et al. 2022, with permission from Elsevier)

Furthermore, sphericity is defined as the cube root of the ratio of the volume of a particle to the volume of its enclosing sphere, while convexity is defined as the ratio of the area of a particle to the convex area of its two-dimensional projection. Meanwhile, the angularity index quantifies the changes in the vertex angles of a particle's two-dimensional silhouette influenced by its sharp corners and smooth edges. As shown in Fig. 2.11, there was a small overall average variation in particle morphology parameters across the entire sample, but there was a substantial overall average variation in large particles, i.e., 45–63 mm particles. For instance, as the number of train loading carriages increases from 100,000 to 500,000, large particles in this size range demonstrate a reduction in their FER from 2.51 to 2.28, an increase in their sphericity from 0.65 to 0.68, an increase in their convexity from 0.9600 to 0.9635, and a decrease in their angularity index from 600 to 550. This is because under dynamic loading conditions, larger ballast particles are more susceptible to edge abrasion and particle attrition compared to smaller particles, resulting in smoother and finer particles over time. During actual train operations, the cumulative number of loading carriages far exceeds 500,000, leading to more pronounced changes in particle morphologies and significant alterations in particle–particle contact dynamics.

2.2.3 Model for Particle Breakage

The process of soil particle breakage progresses gradually during loading, making it essential for determining the evolution pattern of the particle breakage index, B_{r}, throughout this process. Typically, plastic work, expressed in Eq. (2.3), is selected as the variable controlling the evolution of soil particle breakage.

$$W_p = \int p\langle \mathrm{d}\varepsilon_{\mathrm{v}}^{\mathrm{p}} \rangle + q\mathrm{d}\varepsilon_{\mathrm{d}}^{\mathrm{p}} \tag{2.3}$$

where $\langle \mathrm{d}\varepsilon_{\mathrm{v}}^{\mathrm{p}} \rangle$ denotes the MacCauley function, defined as $\langle \mathrm{d}\varepsilon_{\mathrm{v}}^{\mathrm{p}} \rangle = \mathrm{d}\varepsilon_{\mathrm{v}}^{\mathrm{p}}$ when $\mathrm{d}\varepsilon_{\mathrm{v}}^{\mathrm{p}} \geq 0$ and as $\langle \mathrm{d}\varepsilon_{\mathrm{v}}^{\mathrm{p}} \rangle = 0$ when $\mathrm{d}\varepsilon_{\mathrm{v}}^{\mathrm{p}} < 0$. The use of the MacCauley function in this context implies that the influence of plastic work resulting from volumetric expansion on the evolution of particle breakage is disregarded.

However, plastic work fails to describe the different evolution patterns of particle breakage at different shear stages. That is, during shear, a soil undergoes several distinct stages of hardening and softening before reaching its critical state. In these stages, the stress–strain characteristics of soil vary significantly, resulting in distinct particle breakage characteristics. During the hardening stage, the load exerted on the soil continuously increases. This increasing load strengthens external forces and aggravates volume reduction, leading to soil particle compression, which, in turn, results in a significant increase in interaction forces and particle breakage. During the softening stage, the soil load gradually decreases and is accompanied by volume expansion, leading to reduced particle interactions and consequently less particle breakage. When a soil reaches its critical state, it enters a plastic flow state in which the volume is constant and the contact status between soil particles stabilises, which hinders further particle breakage.

Figure 2.12 illustrates particle breakage evolution patterns for three types of sand subjected to varying confining pressures in triaxial compression tests (Yu 2017, 2018; Wang et al. 2020). The horizontal axis of the figure represents the ratio of axial strain (ε_{a}) to the axial strain at peak strength ($\varepsilon_{\mathrm{a-peak}}$). A ratio of less than one indicates the hardening stage, while a ratio exceeding one signifies the softening stage. Evidently, during the hardening stage, the particle breakage indices (B_{r}) of all three sands rapidly increase, indicating significant particle breakage evolution. However, upon entering the softening stage, the rate of increase in B_{r} diminishes significantly, with certain specimens demonstrating almost no change in B_{r}. This indicates that the rate of particle breakage during the softening stage is much slower than that during the hardening stage. Conversely, plastic work, which increases monotonically throughout the shearing process,

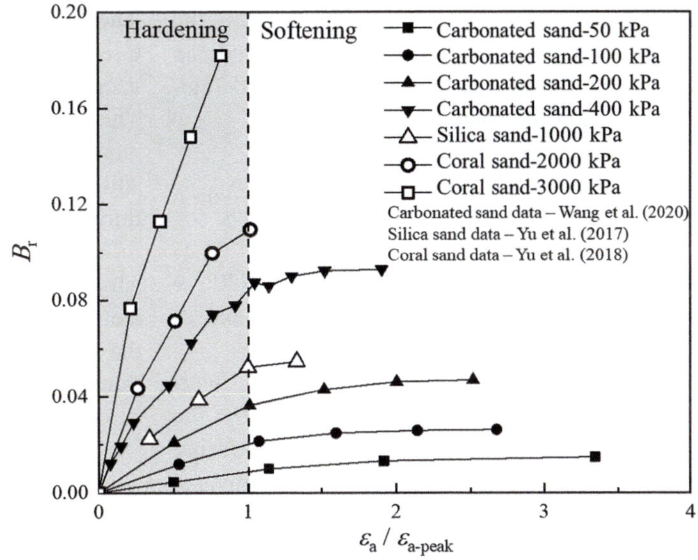

Fig. 2.12 Variation in evolution of particle breakage of different types of sands in different stages of shear

continues to rise in both the softening and critical state stages. Consequently, using plastic work as a control variable for soil particle breakage fails to fundamentally reflect the differences in the evolution patterns of particle breakage at varying shear stages.

To address the shortcomings of utilizing plastic work as a control variable for particle breakage evolution, Ma (2022) introduced a new control variable. As illustrated in Fig. 2.13, under isotropic compression, the state point of soil initially lies at A on the normal consolidation line (NCL). Following the application of an incremental stress dp, if no particle breakage occurs, the postloading state point shifts to B′ on the NCL. Conversely, if particle breakage occurs, the weakening of the soil skeleton causes greater deformation, moving the postloading state point to B below the NCL. Therefore, $\Delta e_{AB'}$ represents loading-induced deformation, while $\Delta e_{B'B}$ denotes particle-breakage-induced deformation. Given that $\Delta e_{B'B}$ reflects the deformation caused by particle breakage, the work done on this deformation is disregarded when evaluating particle breakage evolution. In particular, this analysis explicitly considers the work done on

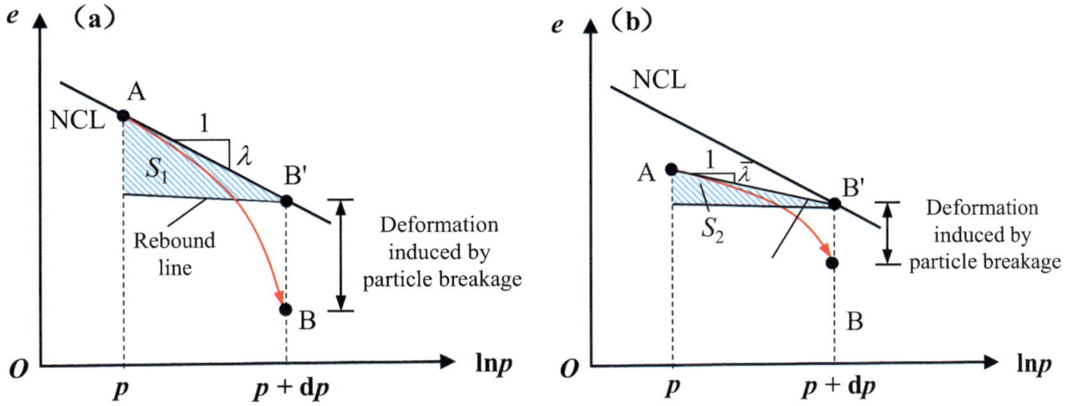

Fig. 2.13 Diagram illustrating the definition of effective plastic work during particle breakage for **a** normally consolidated soil and **b** overly consolidated soil

loading-induced deformation, represented by the shaded area of the triangle displayed in Fig. 2.13a, denoted as S_1. This area can be calculated using the following equation:

$$S_1 = p\mathrm{d}\varepsilon^{\mathrm{p}}_{\mathrm{v-AB'}} = p \cdot \frac{\lambda - \kappa}{1 + e_0} \cdot \frac{1}{p} \mathrm{d}p = \frac{\lambda - \kappa}{1 + e_0} \mathrm{d}p \tag{2.4}$$

where λ denotes the slope of the NCL, κ represents the slope of the rebound line, and e_0 indicates the initial void ratio of the soil.

As illustrated in Fig. 2.13b, for over-consolidated soils, point A is located below the NCL. In this case, the state point of soil during loading is anticipated to approximately traverse the reloading line. Similar to normally consolidated soils, only the work done on loading-induced deformation is considered. This area, denoted by S_2, is represented by the shaded triangular region in Fig. 2.13b and defined as follows:

$$S_2 = p\mathrm{d}\varepsilon^{p}_{\mathrm{v-AB'}} = \frac{\overline{\lambda} - \kappa}{1 + e_0} \mathrm{d}p \tag{2.5}$$

where $\overline{\lambda}$ denotes the slope of the reloading line. According to Yao (2015), the slopes of the NCL, reloading line, and rebound line demonstrate the following relationship:

$$\frac{\lambda - \kappa}{\overline{\lambda} - \kappa} \approx \frac{M_{\mathrm{Y}}^4}{M^4} \tag{2.6}$$

where M_{Y} denotes the current potential strength of over-consolidated soil, as defined by Yao (2015). Combining Eqs. (2.5) and (2.6) yields

$$S_2 = \frac{\lambda - \kappa}{1 + e_0} \frac{M^4}{M_{\mathrm{Y}}^4} \mathrm{d}p \tag{2.7}$$

Based on this derivation, the effective plastic work for soil particle breakage can be defined as follows:

$$W_p' = \int \frac{M^4}{M_{\mathrm{Y}}^4} \langle \mathrm{d}p \rangle \tag{2.8}$$

The MacCauley function is used to prevent the effective plastic work of particle breakage decreasing when the soil is in the softening stage. This non-decreasing nature of effective plastic work is supported by the fact that particles of the soil shown in Fig. 2.12 underwent little breakage during the softening stage.

For the more general case wherein soil is subjected to deviatoric stress and is not under isotropic compression, the overall stress state of the soil is transformed into an equivalent one-dimensional stress (p_{x}) as follows:

$$p_{\mathrm{x}} = p\left(1 + \frac{q^2}{M^2 p^2}\right) \tag{2.9}$$

This one-dimensional equivalent stress (p_{x}) represents the intersection of the yield surface of the modified cam-clay model with the p axis. Consequently, the effective plastic work of soil under normal stress conditions is given as

$$W_p' = \int \frac{M^4}{M_{\mathrm{Y}}^4} \langle \mathrm{d}p_{\mathrm{x}} \rangle \tag{2.10}$$

The effective plastic work of breakage is greater than zero in the hardening stage, while it equals zero in the softening and critical stages. Thus, the different laws of particle breakage in different shear stages of a soil can be reflected by the effective plastic work of breakage, indicating that it is suitable for use as a control variable for the evolution of particle breakage.

Given that the evolution of soil particle breakage slows down gradually during the loading process, a hyperbolic curve is appropriate to describe the relationship between the particle breakage index and the effective plastic work of breakage:

$$B_{\mathrm{r}} = \frac{W_p'}{\chi' + W_p'} \tag{2.11}$$

where χ' denotes a constant (in kPa) reflecting the difficulty of soil particle breakage under the action of the effective plastic work of breakage.

The evolution of B_r during triaxial compression is shown in Figs. 3.15–3.17 of Sect. 3.4.

The particle breakage evolution model established in this section offers a crucial theoretical foundation for the particle breakage constitutive model detailed in Sect. 3.4. This model will be employed to analyze the particle breakage evolution of soil during loading processes, thereby accurately modeling the stress–strain characteristics during soil particle breakage.

2.3 Chemical Degradation

NGHs are cage-like crystalline compounds made up of natural gas (mostly CH_4 gas) and water molecules. These compounds, formed under low temperature and high-pressure conditions, are commonly found in seabed and permafrost regions. Owing to their ice-like appearances, they are often referred to as combustible ice. NGHs are known for their abundant reserves, high energy density, and nonpolluting combustion products, rendering them a promising new clean source of energy. Hydrate reservoirs are geological formations containing solid degradable hydrates, which are dispersed and distributed as skeleton cementing agents, pore fillers, and particle displacers. When these hydrates undergo phase change decomposition, producing water and gas, the strength, deformation, and permeability characteristics of the reservoir are significantly altered (Yuan et al. 2022; Zhu et al. 2022).

2.3.1 Decomposition Equation of Hydrates

Natural gas hydrates are developed artificially primarily by using various methods to alter the temperature and pressure conditions of hydrate reservoirs. These alterations disrupt the phase equilibria of hydrates, thereby inducing their decomposition to form natural gas, which is extracted for utilisation, as illustrated in Fig. 2.14.

The current mainstream methods for hydrate exploitation include depressurization, thermal stimulation, chemical injection, and CO_2 replacement. Among these, depressurization involves extracting pore water from the hydrate reservoir to decrease pore pressure and promote hydrate decomposition. Meanwhile, thermal stimulation involves introducing high-temperature steam or hot brine into the reservoir, thus raising its temperature above the equilibrium temperature of the hydrate phase and

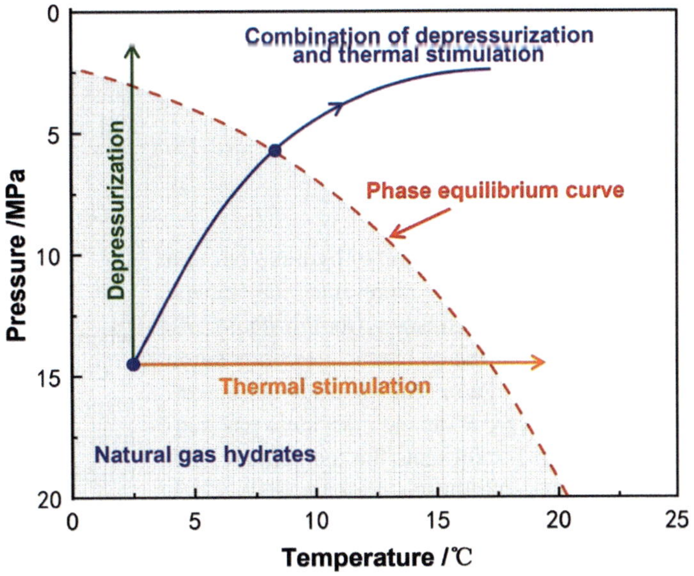

Fig. 2.14 Phase transition and decomposition pathways of hydrates

ultimately inducing its degradation into CH_4 gas and water. Chemical injection involves injecting chemical agents into the reservoir, thus compromising the stability of hydrates. Finally, CO_2 replacement, a recent frontier research hotspot, involves injecting liquid or gaseous CO_2 into the hydrate reservoir. Given the tendency of CO_2 to form hydrates, CO_2 molecules can replace CH_4 molecules within the cage-like hydrate structure, achieving gas exploitation.

Equation (2.12) describes the process of hydrate formation and decomposition, where n denotes the stoichiometric number of hydrates. Typically, under standard conditions, the decomposition of one unit volume of hydrates (for instance, $n = 6$) can generate approximately 164 and 0.8 volumes of natural gas and water, respectively (Sloan and Koh 2008).

$$CH_4 \cdot nH_2O \rightleftharpoons CH_4 + nH_2O \tag{2.12}$$

The decomposition process of pure hydrates can be described using various models, including equilibrium models, quasi-kinetic models, and kinetic models. The classic Kim–Bishnoi kinetic model, a prime example of the final category, describes the kinetic reaction process of hydrates (Kim et al. 1987; Hardwick and Mathias 2018; Sun et al. 2019):

$$\frac{dn_H}{dt} = K_d A_s (p^e - p^g) \tag{2.13}$$

$$K_d = K_0 \exp\left(-\frac{\Delta E}{RT}\right) \tag{2.14}$$

$$p^e = c_e \exp\left(a_e T^{b_e}\right) \tag{2.15}$$

$$A_s = A_{s0} \varphi s_H \tag{2.16}$$

where n_H denotes the amount of hydrates per unit volume (mol/m^3); K_d represents the kinetic decomposition rate constant (mol/m^2/Pa/s); K_0 indicates the intrinsic kinetic decomposition constant (mol/m^2/Pa/s); ΔE signifies the activation energy (J/kg); R denotes the ideal gas constant (J/kg/K); T represents the temperature (K); p^e denotes the equilibrium pressure (Pa); a_e, b_e and c_e (Pa) represent coefficients related to the equilibrium pressure; A_{s0} indicates the initial specific surface area of decomposition (m^{-1}); A_s represents the specific surface area of decomposition; φ denotes porosity; and s_H represents hydrate saturation.

During the decomposition of NGHs, solid hydrates transform into gas and water, accompanied by changes in their energy. The specific mass and energy change rates of hydrates, water, and gas phases per unit volume are described by the following equations:

$$q_H = -M_H \frac{dn_H}{dt} \tag{2.17}$$

$$q_a = M_a \frac{dn_H}{dt} \tag{2.18}$$

$$q_w = nM_w \frac{dn_H}{dt} \tag{2.19}$$

$$q_E = -\Delta H M_h \frac{dn_H}{dt} \tag{2.20}$$

$$\Delta H = c_{H1} + c_{H2} T \tag{2.21}$$

where M_a, M_w, and M_h denote the molar masses of gas, water, and hydrates, respectively (kg/mol); ΔH denotes the enthalpy change of hydrates (J/mol); and c_{H1} (J/mol) and c_{H2} (J/mol/K) represent enthalpy-related coefficients, with values of 44.6120 J/mol and −132.638 J/mol/K, respectively (Sun et al. 2019).

The decomposition of NGHs within reservoirs primarily entails the process of solid mass loss and water and gas production from gas-hydrate-bearing sediments (GHBSs). The degree of solid mass loss of hydrates, defined as D_H, is expressed as

$$D_H = 1 - \frac{m_H^t}{m_H^{t_I}} = 1 + \frac{\rho_H V \int_{t_I}^{t} q_H dt}{m_H^{t_I}} = 1 - \frac{s_H^t}{s_H^{t_I}} \tag{2.22}$$

where m_H^t and $m_H^{t_I}$ denote the masses of solid hydrates in GHBSs at time t and initial time t_I, respectively (kg). D_H denotes the degree of hydrate decomposition. Notably, for undecomposed hydrate reservoirs, $m_H^t = m_H^{t_I}$, indicating $D_H = 0$. Conversely, for fully decomposed GHBSs, m_H^t approaches zero, implying that

$D_H = 1$. Furthermore, s_H^t and $s_H^{t_I}$ signify hydrate saturation in the GHBS at time t and initial time t_I, respectively.

Notably, some parameters appearing in Eqs. (2.13)–(2.16) represent the basic property parameters of pure hydrates and are often treated as constants. In particular, for CH_4 hydrates (Kim et al. 1987; Hardwick and Mathias 2018), these constant parameters have the following values: $K_0 = 3.6 \times 10^4$ mol/m²/Pa/s, $\Delta E = 8.11 \times 10^4$ J/kg, $R = 8.314$ J/kg/K, $a_e = -9{,}459$ K, and $b_e = -1$. Meanwhile, for CO_2 hydrates (Feyzi and Mohebbi 2020; Goel 2006; Merey et al. 2018), these constants have the following values: $K_0 = 1.3 \times 10^{-4}$ mol/m²/Pa/s, $\Delta E = 4.05 \times 10^4$ J/kg, $R = 8.314$ J/kg/K, $a_e = 0.1306 \times 10^{-10}$ K, and $b_e = 1$. The values of other parameters, including c_e and A_{s0}, must be determined experimentally under actual decomposition conditions.

2.3.2 Solid Mass Loss of Hydrates

A previous study comparatively analyzed the theoretical predictions and experimental outcomes of CH_4 hydrate decomposition under triaxial conditions with depressurization (Choi et al. 2020). Considering the temperature and pressure conditions at the onset of hydrate decomposition (T= 274.23 K, p^g= 3.29 MPa), as well as the evolution tendency of the degree of decomposition D_H, the analysis yielded the following parameter values: $c_e = 3.1473 \times$ 1,021 Pa and $A_{s0} = 2.0 \times 10^5$ m^{-1}.

Notably, in the triaxial experiment, hydrate decomposition did not occur initially during depressurization, resulting in the selective production of free gas, as depicted in Fig. 2.15. During this period, the gas production rate from hydrate decomposition was zero. Subsequently,

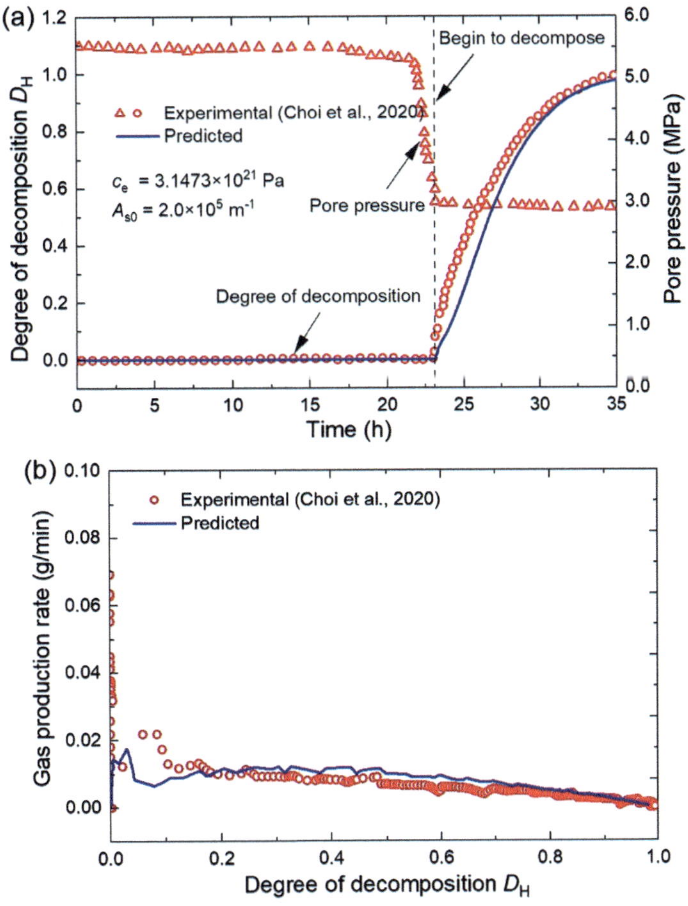

Fig. 2.15 Variations in different parameters during the CH_4 hydrate decomposition experiment: **a** degree of decomposition versus time and **b** gas production rate versus degree of decomposition

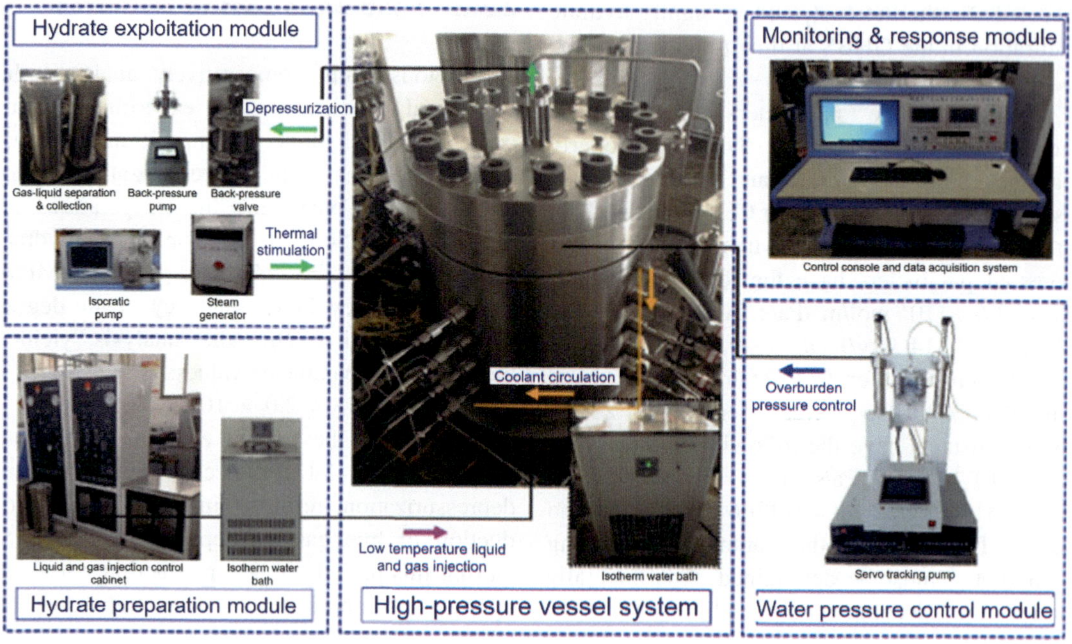

Fig. 2.16 In-flight centrifuge apparatus developed by Zhejiang University to model hydrate exploitation (reprinted from Wang et al. 2024, with permission from the Chinese Journal of Geotechnical Engineering)

during the steady depressurization stage, hydrates gradually decomposed, and the above predicted values effectively reflected the changes in the degree of decomposition and gas production rate observed during the experiment.

Furthermore, experiments have been conducted using the hydrate-extraction simulation device at Zhejiang University (as shown in Fig. 2.16) to explore the relationship between solid-phase mass loss, temperature–pressure variations, and gas production characteristics during CO_2 hydrate depressurisation decomposition (Zhu et al. 2021; Wang et al. 2022a, 2024).

The distribution of monitoring points within the high-pressure vessel is depicted in Fig. 2.17. The vessel features four columns of sensor interfaces, with each column containing four sensors. The sensors are positioned at distances of 30, 95, 160, and 225 mm from the bottom of the vessel and at distances of 35 and 75 mm from the vessel wall. The monitoring points for pore pressure and temperature are denoted as T_i and P_i (i = 1–8), respectively. Furthermore, the vessel has a radius of 11.5 cm and a height of 30 cm. Under normal gravity conditions, pore pressure and temperature are approximately uniform throughout the vessel, and the self-weight stress of soil is much lower than the pore pressure. Hence, the entire vessel can be considered a unit cell test.

Figure 2.18 illustrates the variation in the degree of decomposition and gas production rate of CO_2 hydrates with varying depressurization rates. Considering the temperature and pressure conditions prevalent at the onset of decomposition (T= 279.16 K, p^g= 2.53 MPa), the evolution of the degree of decomposition, and Eqs. (2.13)–(2.16), the following values can be derived: $c_e = 3.7 \times 10^{-10}$ Pa and $A_{s0} = 7.5 \times 10^5$ m^{-1}. The predicted gas production rate reflects the hydrate decomposition rate and the amount of gas flow induced by decomposition within the reservoir (Wang et al. 2022b).

At three different depressurization rates, the peak gas production rates increase from 1.2 to 1.3 L/min and finally to 1.6 L/min. This indicates that as the depressurization rate increases, the driving force behind hydrate decomposition also

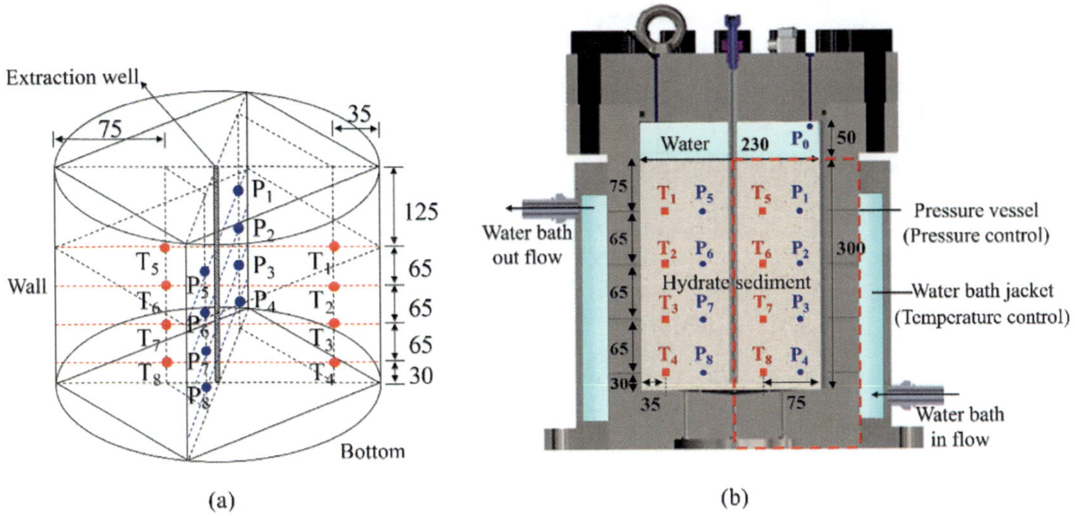

Fig. 2.17 Schematic of the high-pressure vessel: **a** spatial distribution of monitoring points and **b** cross-sectional view of the apparatus (mm) (reprinted from Wang et al. 2024, with permission from the Chinese Journal of Geotechnical Engineering)

increases, leading to gradual increments in peak gas production rates. However, during intense events of gas production and endothermic solid hydrate decomposition, disturbances within the reservoir can cause fine solid particles to migrate. Furthermore, the accompanying temperature drop may lead to the formation of secondary hydrates, suggesting that beyond a certain point, increasing depressurization rates do not necessarily improve the efficiency of hydrate decomposition. The highest overall decomposition efficiency is achieved at an optimal depressurization rate. The gas-production rate and degree of decomposition at each moment are obtained by solving Eqs. (2.13)–(2.16) and Eq. (2.18) simultaneously, using the initial hydrate saturation obtained from the experiments and the outlet pressure and the average temperature at various measurement points as inputs. Remarkably, the predicted values align well with the experimental results. Furthermore, c_e and A_{s0} remain consistent across varying depressurization rates, indicating that these parameters are independent of the depressurization rate.

The values of c_e and A_{s0} are independent of temperature. Figure 2.19 illustrates the decomposition degrees and gas production rates of CO_2 hydrates at varying temperatures. The prevailing temperature and pressure conditions (T= 276.85 K, p^g= 1.87 MPa; T = 279.69 K, p^g= 2.70 MPa, respectively) yield the same values of c_e and A_{s0} as in the previous example. During depressurization at a rate of 0.05 MPa/min, minimal hydrate decomposition is accompanied by a temperature drop. During the initial phase of the steady pressure stage, hydrate decomposition becomes more pronounced, leading to further reduction in temperature, which ultimately stabilizes at a low value. Over time, the decomposition rate decreases, allowing the temperature to gradually rise. This process reflects the dynamic phenomena of energy consumption and replenishment during hydrate decomposition. Furthermore, the gas production rate initially increases and subsequently decreases gradually with hydrate decomposition. This model effectively describes the dynamics of both hydrate decomposition and gas production rates during depressurization-induced decomposition under varying temperature conditions.

The coefficient c_e related to equilibrium pressure and the initial specific surface area of decomposition (A_{s0}) can be derived using the Kim–Bishnoi kinetic decomposition equations

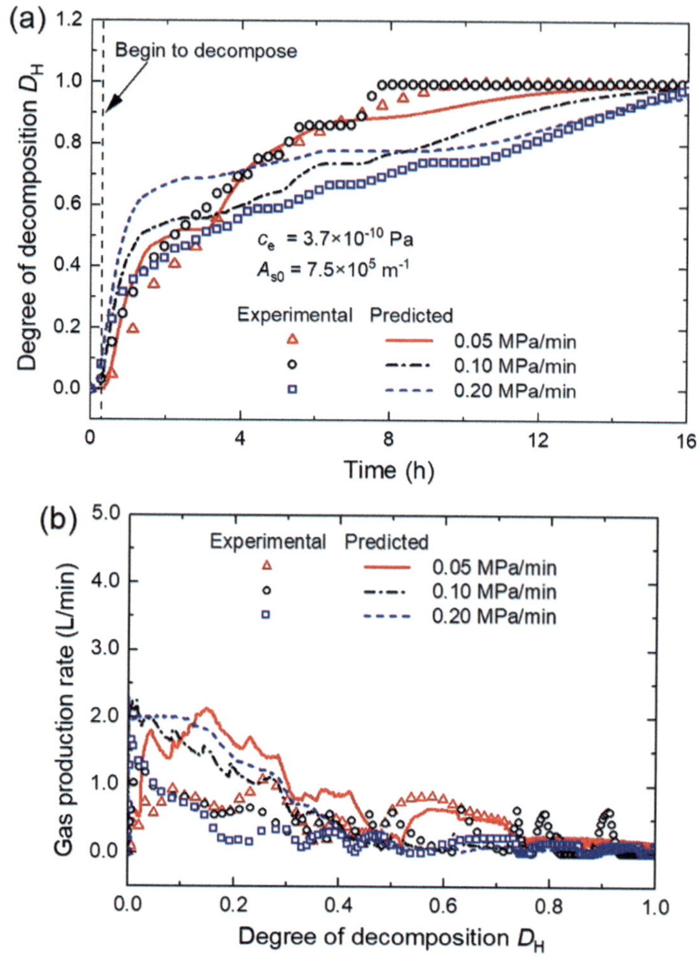

Fig. 2.18 Experimental and predicted variations in CO_2 hydrate decomposition at different depressurization rates: **a** degree of decomposition versus time and **b** gas production rate versus degree of decomposition

for hydrates—Eqs. (2.13)–(2.16)—and triaxial or normal gravity modeling test data. These parameters can then be used to determine the kinetic reaction rate of hydrates, thus providing a crucial theoretical foundation and the values of several decomposition-related parameters for the constitutive model of HBSs.

2.4 Biochemical Degradation

2.4.1 Unified Equations

The skeletons of skeleton-degradable soils comprise both active and inert components. Biological and chemical processes in these soils involve the generation and consumption of solid-, liquid-, and gas-phase substances. Each biological reaction group has a limiting substrate, which directly serves as the reactant for microbial action, providing essential nutrients (such as carbon and nitrogen) for microbial growth, or is acted upon by microbial enzymes to expedite reactions with other reactants. Hence, this limiting substrate dictates the rate of the reaction process within that specific reaction group. For chemical reactions not influenced by microorganisms, the limiting substrate is defined as the reactant that is most easily consumed (Voet and Voet 2011). The following paragraph describes the mass changes of various substances during biochemical reactions.

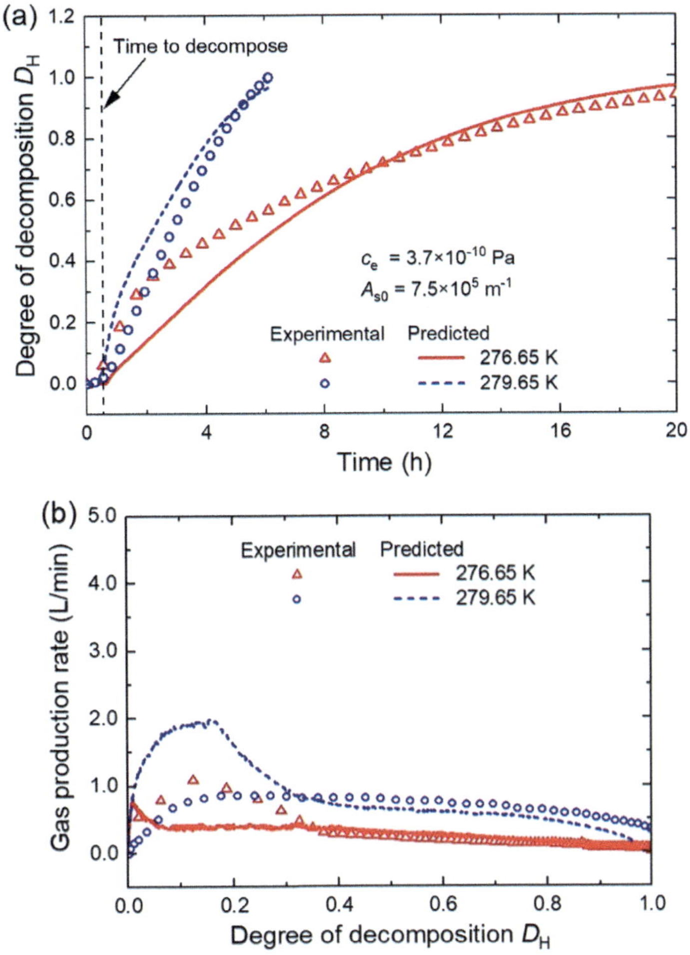

Fig. 2.19 Experimental and predicted variations in CO_2 hydrate decomposition at different temperatures: **a** degree of decomposition versus time and **b** gas production rate versus degree of decomposition

First, biochemical reaction processes must adhere to the principle of mass conservation, typically represented by chemical equilibrium equations. Notably, these equations capture the proportional relationships among the mass changes of reactants and products involved in biochemical reactions. If m biochemical reaction processes involving n chemical substances are assumed to occur in any soil, the m biochemical reaction processes can be uniformly represented using mass conservation equations involving these n chemical substances, as illustrated in Fig. 2.20. Considering the limiting substrate as a reference, the conversion coefficients of individual substances in the reaction process, i.e., the ratio of the mass change of each substance to the mass change of the limiting substrate, can be derived as follows:

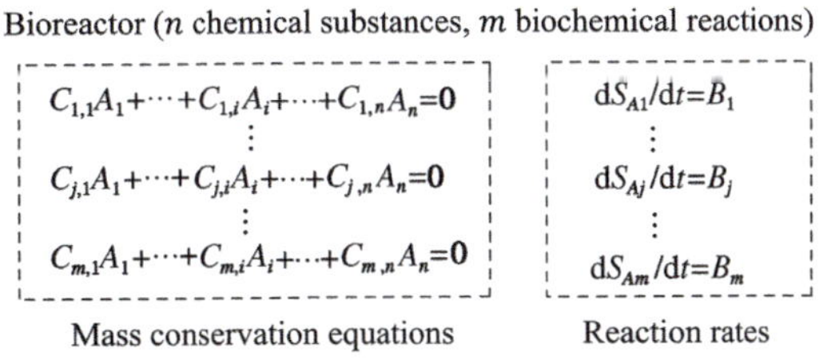

Fig. 2.20 Framework of biochemical reactions (reprinted from Chen 2014, with permission from the Chinese Journal of Geotechnical Engineering)

$$C = C_c \cdot C_m \tag{2.23}$$

$$C_c = \begin{bmatrix} c_{1,1} & \cdots & c_{1,i} & c_{1,n} \\ \vdots & \cdots & \cdots & \vdots \\ c_{j,1} & \cdots & c_{j,i} & c_{j,n} \\ c_{m,1} & \cdots & c_{m,i} & c_{m,n} \end{bmatrix} \tag{2.24}$$

$$C_m = \begin{bmatrix} m_{1,1} & \cdots & m_{1,i} & m_{1,n} \\ \vdots & \cdots & \cdots & \vdots \\ m_{j,1} & \cdots & m_{j,i} & m_{j,n} \\ m_{m,1} & \cdots & m_{m,i} & m_{m,n} \end{bmatrix} \tag{2.25}$$

where C_c and C_m represent the stoichiometric coefficient matrices of the mass conservation equations and the molar mass ratio coefficient matrices, respectively. Furthermore, $c_{j,i}$ denotes the stoichiometric coefficient of the ith chemical species in the jth chemical reaction (mol/mol). Notably, chemical species that do not participate in the reaction have a stoichiometric coefficient of zero, while the limiting substrate has a stoichiometric coefficient of -1 in each reaction. Moreover, $m_{j,i}$ denotes the molar mass ratio of the ith chemical substance to the limiting substrate in the jth chemical reaction (kg/mol/(kg/mol).

Furthermore, the kinetics of biochemical reactions follow specific dynamic laws, and the content change rate of each substance can be expressed as

$$\frac{dS}{dt} = B \tag{2.26}$$

$$S = \{ S_1 \quad \cdots \quad S_i \quad S_n \}^T \tag{2.27}$$

$$B = \{ B_1 \quad \cdots \quad B_i \quad B_n \}^T \tag{2.28}$$

where S denotes the concentration matrix of chemical substances, with S_i representing the concentration of the ith chemical substance (kg/m^3), indicating the mass of the substance per unit volume of soil. Considering that changes in solute concentrations in the liquid phase of soil typically correspond to the mass changes of the solute per unit volume of the solvent, the concentration of solute substances within the liquid phase is defined as the mass of the solute per unit volume of pore water. Furthermore, B represents the matrix of reaction rate coefficients, with B_i representing the reaction rate coefficient of the ith chemical substance (kg/m^3/s), indicating the concentration change rate for the ith chemical substance. The value of B_i can be obtained by summing the reaction rate coefficients of the given chemical substance in all biochemical reactions:

$$B_i = \beta_i \sum_{j=1}^{m} B_{j,i} (i = 1, 2, \ldots, n) \tag{2.29}$$

where $B_{j,i}$ denotes the reaction rate coefficient of the ith chemical substance in the jth biochemical reaction, and β_i represents the conversion coefficient of chemical substance concentration units. For solute substances in the liquid phase, β_i equals $1/\theta$, while for other substances, its value is one. Here, θ represents the volumetric water content.

Considering the limiting substrate as a reference and ensuring mass conservation in each reaction, $B_{j,i}$ can be expressed as the product of the conversion ratio coefficient, $C_{j,i}$ ($C_{j,i} = c_{j,i} m_{j,i}$), of the substance and the reaction rate, R_j, of the limiting substrate:

$$B_{j,i} = C_{j,i} R_j (i = 1, 2, \ldots, n) \tag{2.30}$$

Here, the value of R_j, representing the reaction rate of the limiting substrate in the jth chemical reaction, is predominantly influenced by two factors: ① the reaction rate when the chemical reaction occurs independently and ② the influence of other simultaneous chemical reactions on the current reaction process. This analysis exclusively focuses on the competitive influence of other chemical reactions that use the same limiting substrate as the current chemical reaction. The specific expression for R_j is as follows:

$$R_j = \phi_j \varphi_j \tag{2.31}$$

$$\varphi_j = f(S, T, \text{ pH}, w, \ldots) \tag{2.32}$$

where φ_j denotes the reaction rate of the limiting substrate when the jth chemical reaction occurs independently. Furthermore, φ_j is related to factors such as substance concentration (S), temperature (T), pH value, and water content (w). Moreover, ϕ_j denotes the competitive allocation factor for biochemical reactions utilizing the same limiting substrate. In simple terms, this factor can be expressed as the ratio of φ_j to the sum of the reaction rates of the limiting substrate in all independently occurring biochemical reactions utilizing the same limiting substrate (Haarstrick et al. 2001).

The overall mass changes of the solid, liquid, and gas phases induced by biochemical reactions can be expressed as

$$f_{\text{sd}} = \sum_{i=1}^{n_1} \frac{dS_{\text{s}i}}{dt} \tag{2.33}$$

$$f_{\text{ld}} = \theta \sum_{j=1}^{n_2 - 1} \frac{dS_{\text{l}j}}{dt} + \frac{dS_{\text{w}}}{dt} \tag{2.34}$$

$$f_{\text{ad}} = \sum_{k=1}^{n_3} \frac{dS_{\text{g}k}}{dt} \tag{2.35}$$

where f_{sd}, f_{ld}, and f_{ad} denote the mass change rates of solid-, liquid-, and gas-phase substances per unit volume of soil at time t, respectively. Furthermore, $S_{\text{s}i}$, $S_{\text{l}j}$, and $S_{\text{g}k}$ represent the concentrations (kg/m^3) of the ith solid phase substance (out of n_1 types), jth liquid phase substance (out of n_2 types, with one being pore water and the remaining $n_2 - 1$ being liquid phase solutes), and kth gas-phase substance (out of n_3 types) participating in biochemical reactions.

Under natural conditions, soil structure degradation can negatively affect practical engineering projects. However, these adverse effects can be mitigated through targeted human interventions aimed at modifying the degradation processes and modes of soil, thereby enhancing engineering efficiency. For instance, in MSW landfills, anaerobic degradation predominates under natural conditions, leading to slow rates of degradation, high concentrations of leachate pollutants, generation of abundant greenhouse gases such as CH_4, and high concentrations of odorous gases such as hydrogen sulfide.

Conventionally, two primary control measures are adopted to improve degradation efficiency and mitigate environmental impacts: (1) Leachate recirculation involves the reintroduction of leachates into waste to increase their microorganism concentration and alkalinity. This approach alleviates acid inhibition, enhances anaerobic degradation rates, increases CH_4 production, and improves power generation efficiency. However, it does not alter the degradation mode, leaving the types of degradation products unchanged. (2) Conversely, air injection involves the introduction of air into a landfill to shift the degradation mode from anaerobic to aerobic, thus significantly increasing degradation rates and reducing both liquid and gas pollution loads. Interestingly, aerobic degradation generates CO_2 instead of conventional degradation products, thus reducing greenhouse gas emissions and eliminating odor issues. Additionally, the exploitation of NGHs can be managed by regulating pressure and temperature conditions to ensure safe and efficient extraction. Different engineering projects can adopt diverse degradation control measures tailored to their specific requirements, exploiting all known mass transformation relationships during degradation to achieve green, efficient, and sustainable development.

2.4.2 Anaerobic Degradation of Municipal Solid Waste

2.4.2.1 Anaerobic Degradation Equations

The primary components of degradable organic matter in MSW comprise total sugars, proteins, lipids, and cellulose (including hemicellulose). Among these, the total sugars, proteins, and lipids predominantly originate from food waste,

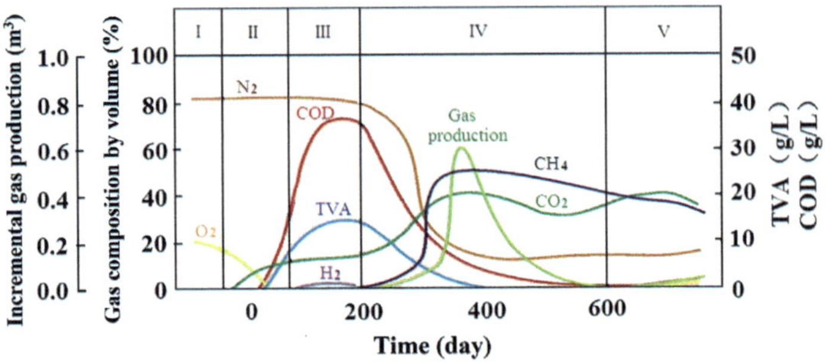

Fig. 2.21 Anaerobic degradation of MSW (ITRC 2006)

while cellulose is primarily sourced from paper waste, yard waste, wood, and textile waste. Notably, food wastes, including fruits and vegetables, contain numerous intact cells. Cellular water is sometimes trapped within these cells by their cell walls. This water differs from the free water found in the voids of MSW and is referred to as intracellular water.

The process of MSW degradation occurs in five distinct stages, as illustrated in Fig. 2.21.

Stage I, representing the initial sealing stage of landfilling, involves the aerobic biochemical degradation of easily degradable waste components using oxygen from the landfill body, producing CO_2 and water and releasing heat.

Stage II, denoting the transition phase, commences as ongoing aerobic reactions lead to oxygen depletion within the landfill body owing to the sealing effect of the cover system, which prevents the entry of atmospheric oxygen. This leads to a significant decline in aerobic microbial activity. Consequently, anaerobic microorganisms begin to dominate, albeit in smaller numbers, causing the landfill environment to gradually transition from aerobic to anaerobic conditions.

Stage III, signifying the acidification phase, is characterized by the rapid proliferation of anaerobic microorganisms. During this stage, degradable waste components undergo hydrolysis reactions under anaerobic microbial action, producing volatile fatty acids (VFAs). This leads to a reduction in the pH of the leachate, along with a significant increase in its chemical oxygen demand (COD), reaching its peak. Furthermore, chemical solutes, such as metal ions, undergo precipitation.

Stage IV, representing the CH_4 production phase, initiates later in the acidification phase when methanogens proliferate using VFAs as substrates. CH_4 and CO_2 predominantly constitute the LFG generated during this phase. Furthermore, the rate of volatile acid generation is significantly lower than its rate of consumption, leading to a gradual decline in VFA concentrations and COD values within the leachate, along with a subsequent rise in leachate pH. In particular, this stage harnesses significant amounts of valuable LFGs.

Stage V, denoting the stable mineralization phase, signifies the stabilization of the landfill body following its biochemical degradation in the preceding stages. At this stage, the landfill body stabilizes as most degradable waste components are consumed.

Typically, Stages I and II have short durations, often lasting a few weeks, and are usually integrated into the aerobic stage. Conversely, Stages III to V constitute the anaerobic stage, dominating biochemical reactions in landfills (Barlaz et al. 1989). The process of anaerobic MSW degradation occurs in two distinct stages: hydrolysis and methanogenesis (illustrated in Fig. 2.22). During the hydrolysis stage, the four major types of solid phase materials—total sugars, proteins, lipids, and cellulose—are hydrolyzed by microorganisms into species such as soluble VFAs, hydrogen, and CO_2, leading to a loss of solid phase mass. In particular, the hydrolysis of cellulose, a primary component of cell walls, ruptures cell walls and releases intracellular water, producing leachates during degradation. Meanwhile, during the methanogenesis stage, intermediate hydrolysis products

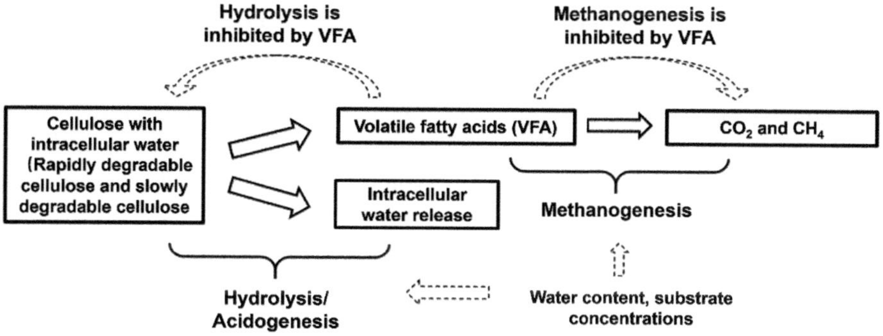

Fig. 2.22 Diagram of the two-stage anaerobic degradation model of MSW

are subsequently converted into CH_4 and CO_2 by methanogens, producing gas during degradation. Notably, methanogens, similar to other organisms, possess limited lifecycles and undergo programmed decay.

Equations modeling the anaerobic MSW degradation process are as follows: Eq. (2.36) represents the release of intracellular water due to cell wall rupture, where λ_w denotes the mass ratio of released intracellular water to degraded cellulose. Furthermore, $C_6H_{10}O_5$ in Eqs. (2.37) and (2.41) represents total sugars.

$$C_6H_{10}O_5 \cdot \lambda_w H_2O + 5H_2O \rightarrow CH_3COOH + 8H_2 + 4CO_2 + \lambda_w H_2O \tag{2.36}$$

$$C_6H_{10}O_5 + 5H_2O \rightarrow CH_3COOH + 8H_2 + 4CO_2 \tag{2.37}$$

$$C_{46}H_{77}O_{17}N_{12}S + 59.26H_2O \rightarrow 7.88CH_3COOH + 30CO_2 + 63H_2 + 12NH_3 + H_2S \tag{2.38}$$

$$C_{55}H_{104}O_6 + 78H_2O \rightarrow 13CH_3COOH + 29CO_2 + 104H_2 \tag{2.39}$$

$$CH_3COOH + 7.43H_2 + 0.93CO_2 + 0.029NH_3 \rightarrow 2.79CH_4 + 3.8H_2O + 0.029C_5H_7NO_2 \tag{2.40}$$

$$C_5H_7NO_2 + 2H_2O \rightarrow 0.83C_6H_{10}O_5 + NH_3 \tag{2.41}$$

The above reactions involve 12 substances, namely cellulose ($C_6H_{10}O_5$), total sugars ($C_6H_{10}O_5$), proteins ($C_{46}H_{77}O_{17}N_{12}S$), lipids ($C_{55}H_{104}O_6$), water ($H_2O$), VFAs (represented by CH_3COOH), ammonia (NH_3), methanogens ($C_5H_7NO_2$), CO_2, hydrogen sulfide (H_2S), CH_4, and hydrogen gas (H_2). These substances are denoted as $i = 1, 2, \ldots, 12$. Furthermore, the reactions encompass four hydrolysis reactions for different degradable substances (Eqs. 2.36–2.39), a methanogenesis reaction for intermediate products (Eq. 2.40), and a decay reaction for methanogens (Eq. 2.41). The corresponding limiting substrates include cellulose, total sugars, proteins, lipds, VFAs, and methanogens.

The stoichiometric coefficient matrix ($\boldsymbol{C}_{c}$) and matrix of molar mass ratios (C_m) for Eqs. (2.36)–(2.41) are presented below. Further, the chemical reaction matrix ($\boldsymbol{C}$) is obtained by the element-wise multiplication of $\boldsymbol{C}_{c}$ and $\boldsymbol{C}_{m}$.

$$\boldsymbol{C}_{c} = \begin{pmatrix} -1 & 0 & 0 & 0 & \lambda_w - 5 & 1 & 0 & 0 & 4 & 0 & 0 & 8 \\ 0 & -1 & 0 & 0 & -5 & 1 & 0 & 0 & 4 & 0 & 0 & 8 \\ 0 & 0 & -1 & 0 & -59.26 & 7.88 & 12 & 0 & 30 & 1 & 0 & 63 \\ 0 & 0 & 0 & -1 & -78 & 13 & 0 & 0 & 29 & 0 & 0 & 104 \\ 0 & 0 & 0 & 0 & 3.8 & -1 & -0.029 & 0.029 & -0.93 & 0 & 2.79 & -7.43 \\ 0 & 0.83 & 0 & 0 & -2 & 0 & 1 & -1 & 0 & 0 & 0 & 0 \end{pmatrix} \quad (2.42)$$

$$\boldsymbol{C}_{m} = \begin{pmatrix} 1 & 1 & 6.80 & 5.31 & 0.11 & 0.37 & 0.10 & 0.70 & 0.27 & 0.21 & 0.099 & 0.012 \\ 1 & 1 & 6.80 & 5.31 & 0.11 & 0.37 & 0.10 & 0.70 & 0.27 & 0.21 & 0.099 & 0.012 \\ 0.15 & 0.15 & 1 & 0.78 & 0.016 & 0.054 & 0.015 & 0.10 & 0.040 & 0.031 & 0.015 & 0.002 \\ 0.19 & 0.19 & 1.28 & 1 & 0.021 & 0.070 & 0.020 & 0.13 & 0.051 & 0.040 & 0.019 & 0.002 \\ 2.7 & 2.7 & 18.35 & 14.33 & 0.30 & 1 & 0.28 & 1.88 & 0.73 & 0.57 & 0.27 & 0.033 \\ 1.43 & 1.43 & 9.74 & 7.61 & 0.16 & 0.53 & 0.15 & 1 & 0.39 & 0.30 & 0.14 & 0.018 \end{pmatrix} \quad (2.43)$$

$$\boldsymbol{C} = \begin{pmatrix} -1 & 0 & 0 & 0 & 0.11\lambda_w - 0.56 & 0.37 & 0 & 0 & 1.08 & 0 & 0 & 0.10 \\ 0 & -1 & 0 & 0 & -0.56 & 0.37 & 0 & 0 & 1.08 & 0 & 0 & 0.10 \\ 0 & 0 & -1 & 0 & -0.97 & 0.43 & 0.19 & 0 & 1.20 & 0.03 & 0 & 0.11 \\ 0 & 0 & 0 & -1 & -1.63 & 0.91 & 0 & 0 & 1.48 & 0 & 0 & 0.24 \\ 0 & 0 & 0 & 0 & 1.14 & -1 & -0.008 & 0.055 & -0.68 & 0 & 0.74 & -0.25 \\ 0 & 1.19 & 0 & 0 & -0.32 & 0 & 0.15 & -1 & 0 & 0 & 0 & 0 \end{pmatrix} \quad (2.44)$$

As each reaction involves different limiting substrates, the competitive allocation factor (ϕ_j) is 1.0 or all reactions describing the degradation process (Eqs. 2.36–2.41). Meanwhile, the reaction rates for the four hydrolysis reactions (Eqs. 2.36–2.39) can be derived from first-order kinetics equations (Vavilin et al. 2003), as follows:

$$R_j = -k_{Hj} f_w f_{Ih} S_j \quad (2.45)$$

$$f_w(\theta) = \begin{cases} 0 & (\theta < \theta_{min}) \\ \frac{\theta - \theta_{min}}{\theta_{max} - \theta_{min}} & (\theta_{min} \le \theta \le \theta_{max}) \\ 1 & (\theta > \theta_{max}) \end{cases} \quad (2.46)$$

$$f_{Ih}(S_6) = \frac{1}{1 + (S_6/K_h)^{n_h}} \quad (2.47)$$

where R_j denotes the reaction rate of the limiting substrate in each hydrolysis reaction (kg/m^3/d), with j = 1, 2, 3, and 4; t represents time (d); k_{Hj} indicates the hydrolysis rate constant for limiting substrate j (d^{-1}); f_w signifies the water content influence function; f_{Ih} denotes the inhibition function of VFAs on hydrolysis; θ represents the volumetric water content of MSW, indicating volume of water per unit volume of MSW; θ_{min} denotes the minimum volumetric water content required for hydrolysis; θ_{max} represents the upper water content limit affecting hydrolysis, beyond which moisture no longer limits hydrolysis; f_{Ih} denotes an improved form of the noncompetitive inhibition function, with K_h corresponding to the hydrolysis inhibition constant (kg/m^3); and n_h represents the hydrolysis inhibition index.

Further, the reaction rate of the methanogenesis reaction using VFA substrates for proliferation (Eq. 2.40) can be described using the Monod equation (White et al. 2004), as follows:

$$R_5 = \theta \frac{1}{Y} \frac{k_{mmax} S_8}{K_s + c} f_{Im} S_6 \quad (2.48)$$

$$f_{Im}(S_6) = \frac{1}{1 + (S_6/K_m)^{n_m}} \quad (2.49)$$

where k_{mmax} denotes the maximum growth rate of methanogens (d^{-1}); K_s represents the half-saturation constant for methanogen growth (kg/m^3); and f_{Im} indicates the inhibition function of VFAs on methanogen proliferation, reflecting the impact of VFAs (pH) on methanation. Notably, the form of f_{Im} is similar to that of the inhibition function of VFAs on hydrolysis. Additionally, K_m represents the methanation inhibition constant (kg/m^3), while n_m denotes the methanation inhibition index.

Finally, the reaction rate of the methanogen decay process (Eq. 2.41) can be derived from a first-order kinetic equation, as follows:

$$R_6 = \theta k_d S_8 \quad (2.50)$$

where k_d denotes the decay rate of methanogens (d^{-1}).

Combining the above equations yields the governing equation for anaerobic MSW degradation:

$$\frac{dS_i}{dt} = \beta_i \sum_{j=1}^{6} (C_{j,i} R_j) \quad (i = 1, 2, \ldots, 12) \quad (2.51)$$

Notably, this equation describes the generation and consumption of substances within the reactor. Its control variables include the concentrations of the 12 chemical substances involved in biochemical degradation. Interestingly, compared to existing biochemical degradation models, this model more comprehensively classifies MSW according to its composition, considering the distinct degradation behaviors of total sugars and cellulose, intracellular water release during waste degradation, and simultaneous influence of VFAs (pH) on both hydrolysis and methanogenesis.

The solution to Eq. (2.51) yields the following mass change rates:

(1) Mass change rates for solid, liquid, and gas phases per unit volume of MSW:

$$f_{sd} = \sum_{i=1}^{4} \frac{dS_i}{dt} \quad (2.52)$$

$$f_{wd} = \frac{dS_5}{dt} \quad (2.53)$$

$$f_{gd} = \sum_{i=9}^{12} \frac{dS_i}{dt} \quad (2.54)$$

(2) Mass change rate of pollutant VFAs per unit volume of the liquid phase:

$$f_{ad} = dS_6/dt \quad (2.55)$$

(3) Mass change rates of CO_2, hydrogen sulfide, and CH_4 per unit volume of MSW:

$$f_{cd} = dS_9/dt \quad (2.56)$$

$$f_{hd} = dS_{10}/dt \quad (2.57)$$

$$f_{md} = dS_{11}/dt \quad (2.58)$$

According to McDougall (2007) and Meima et al. (2008), certain parameters in the governing equation (Eq. 2.51) are constants with the following values: θ_{min} = 12%, θ_{max} = 45%, K_h = 10 kg/m^3, n_h = 4, K_m = 5 kg/m^3, n_m = 3, and Y= 0.08. Meanwhile, the values of other parameters in the governing equation must be experimentally determined under real degradation conditions. Among these parameters, the water hydrolysis rate constant (k_{Hi}) is significantly influenced by the dimensions and morphologies of degradable materials. Specifically, fine particles typically exhibit rapid hydrolysis. Furthermore, the maximum reproduction rate of methanogens (k_{mmax}) and the decay rate (k_d) are impacted by the composition of the MSW sample, which may vary owing to differences in microbial populations. Overall, these parameters can be determined through CH_4 production rate experiments (Li et al. 2005).

2.4.2.2 Anaerobic Degradation Behavior

Tables 2.2 and 2.3 detail the physical compositions and water contents of typical MSW types found in China and America, respectively. In particular, Chinese MSW is characterized by a high food waste content (HFWC), while American MSW comprises low food waste content (LFWC). Previous studies conducted indoor column degradation experiments on Chinese and American MSWs (Hossain et al. 2003; He et al. 2007), denoting them as Column 1 and Column 2, respectively. Notably, both columns were isolated degradation units, not subjected to any external

Table 2.2 Compositions of Chinese and American MSWs (%) (reprinted from Chen 2014, with permission from the Chinese Journal of Geotechnical Engineering)

Composition	Food waste	Paper	Plastic	Textiles	Others
China	53.0	22.0	16.0	3.0	6.0
America	5.8	60.0	15.6	5.0	13.6

Table 2.3 Water content of Chinese and American MSWs (%) (reprinted from Chen 2014, with permission from the Chinese Journal of Geotechnical Engineering)

Composition	Food waste	Paper	Plastic	Textiles	Others
China	85	20	5	15	20
America	70	5	2	10	8

Table 2.4 Initial degrading conditions of the two MSW columns (kg/m^3) (reprinted from Chen 2014, with permission from the Chinese Journal of Geotechnical Engineering)

Initial conditions	S_1	S_2	S_3	S_4	S_5	S_8
Column 1	30	65	8	8	0.30	0.01
Column 2	90	8	5	5	0.36	0.01

loads and lacking matter exchange with the external environment. Furthermore, mass changes for each substance during biochemical degradation in Columns 1 and 2 were analyzed using Eq. (2.33). The initial degradation conditions were as listed in Table 2.4, with other relevant parameter values set as follows: $k_{H1} = 0.03$ d^{-1}, $k_{H2} = 0.03$ d^{-1}, $k_{H3} = 0.003$ d^{-1}, $K_H = 0.003$ d^{-1}, $k_{mmax} = 0.05$ d^{-1}, $k_d = 0.005$ d^{-1}, and $\lambda_w = 9.9/0.93$ (release coefficients of intracellular water for Chinese and American MSWs, respectively). Additionally, the initial density of Column 1 was 8 kN/m^3, while that of Column 2 was 6 kN/m^3.

The experimental and simulated results for solid mass loss, intracellular water production, VFA concentration in the liquid phase, pH value, CH_4 production, and H_2S production are displayed in Figs. 2.23, 2.24, 2.25, 2.26, 2.27 and 2.28. Notably, both experimental and simulated results indicate that during the initial stages of degradation, easily degradable substances undergo rapid hydrolysis, generating substantial amounts of VFAs. However, the methanogenic environment is yet to be established at this stage, hindering the prompt conversion of VFAs. Consequently, VFAs accumulate, inhibiting reaction progress, known as acidification inhibition.

Chinese MSW typically contains significantly higher proportions of food waste compared to MSWs from western developed countries, thus containing substantial amounts of easily degradable substances. Consequently, acidification in

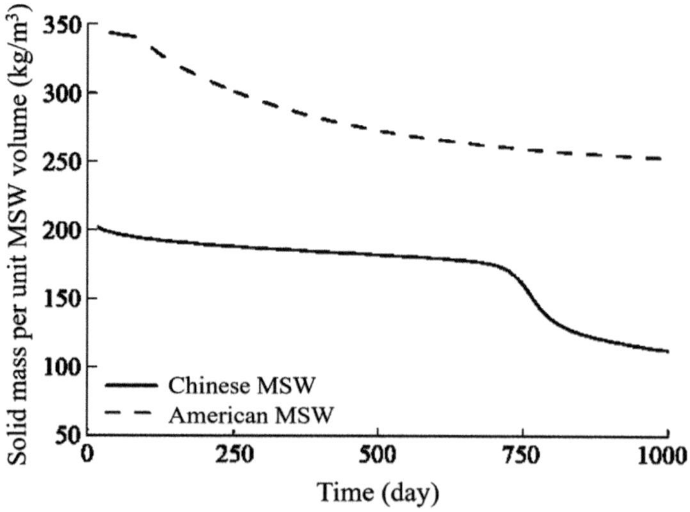

Fig. 2.23 Variations in solid mass per unit MSW volume based on typical Chinese and American MSWs (reprinted from Chen 2014, with permission from the Chinese Journal of Geotechnical Engineering)

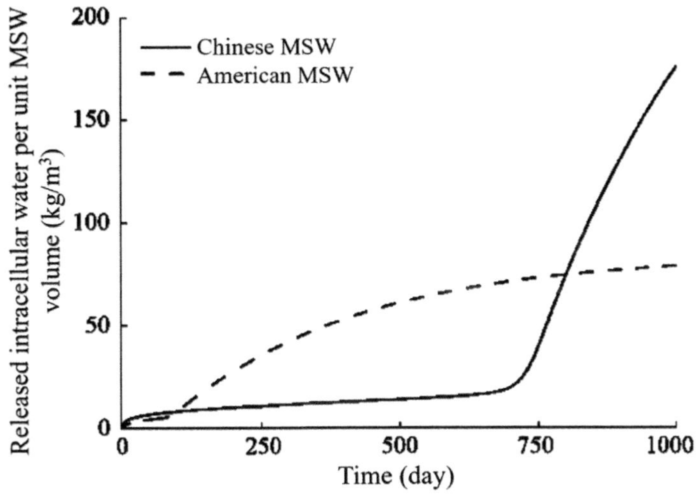

Fig. 2.24 Variaitons in released intracellular water per unit MSW volume based on typical Chinese and American MSWs (reprinted from Chen 2014, with permission from the Chinese Journal of Geotechnical Engineering)

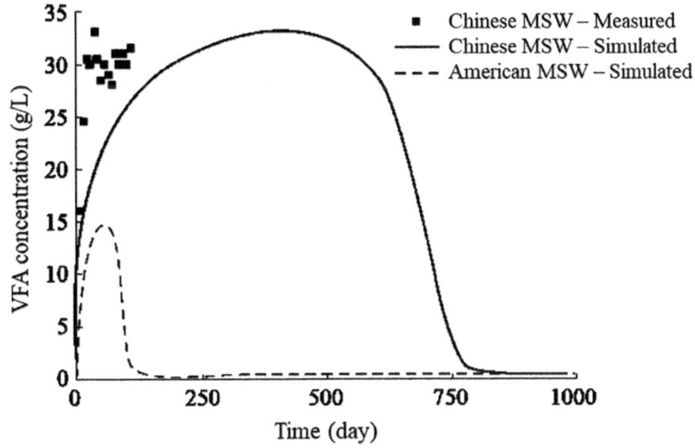

Fig. 2.25 Variations in VFA concentration in the leachate based on typical Chinese and American MSWs (reprinted from Chen 2014, with permission from the Chinese Journal of Geotechnical Engineering)

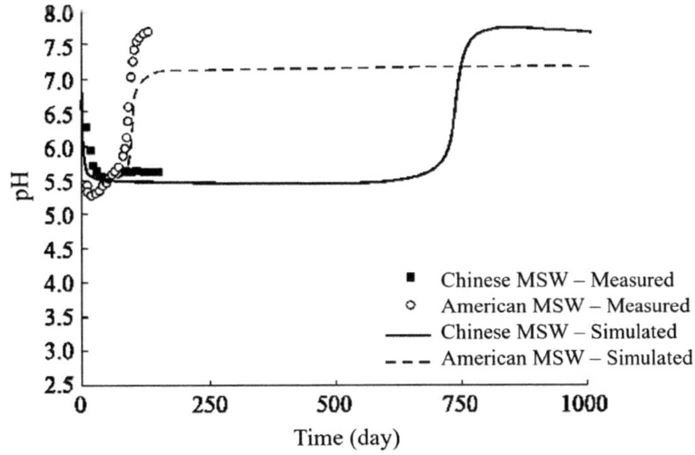

Fig. 2.26 Variations in pH of the leachate based on typical Chinese and American MSWs (reprinted from Chen 2014, with permission from the Chinese Journal of Geotechnical Engineering)

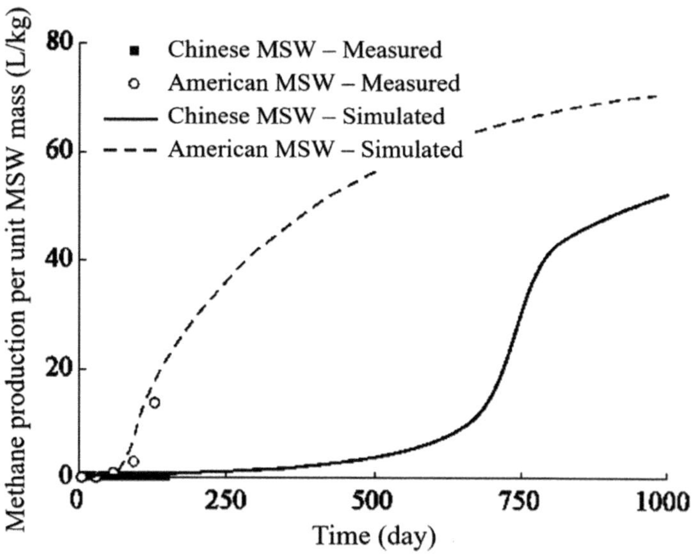

Fig. 2.27 Variations in CH_4 production per unit MSW mass based on typical Chinese and American MSWs (reprinted from Chen 2014, with permission from the Chinese Journal of Geotechnical Engineering)

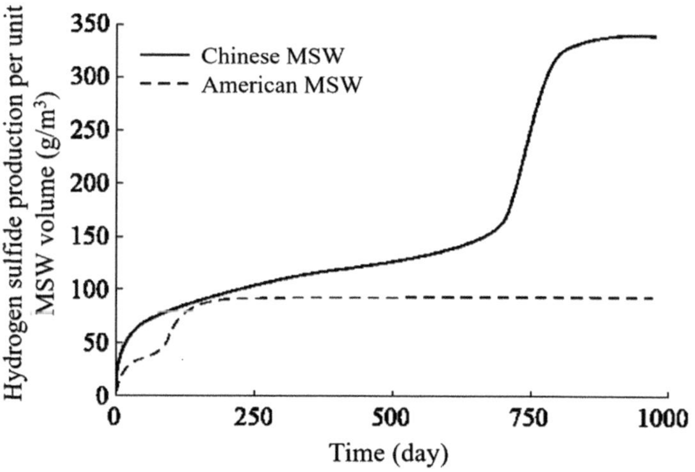

Fig. 2.28 Variations in hydrogen sulfide production per unit MSW volume based on typical Chinese and American MSWs (reprinted from Chen 2014, with permission from the Chinese Journal of Geotechnical Engineering)

Chinese MSW appears more pronounced and thus extends the hydrolysis acidification stage of the degradation process, preventing its smooth transition into methanogenesis. This is evidenced by the near stagnation of degradation, high VFA concentration in the liquid phase (accompanied by a low pH), and almost negligible gas production.

The typical duration of acidification inhibition for Chinese MSW can extend up to 600 days, whereas for typical American MSW, this duration is less than 100 days, with significantly less inhibition. This model effectively captures the degradation behavior of MSW with low food waste content while also reflecting the severe acidification inhibition phenomenon observed during the degradation of Chinese MSW featuring high food waste contents.

The above-described examples of degradation of Chinese MSW and U.S. MSW indicate that significant acidification is the main obstacle to the degradation of waste with a high content of food waste. Such acidification occurs due to a low concentration of methanogens, as this prevents produced VFAs from being consumed promptly, resulting in acidification inhibition. The injection of high concentrations of methanogens and highly alkaline leachate into landfills is an effective measure for preventing and alleviating acidification inhibition, thereby accelerating waste degradation. For example, the high proportion of food waste in Tianziling Landfill in Hangzhou led to a significant acidification inhibition phenomenon in this landfill. Therefore, a landfill unit (1×10^5 m^3) was selected, and a biologically pretreated leachate was injected into the landfill body to increase its concentration of methanogens and the alkalinity of its leachate, thereby accelerating landfill degradation (Tongji University 2008; Liu 2016). Furthermore, adopting an anaerobic degradation model, simulations were conducted under both natural and controlled conditions. Specifically, initial conditions for degradation were as follows: S_1 = 61.6 kg/m^3, S_2 = 9 kg/m^3, S_3 = 7.2 kg/m^3, S_4 = 60 kg/m^3, S_5 = 0.40 kg/m^3, S_8 = 0.5/0.01 kg/m^3 (0.5 and 0.01 represent CH_4 concentrations under controlled and natural conditions, respectively). Meanwhile, kinetic parameters were determined based on waste sampling tests: K_{H1} = 0.02 d^{-1}, K_{H2} = 0.02 d^{-1}, K_{H3} = 0.009 d^{-1}, K_{H4} = 0.009 d^{-1}, k_{mmax} = 0.1 d^{-1}, and k_d = 0.01 d^{-1}. Parameters not listed here were assumed to have the same values as previously described.

The obtained experimental and simulated results are illustrated in Figs. 2.29, 2.30, 2.31 and 2.32. There is good agreement between the simulated concentrations of VFAs in the leachate

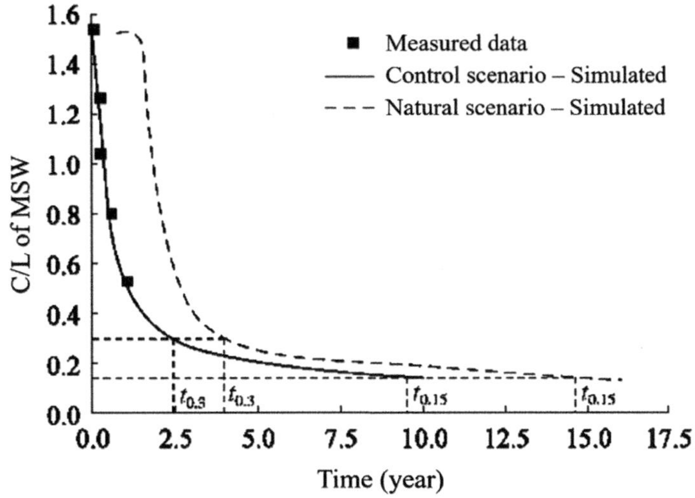

Fig. 2.29 Ratio of cellulose to lignin (C/L) in MSW during degradation (reprinted from Chen 2014, with permission from the Chinese Journal of Geotechnical Engineering)

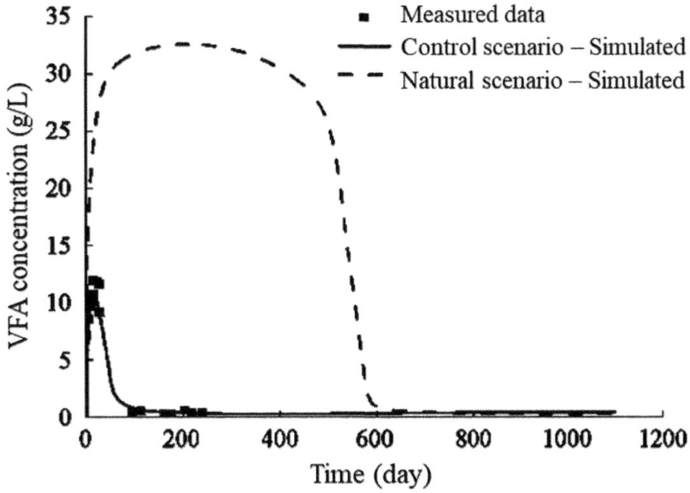

Fig. 2.30 VFA concentration in the leachate (reprinted from Chen 2014, with permission from the Chinese Journal of Geotechnical Engineering)

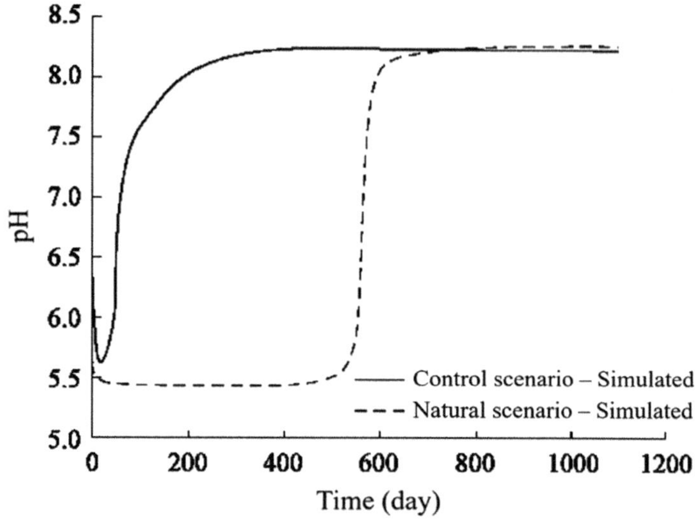

Fig. 2.31 pH of the leachate (reprinted from Chen 2014, with permission from the Chinese Journal of Geotechnical Engineering)

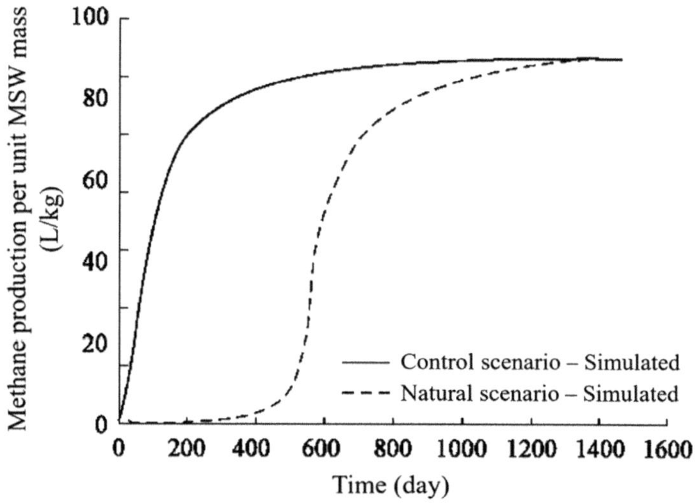

Fig. 2.32 CH_4 production per unit MSW mass (reprinted from Chen 2014, with permission from the Chinese Journal of Geotechnical Engineering)

after the control treatment and the measured results after the control treatment. Simulation predictions indicated that the time for basic stabilization ($t_{0.3}$) was 900 days, while for complete stabilization ($t_{0.15}$), it was 9.4 years. Gas production essentially stopped around 920 days, aligning with the predicted time for basic stabilization and thus validating the feasibility of the degradation assessment and prediction method. Meanwhile, the simulations for natural degradation indicated the presence of a severe acidification inhibition stage lasting approximately 500 days, with high pollutants levels in the leachate. The time for basic stabilization ($t_{0.3}$) in this case was 3.9 years, while that for complete stabilization ($t_{0.15}$) was 15 years, significantly longer than the complete stabilization duration of 9.4 years under controlled conditions. This finding highlights the effectiveness of leachate recirculation in alleviating acidification inhibition and expediting waste degradation.

2.4.3 Aerobic Degradation of Municipal Solid Waste

2.4.3.1 Aerobic Degradation Equations

Under anaerobic degradation conditions, the stabilization duration of MSW can extend up to decades. However, introducing air into a landfill allows aerobic degradation to occur, which can shorten the degradation time of organic matter by 5–30 times, thereby significantly reducing the liquid and gas pollution load. During aerobic MSW degradation, solid phase organic compounds rich in carbon and nitrogen undergo aerobic hydrolysis facilitated by aerobic microorganisms. This process converts these compounds into small, water-soluble molecules and gases such as CO_2 and N_2. Xiao et al. (2021) developed an aerobic degradation model based on the principles of electron charge and mass conservation. This model assumed organic hydrolysis as the rate-limiting step, with various organic compounds functioning as electron donors and O_2 acting as the electron acceptor. Meanwhile, rapid degradation cellulose (RDC: $C_6H_{10}O_5$), slow degradation cellulose (SDC: $C_6H_{10}O_5$), total sugars ($C_6H_{10}O_5$), lipids ($C_{55}H_{104}O_6$), proteins ($C_{46}H_{77}O_{17}N_{12}S$), ammonia ($NH_3$), and microorganisms (specifically aerobic bacteria, nitrifying bacteria, and denitrifying bacteria) were selected as the limiting substrates.

(1) Aerobic hydrolysis of organic compounds

RDC:

$$C_6H_{10}O_5 + 4O_2 + \frac{2}{5}NH_3 = \frac{2}{5}C_5H_7O_2N + 4CO_2 + \frac{21}{5}H_2O + \Delta Q_1 \tag{2.59}$$

SDC:

$$C_6H_{10}O_5 + 4O_2 + \frac{2}{5}NH_3 = \frac{2}{5}C_5H_7O_2N + 4CO_2 + \frac{21}{5}H_2O + \Delta Q_2 \tag{2.60}$$

Total sugars:

$$C_6H_{10}O_5 + 4O_2 + \frac{2}{5}NH_3 = \frac{2}{5}C_5H_7O_2N + 4CO_2 + \frac{21}{5}H_2O + \Delta Q_3 \tag{2.61}$$

Lipids:

$$C_{55}H_{104}O_6 + 58O_2 + 4NH_3 = 4C_5H_7O_2N + 35CO_2 + 44H_2O + \Delta Q_4 \tag{2.62}$$

Proteins:

$$C_{46}H_{77}O_{17}N_{12}S + 34.25O_2 = 3C_5H_7O_2N + 31CO_2 + 9NH_3 + H_2SO_4 + 13.5H_2O + \Delta Q_5 \tag{2.63}$$

In the above equations, ΔQ represents the heat released by the biochemical reactions of various organic substances. In particular, the ΔQ values for RDC, SDC, total sugars, lipids, and proteins were 2,575.8 kJ/mol, 2,575.8 kJ/mol, 2,575.8 kJ/mol, 30,616 kJ/mol, and 20,698.8 kJ/mol, respectively.

(2) Synchronous nitrification and denitrification

Aged MSW has been under anaerobic conditions for a long time and thus has accumulated a large amount of ammonia nitrogen. When anaerobic conditions shift to aerobic conditions, a process known as synchronous nitrification and denitrification occurs, removing ammonia nitrogen. The following equation describes this process. However, for simplicity, complex intermediate products have been omitted, exclusively retaining the initial substances and final products.

$$NH_3 + 0.75O_2 = 0.5N_2 + 1.5H_2O \tag{2.64}$$

(3) Microbial decay

Low concentrations of organic matter prove fatal to microorganisms that depend on it. Under aerobic conditions, these dead microorganisms decompose into small molecules, CO_2, ammonia, and water. The following chemical equation captures this aerobic oxidation reaction:

$$C_5H_7O_2N + 5O_2 = 5CO_2 + NH_3 + 2H_2O \quad (2.65)$$

Furthermore, the rate at which these aerobic oxidation reactions occur for different limiting substrates can be expressed using the following equation:

$$R_j = -k_{mTj} \cdot f(T) \cdot f(w) \cdot f(C_{O2}) \cdot f(C_{C/N}) \cdot S_j \quad (j = 1, 2, 3, 4, \text{ and } 5) \quad (2.66)$$

where k_{mTj} denotes the first-order hydrolysis rate constant for the aerobic hydrolysis of the jth limiting substrate (d^{-1}), R_j represents the reaction rate of the jth limiting substrate, S_j indicates the concentration of the jth limiting substrate, $f(T)$ signifies the temperature influence function (dimensionless), $f(w)$ represents the water content influence function (dimensionless), $f(C_{O2})$ denotes the oxygen content influence function, $f(C_{C/N})$ represents the initial carbon-to-nitrogen ratio influence function (dimensionless), w is the water content, C_{O2} represents the oxygen volumetric content, and $C_{C/N}$ indicates the initial carbon-to-nitrogen ratio.

Notably, microbial growth under aerobic conditions is optimal within a specific temperature range. Rosso et al. (1993) proposed the following expression for the temperature influence function:

$$f(T) = \frac{(T - T_{\max})(T - T_{\min})^2}{(T_{opt} - T_{\min})[(T_{opt} - T_{\min})(T - T_{opt}) - (T_{opt} - T_{\max})(T_{opt} - T_{\min} - 2T)]} \quad (2.67)$$

where $T_{\min}$, T_{opt}, and $T_{\max}$ denote the minimum, optimal, and maximum temperatures, set to 0 °C, 50 °C, and 65 °C, respectively.

The influence of water content on microbial nutrient uptake and aeration space is represented by the following equation (Haug 2018):

$$f(w) = \frac{1}{\exp^{(-17.684w + 7.0622)} + 1} \quad (2.68)$$

Further, the effect of oxygen content on the aerobic oxidation reaction rate is described by the following equation:

$$f(C_{O2}) = \frac{C_{O2}}{C_{O2} + 2\%} \quad (2.69)$$

The influence of the initial carbon-to-nitrogen ratio on the aerobic oxidation reaction rate is modeled as follows (Wang et al. 2014):

$$f(C_{C/N}) = \exp\left[-\left(\frac{C_{C/N} - 28}{14}\right)^2\right] \quad (2.70)$$

Moreover, the rate of the nitrification–denitrification reaction is described by the following first-order kinetic equation:

$$R_6 = \theta k_N \cdot f(T) \cdot f(w) \cdot f(C_{O2}) \cdot f(C_{C/N}) \cdot S_j \quad (2.71)$$

where k_N denotes the first-order kinetic reaction rate constant for synchronous nitrification and denitrification, θ represents the volumetric water content, and the definitions of other influencing functions are identical to those for oxidation reactions.

The rate of bacterial decay is described by the following first-order kinetic equation:

$$R_7 = \theta K_d f_{Td} S_6 \quad (2.72)$$

$$f_{Td} = \begin{cases} \exp\left[-\lambda_d (T - T_s)^2\right] & (T < T_s) \\ 1 & (T \geq T_s) \end{cases} \quad (2.73)$$

where K_d denotes the bacterial decay rate constant, f_{Td} represents the effect of temperature on the microbial decay rate, and λ_d denotes the temperature influence parameter with a default value of 0.004 °C^{-2}. Furthermore, T_s is set to 65 °C.

The heat change during aerobic degradation can be expressed as

$$\frac{dQ}{dt} = \sum_{j=1}^{5} \Delta Q_j R_j \tag{2.74}$$

For accuracy validation, Xiao et al. (2021) adopted the above aerobic model to simulate the aerobic degradation experiments conducted by Wu et al. (2016) using column tests. Experimental waste samples comprising ceramic stones (28.6%), glass (6.7%), wood bamboo straw (6.6%), food waste (6.0%), plastics (4.7%), and fine soil (43%) were procured from the Laogang Landfill in Shanghai, aged between 5 and 8 years. These samples were packed into two pilot-scale column units (lysimeters), each with a diameter and height of 1 m and 3 m, respectively. The initial concentration of degradable organic matter corresponded to the sum of the SDC concentrations in the solid and liquid phases, with values of 10.8 kg/m^3 and 12.0 kg/m^3 in lysimeters A and B, respectively. Meanwhile, the concentrations of total nitrogen compounds in the solid and liquid phases, represented by NH_3, were 3.1 kg/m^3 and 2.6 kg/m^3 in lysimeters A and B, respectively. Moreover, the carbon-to-nitrogen ratio was set to 15 for both lysimeters.

Figures 2.33 and 2.34 illustrate the variations in SDC and NH_3 concentrations during aerobic degradation, respectively. Remarkably, the simulation results closely align with the experimental data. Specifically, during the initial 100 days of aerobic degradation, SDC degradation was particularly rapid, facilitated by favorable temperatures and high organic matter concentrations. Subsequently, from Day 100 to Day 200, microbial activity slowed as the environmental temperature decreased, leading to lower SDC degradation rates. As the temperature increased again after 200 days, the SDC degradation rate increased. By Day 400, SDC concentrations reduced significantly, with notable reduction in degradation. Under consistent oxygen supply, SDC concentrations approached zero by Day 800. NH_3 concentrations mirrored the variation trend of SDC concentrations. Specifically, NH_3 levels decreased rapidly in the first 100 days. The degradation rate slowed down between Day 100 and Day 200 owing to reduced temperatures, then increased after 200 days, and finally approached zero by Day 400.

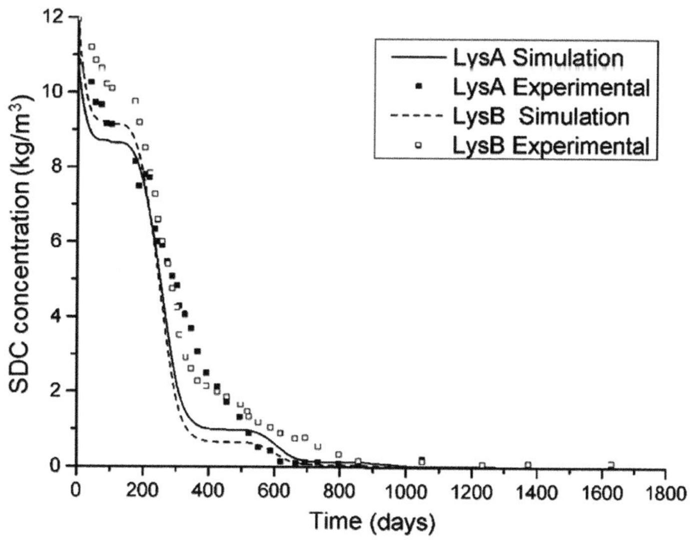

Fig. 2.33 Time profile of SDC degradation (Xiao et al. 2021)

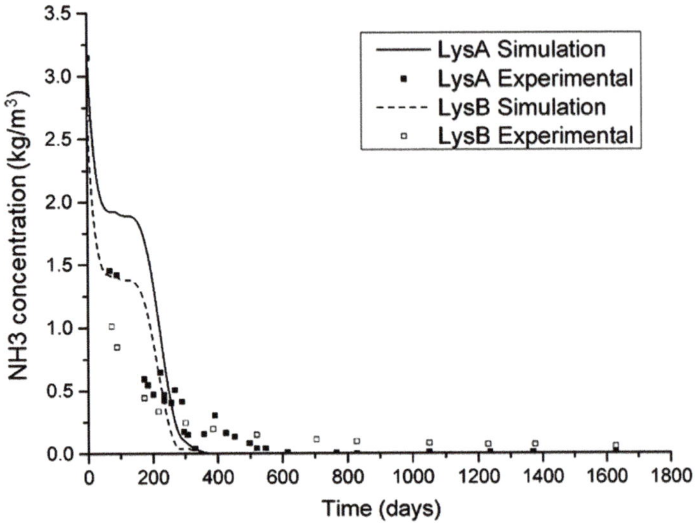

Fig. 2.34 Time profile of NH_3 degradation (Xiao et al. 2021)

2.4.3.2 Evaluation of Aerobic Degradation

Xiao et al. (2022) conducted indoor experiments on MSW samples, aged between 3 and 4 years, procured from the Shanghai Laogang Landfill to compare anaerobic and aerobic degradation. The sampled MSW comprised residual waste (27.8%), rubber and plastic (64.5%), bamboo and wood (1.8%), stones (1.3%), metals (1.7%), and glass (2.9%). The experimental setup consisted of a test cylindrical test reactor with an inner diameter of 0.4 m and a total height of 1.0 m. The MSW layer was 0.9 m thick, containing 115 kg of MSW.

As illustrated in Fig. 2.35, the reactor design featured several layers arranged from bottom to

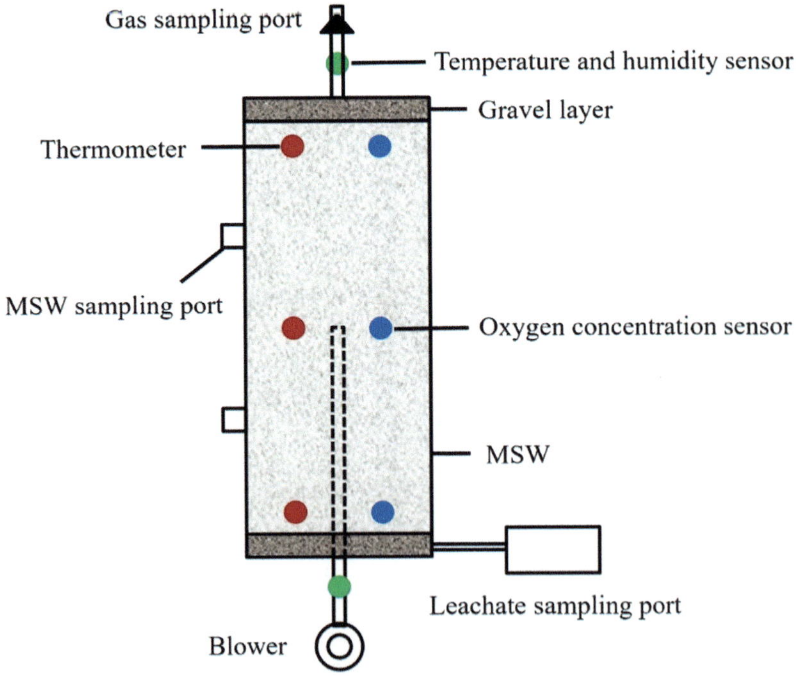

Fig. 2.35 Indoor forced ventilation test device (reprinted from Xiao et al. 2022, with permission from China Environmental Science)

top as follows: a 0.05 m gravel layer, a nonwoven geotextile filter, the MSW layer, another nonwoven geotextile filter, and a second 0.05 m gravel layer. Two MSW sampling ports were installed on the column wall at depths of 25 and 65 cm below the surface, each with a diameter of 5 cm. The reactor was equipped with a forced ventilation system and a leachate drainage system. A gas distribution pipe with a diameter of 25 mm and an opening ratio of 12.5% is installed at the centre of the column and inserted halfway into the MSW layer. This pipe facilitated forced ventilation from the bottom. Meanwhile, a gas sampling port was set at the top to monitor gas flow, humidity, and composition. Leachate was drained from the bottom of the reactor, and samples were regularly collected for testing. Changes in MSW mass were recorded using a bottom load sensor. Additionally, temperature and oxygen concentration sensors were buried at depths of 5, 45, and 85 cm below the surface of the MSW body. Apart from aeration conditions, all other test settings for the aerobic and anaerobic columns were identical.

Figures 2.36, 2.37, and 2.38 respectively compare the temporal variations of various quantities under aerobic and anaerobic conditions: solid phase C/L ratio, liquid phase NH_4-N concentration, and gas-phase CH_4 production. Under anaerobic conditions, the C/L ratio, reflecting the degree of waste degradation, decreased from 0.8 to 0.65 in the initial stages of the experiment, remaining almost constant thereafter. Conversely, under aerobic conditions, the C/L ratio consistently decreased, reaching 0.35 at the end of aeration period. This ongoing reduction is attributed to the rapid decomposition of slowly degradable cellulose under aerobic conditions, further enhancing waste stabilization. Under anaerobic conditions, the concentration of NH_4-N, a typical pollutant load indicator in leachate, increased, whereas under aerobic conditions, it decreased to zero owing to nitrification–denitrification reactions. Correspondingly, the concentration of CH_4, a greenhouse gas, continuously increased in volume fractions to 50% under anaerobic conditions. Conversely, no CH_4 production was observed during aerobic degradation, maintaining its concentration at zero. These results clearly underscore the superior capability of aerobic reactions to accelerate solid phase waste degradation and mitigate liquid and gas pollution loads.

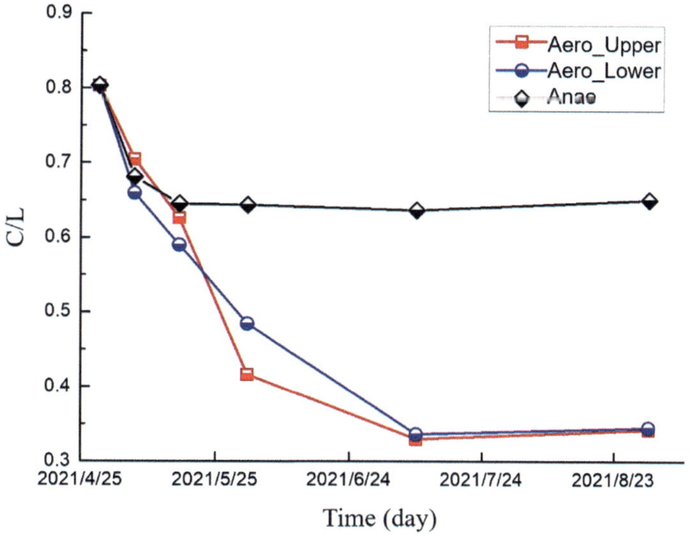

Fig. 2.36 Variation of C/L with time (Xiao 2022)

Fig. 2.37 Variation of NH_4-N concentration with time (Xiao 2022)

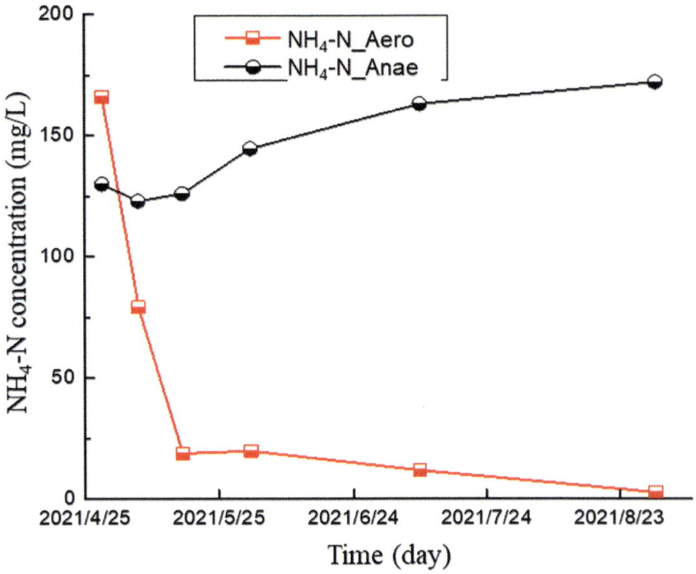

Fig. 2.38 Variation of CH_4 volume fraction with time (Xiao 2022)

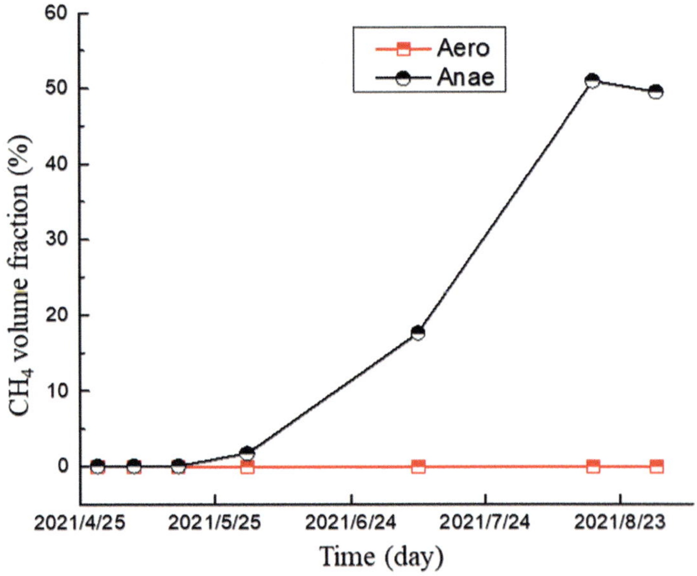

References

Barlaz MA, Ham RK, Schaefer DM (1989) Mass-balance analysis of anaerobically decomposed refuse. J Environ Eng 115(6):1088–1102

Bian XC, Shi KH, Li W, Luo X, Tutumluer E, Chen YM (2021) Quantification of railway ballast degradation by abrasion testing and computer-aided morphology analysis. J Mater Civ Eng 33(1):4020411

Blum WEH (2011) Physical degradation of soils, risks and threats. In: Encyclopedia of earth sciences series. Springer, Dordrecht

Cai ZZ, Ke H, Zhang PZ, Lan JW, Ma PC, Hu J, Xiao B, Chen YM (2024) Strength behaviour of stacked phosphogypsum incorporating dissolution–recrystallisation equilibrium. Géotechnique:1–13

Chen YM (2014) A fundamental theory of environmental geotechnics and its application. Chin J Geotech Eng 36(1):1–46 (in Chinese)

Chen YM, Guo RY, Li YC, Liu HL, Zhan TL (2016) A degradation model for high kitchen waste content municipal solid waste. Waste Manag 58:376–385

Chen YM, Xu WJ, Ling DS, Zhan LT, Gao W (2020) A degradation–consolidation model for the stabilization behavior of landfilled municipal solid waste. Comput Geotech 118:103341

Choi JH, Lin JS, Dai S, Lei L, Seol Y (2020) Triaxial compression of hydrate-bearing sediments undergoing hydrate dissociation by depressurization. Geomech Energy Environ 23:100187

Dai S, Santamarina JC, Waite WF, Kneafsey TJ (2012) Hydrate morphology: physical properties of sands with patchy hydrate saturation. J Geophys Res: Solid Earth 117(B11):B11205

Daouadji A, Hicher PY, Rahma A (2001) An elastoplastic model for granular materials taking into account grain breakage. Eur J Mech A/Solids 20(1):113–137

Einav I (2007) Breakage mechanics—Part I: Theory. J Mech Phys Solids 55(6):1274–1297

Feyzi V, Mohebbi V (2020) Experimental study and modeling of the kinetics of carbon-dioxide hydrate formation and dissociation: a mass transfer limited kinetic approach. J Nat Gas Sci Eng 77:103273

Goel N (2006) In situ methane hydrate dissociation with carbon dioxide sequestration: current knowledge and issues. J Petrol Sci Eng 51(3–4):169–184

Gu QS (2022) Trackbed deterioration and controlling measures of ballasted track under long-term train loads. Ph.D. thesis, Zhejiang University, Hangzhou, China (in Chinese)

Gu QS, Zhao C, Bian XC, Morrissey JP, Ooi JY (2022) Trackbed settlement and associated ballast degradation due to repeated train moving loads. Soil Dyn Earthq Eng 153:107109

Guo RY (2017) Research on the biochemical and mechanical behaviors of high food content MSW. Ph.D. thesis, Zhejiang University, Hangzhou, China (in Chinese)

Haarstrick A, Hempel DC, Ostermann L, Ahrens H, Dinkler D (2001) Modelling of the biodegradation of organic matter in municipal landfills. Waste Manag Res 19(4):320–331

Hardin BO (1985) Crushing of soil particles. J Geotech Eng 111(10):1177–1192

Hardwick JS, Mathias SA (2018) Masuda's sandstone core hydrate dissociation experiment revisited. Chem Eng Sci 175:98–109

Haug RT (2018) The practical handbook of compost engineering. Lewis Publishers, Chicago

He PJ, Qu X, Shao LM, Li GJ, Lee DJ (2007) Leachate pretreatment for enhancing organic matter conversion in landfill bioreactor. J Hazard Mater 142(1–2):288–296

Hossain MS, Gabr MA, Barlaz MA (2003) Relationship of compressibility parameters to municipal solid waste decomposition. J Geotech Geoenviron Eng 129(12):1151–1158

ITRC (The Interstate Technology & Regulatory Council) (2006) Characterization, design, construction, and monitoring of bioreactor landfills. ALT-3. Interstate Technology & Regulatory Council, Alternative Landfill Technologies Team, Washington DC

Kong XJ, Liu JM, Zou DG, Fu M (2014) Experimental study of particle breakage of Zipingpu rock fill material. Rock Soil Mech 35(1):35–40 (in Chinese)

Kim HC, Bishnoi PR, Heidemann RA, Rizvi SS (1987) Kinetics of methane hydrate decomposition. Chem Eng Sci 42(7):1645–1653

Li QB, Liu D, Ouyang F (2005) Hydrolysis rate of solid waste in bioreactor landfill. J Southwest Jiaotong Univ 40(1):126–130 (in Chinese)

Liu HL (2016) Research on biodegradation-consolidation-solute migration coupled behaviors of municipal solid waste and landfill stabilization. Ph.D. thesis, Zhejiang University, Hangzhou, China (in Chinese)

Ma PC (2022) Study on constitutive models for soils with degradable solid skeleton and their applications. Ph.D. thesis, Zhejiang University, Hangzhou, China (in Chinese)

McDougall J (2007) A hydro-bio-mechanical model for settlement and other behaviour in landfilled waste. Comput Geotech 34(4):229–246

Meima JA, Naranjo NM, Haarstrick A (2008) Sensitivity analysis and literature review of parameters controlling local biodegradation processes in municipal solid waste landfills. Waste Manag 28(5):904–918

Merey S, Al-Raoush RI, Jung J, Alshibli KA (2018) Comprehensive literature review on CH_4-CO_2 replacement in microscale porous media. J Petrol Sci Eng 171:48–62

Rosso L, Lobry JR, Flandrois JP (1993) An unexpected correlation between cardinal temperatures of microbial growth highlighted by a new model. J Theor Biol 162(4):447–463

Shi KH (2020) Quantitative analysis and application of ballast particle morphology based on computer vision. Master thesis, Zhejiang University, Hangzhou, China (in Chinese)

Sloan ED Jr, Koh CA (2008) Clathrate hydrates of natural gases. CRC Press, Florida
Sun X, Li YH, Liu Y, Song YC (2019) The effects of compressibility of natural gas hydrate-bearing sediments on gas production using depressurization. Energy 185:837–846
Tongji University (2008) Integrated technology and demonstration of anaerobic bioreactor landfill for municipal solid waste (in Chinese)
Turcotte DL (1986) Fractals and fragmentation. J Geophys Res: Solid Earth 91(B2):1921–1926
Vavilin VA, Rytov SV, Lokshina LY, Pavlostathis SG, Barlaz MA (2003) Distributed model of solid waste anaerobic digestion: effects of leachate recirculation and pH adjustment. Biotechnol Bioeng 81(1):66–73
Voet D, Voet JG (2011) Biochemistry. Wiley, New Jersey
Wang LJ, Wang P, Wang XB, Zhu B, Chen YM (2022a) Centrifuge modelling on behaviour of hydrate bearing sediments during gas production by depressurization. In: 10th International conference on physical modelling in geotechnics, Daejeon Korea, pp 495–499
Wang LJ, Wang P, Zhu B, Wang XB, Yang SQ, Chen YM (2024) Development and application of in-flight centrifuge apparatus for modelling hydrate exploitation. Chin J Geotech Eng 46(2):316–324 (in Chinese)
Wang X, Lu X, Li F, Yang G (2014) Effects of temperature and carbon-nitrogen (C/N) ratio on the performance of anaerobic co-digestion of dairy manure, chicken manure and rice straw: focusing on ammonia inhibition. PLoS ONE 9(5):e97265
Wang XB, Wang LJ, Zhu B, Wang P, Yuan SM, Chen YM (2022b) Experimental study of behavior of hydrate-bearing sediments during servo depressurization. Rock Soil Mech 43(9):2360–2370 (in Chinese)
Wang ZN, Wang G, Ye QG (2020) A constitutive model for crushable sands involving compression and shear induced particle breakage. Comput Geotech 126:103757
Wei HY (2007) Experimental and numerical study on gas migration in landfill of municipal solid waste. Ph.D. thesis, Zhejiang University, Hangzhou, China (in Chinese)
White J, Robinson J, Ren QC (2004) Modelling the biochemical degradation of solid waste in landfills. Waste Manag 24(3):227–240
Wu CF, Shimaoka T, Nakayama H, Komiya T, Chai XL (2016) Stimulation of waste decomposition in an old landfill by air injection. Biores Technol 222:66–74
Xiao DK (2022) Aerobic degradation-stabilization model in municipal solid waste landfill and its application. Ph.D. thesis, Zhejiang University, Hangzhou, China (in Chinese)
Xiao DK, Chen YM, Xu WJ, Zhan LT (2021) An aerobic degradation model for landfilled municipal solid waste. Appl Sci 11(16):7557
Xiao DK, Chen YM, Xu WJ, Zhan LT, Ke H, He HJ, Liu YF (2022) Experimental study on aerobic accelerated stabilization and carbon and nitrogen migration of municipal solid waste. China Environ Sci 42 (5):2204–2212 (in Chinese)
Xiao Y, Desai CS, Daouadji A, Stuedlein AW, Liu H, Abuel-Naga H (2020) Grain crushing in geoscience materials—key issues on crushing response, measurement and modeling: review and preface. Geosci Front 11(2):363–374
Yao YP (2015) Advanced UH models for soils. Chin J Geotech Eng 37(2):193–217 (in Chinese)
Yu FW (2017) Characteristics of particle breakage of sand in triaxial shear. Powder Technol 320:656–667
Yu FW (2018) Particle breakage in triaxial shear of a coral sand. Soils Found 58(4):866–880
Yuan SM, Wang LJ, Zhu B, Chen YM (2022) Volumetric strain analysis model for gas hydrate-bearing sediment considering effects of hydrate dissociation. Chin J Geotech Eng 44(6):1044–1052 (in Chinese)
Zhang JR, Hu Y, Zhang BW, Liu YZ (2015) Fractal behavior of particle-size distribution during particle crushing of quartz sand and gravel. Chin J Geotech Eng 37(5):784–791 (in Chinese)
Zhu B, Wang LJ, Yang SQ, Chen YM, Kong DQ, Lin WA (2021) Pressure-control temperature-control hypergravity experimental device for simulating deep-sea seabed responses. U.S. Patent, US11,187,691 B2. U.S. Patent and Trademark Office, Washington DC
Zhu B, Wang LJ, Kong DQ, Chen YM (2022) Physical modelling of multiphase interaction problems in marine sediment. In: 10th International conference on physical modelling in Geotechnics, Daejeon Korea, pp 97–104

Constitutive Models for Skeleton-Degradable Soils

3

Abstract

This chapter focuses on developing stress–strain constitutive models specifically tailored to skeleton-degradable soils, including soils subjected to particle breakage, NGHs, and MSW. The discussion begins with an overview of the classical Modified Cam-Clay model for normally consolidated soils and the unified hardening (UH) model for overly consolidated soils. Considering soil phenomena such as compressive hardening, friction, and shear contraction and dilation, the above classical models are then adapted to analyze the complex behaviors of skeleton-degradable soils. Furthermore, new state variables are introduced to facilitate the development of stress–strain constitutive models for skeleton-degradable soils. In particular, the constitutive model for soils subjected to particle breakage emphasizes the influence of particle size distribution variations on the stress–strain relationship. Meanwhile, the constitutive model for NGHs highlights the significant influence of depressurization and hydrate decomposition rates on volumetric strain behavior. Finally, the constitutive model for MSW considers the influence of degradation on waste hardening behavior. Additionally, parameter sensitivity analyses are conducted to evaluate model variations, and simulated results are rigorously compared with experimental data to validate the effectiveness of the newly developed constitutive models.

Keywords

Skeleton-degradable soil · Particle breakage · Hydrates · Municipal solid waste · Constitutive model · Modified Cam-clay model · Unified hardening model

3.1 General Introduction

A constitutive model is a mathematical expression describing the stress–strain relationship of soil that forms the basis for analytical calculations and numerical analysis in geotechnical engineering. Therefore, to accurately assess the impact of soil skeleton degradation on geotechnical structures, it is necessary to study the stress–strain relationship of degradable soil skeleton and establish corresponding constitutive models. Since Terzaghi established the discipline of soil mechanics in 1923, the constitutive theory of soil has undergone two main stages: the traditional stage of soil mechanics, which considers strength and deformation problems separately, and the modern stage of soil mechanics, based on critical state theory and represented by the Cam-clay models, which unify the analysis of strength and deformation. These models are extensively applied in geotechnical engineering research.

In 1923, Terzaghi introduced the concept of effective stress in soil and the theory of one-dimensional consolidation, marking the beginning of soil mechanics as an independent

Y. Chen, *Soil Degradation-Consolidation Theory and Its Applications*,
https://doi.org/10.1007/978-981-97-7985-7_3

discipline. This traditional or classical stage of soil mechanics was characterised by a focus on the principles of effective stress, deformation analysis based on elasticity, and failure analysis based on rigid-plastic models (Shen 2000). For stress and deformation problems, traditional soil mechanics employs analytical methods from elasticity theory, such as elastic solutions for stress analysis in foundations, elastic foundation beam theory, and consolidation theory, all of which treat soil as an ideal elastic material. When analyzing soil failure, traditional soil mechanics utilizes the limit equilibrium theory based on rigid-plastic models, transforming the failure problem into the determination of the limit equilibrium state of rigid-plastic bodies. This approach is exemplified by widely used theories like Terzaghi's ultimate bearing capacity theory for foundations and the sliding circle method for slope stability analysis. Traditional soil mechanics is convenient for practical applications and remains the primary analytical method used in geotechnical engineering design. However, it has two inherent limitations: firstly, it fails to establish a relationship between soil deformation and strength, making it difficult to describe the gradual development of deformation leading to failure; secondly, it simplifies soil behavior to elastic and rigid-plastic models, neglecting essential soil characteristics such as compressive hardening, friction, and dilatancy (Yao et al. 2012), resulting in computational inaccuracies that often fail to meet the requirements of quantitative analysis (Chen et al. 2020).

To enhance computational accuracy, many scholars have devoted to develop elastoplastic constitutive models for soils. Roscoe et al. (1963) proposed the Cam-clay model based on the critical state theory, which marked a milestone in soil mechanics by being the first elastoplastic model to comprehensively reflect the basic characteristics of saturated normally consolidated remolded soils, ushering soil mechanics into a new stage of development. Subsequently, Roscoe and Burland (1968) modified the yield surface of the Cam-clay model from elliptical to ellipsoidal, introducing the Modified Cam-clay model, which demonstrated better predictive performance for the stress–strain behavior of soils. The Modified Cam-clay model becomes one of the most widely used constitutive models in geotechnical engineering. However, the Modified Cam-clay model is applicable only to normally consolidated soils, and thus many scholars have focused on extending the Modified Cam-clay model to afford models with greater applicability. Yao et al. (2008) proposed the unified hardening model (UH model), which retains the advantages of the Modified Cam-clay model while further developing it into a unified mechanical model that reflects both hardening and softening, as well as shear contraction and shear dilation behaviors of soils. This advancement enables the model to accurately describe the stress–strain relationship in over-consolidated soils.

However, conventional soil constitutive theories do not account for changes in soil skeleton. For skeleton-degradable soils, the aforementioned models fail to accurately describe their stress–strain relationship. Therefore, there is a need to develop constitutive theories that consider soil skeleton degradation. This chapter first introduces the classical Modified Cam-clay model and the UH model. Subsequently, novel constitutive models for breakable soils and degradable soils (taking municipal solid waste as an example) are established based on UH model and Modified Cam-clay model, respectively. In addition, the chapter also includes a constitutive model for hydrate-bearing sediment with chemical degradation.

3.2 Modified Cam-Clay Model

The Modified Cam-clay model is based on experimental results obtained from normally consolidated remolded clays. "Normally consolidated" refers to a condition where the current consolidation pressure of the soil is equal to the maximum consolidation pressure it has experienced in its history. Roscoe et al. (1963) referred to normally consolidated soils as "wet clay," in contrast to over-consolidated soils, which have experienced higher consolidation pressures in the past. Consequently, over-consolidated soils tend to be denser and have a lower moisture content than normally consolidated soils at the same stress levels. The following sections will discuss the fundamental characteristics of normally consolidated soils and their representation in the Modified Cam-clay model.

3.2.1 Compressive Hardening

The normalized compression line (NCL) represents the compressive hardening of normally consolidated soils. As shown in Fig. 3.1, the NCL is approximately linear in the e−lnp coordinates, and the Modified Cam-clay model describes it with the following equation:

$$e = N - \lambda lnp \tag{3.1}$$

in which, N represents the intercept of NCL, which physically corresponds to the void ratio of the soil at $p = 1$ kPa; λ is the slope of NCL. Similarly, the volumetric strain generated in isotropic compression tests is:

$$\varepsilon_v = \frac{\lambda}{1 + e_0} \ln \frac{p}{p_0} \tag{3.2}$$

where e_0 and p_0 represent the initial void ratio and initial mean stress, respectively. The

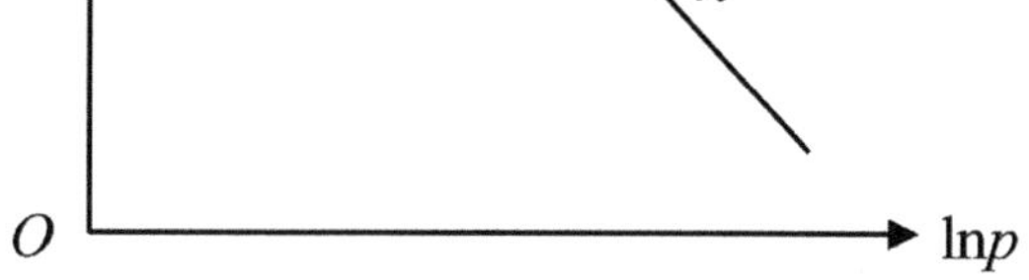

Fig. 3.1 Normal consolidation line (NCL) and swelling line of normally consolidated soil in the $e - \ln p$ coordinates (reprinted from Chen et al. 2020, with permission of Chinese Journal of Theoretical and Applied Mechanics)

Modified Cam-clay model employs Eq. (3.2) to describe the soil's compressive hardening. During unloading, the soil experiences swelling, and the swelling line also approximates a straight line, hence the elastic strain is:

$$\varepsilon_v^e = \frac{\kappa}{1 + e_0} \ln \frac{p}{p_0} \tag{3.3}$$

Therefore, the plastic strain generated by normally consolidated soil in isotropic compression tests is:

$$\varepsilon_v^p = c_p \ln \frac{p}{p_0} \tag{3.4}$$

where $c_p = (\lambda - \kappa)/(1 + e_0)$.

3.2.2 Friction

The friction of soil is reflected by the fact that the shear strength q_f is proportional to the mean stress p. Experimental evidence indicates that the aforementioned parameters have an approximately linear relationship (see Fig. 2a). Therefore, the Modified Cam-clay model describes the soil's frictional behavior using the following expression:

$$q_f = Mp \tag{3.5}$$

Where M is the critical stress ratio. In the critical state soil mechanics, the soil's failure state is referred to as the critical state. Therefore, M is termed as the critical state stress ratio.

When a soil is in its critical state, its stress state and volume (void ratio) remain unchanged but its shear strain develops rapidly, indicating a state of plastic flow (Schofield and Wroth 1968). Experiments have shown that the stress state and volume state of normally consolidated soil are uniquely correlated. The curve describing the unique correspondence between the stress state and volume state in the $p - q - e$ space at the critical state is referred to as the Critical State Line (CSL). Figures 2a and 2b respectively depict the projection of the CSL in the $p - q$ and $e - p$ coordinates. In the $p - q$ coordinates, the

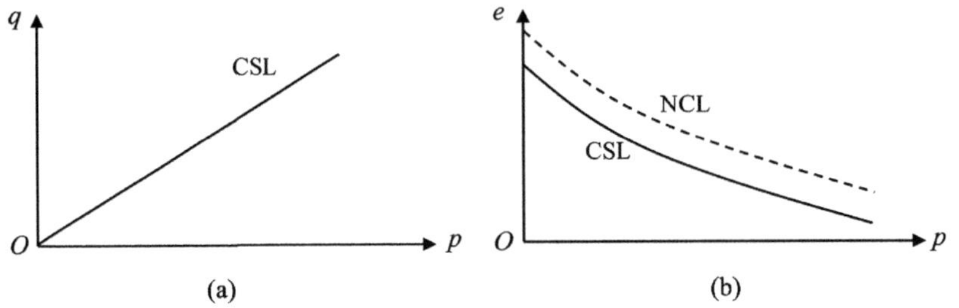

Fig. 3.2 Critical state line (CSL): **a** $p - q$ coordinate; **b** $e - p$ coorindate (reprinted from Chen et al. 2020 with permission of Chinese Journal of Theoretical and Applied Mechanics)

projection of CSL is the strength envelope described by Eq. (3.5); in the $e - p$ coordinates, the projection of CSL is a curve parallel to the NCL. The critical state reveals the stress state and deformation characteristics of soil failure, establishing a connection between soil deformation and strength, serving as the theoretical basis for their unified analysis.

3.2.3 Dilatancy

The dilatancy of soil reflects the influence of shear stress q variations on the volume of the soil. To describe the dilation of soil, the following expression is commonly used (Rowe 1962; Li and Dafalias 2000):

$$\frac{d\varepsilon_{\mathrm{v}}^{\mathrm{p}}}{d\varepsilon_{\mathrm{d}}^{\mathrm{p}}} = f(\eta, C) \tag{3.6}$$

Where $d\varepsilon_{\mathrm{v}}^{\mathrm{p}}$ is the increment of plastic volumetric strain, $d\varepsilon_{\mathrm{d}}^{\mathrm{p}}$ is the increment of plastic shear strain, η is the stress ratio ($\eta = q/p$), and C is a constant reflecting the properties of the soil. Equation (3.6) is commonly referred to as the shear dilation equation.

Roscoe and Burland (1968) derived the shear dilation equation used in the Modified Cam-clay model based on the energy equation. The general expression for the plastic work of a soil element is:

$$\mathrm{d}W_{\mathrm{P}} = p\mathrm{d}\varepsilon_{\mathrm{v}}^{\mathrm{p}} + q\mathrm{d}\varepsilon_{\mathrm{d}}^{\mathrm{p}} \tag{3.7}$$

Taking two specific points in Fig. 3.3 for further analysis. Point A is at critical state, where only plastic shear strain is generated; Point B represents isotropic compression state, where only plastic volumetric strain is generated. Therefore, their expressions for plastic work are respectively:

$$\begin{cases} dW_{\mathrm{P-A}} = Mpd\varepsilon_{\mathrm{d}}^{\mathrm{p}} \\ dW_{\mathrm{P-B}} = pd\varepsilon_{\mathrm{v}}^{\mathrm{p}} \end{cases} \tag{3.8}$$

Based on Eq. (3.8), a special expression for plastic work is constructed:

Fig. 3.3 Schematic diagram of the yield surface of modified Cam-clay model (reprinted from Chen et al. 2020, with permission of Chinese journal of theoretical and applied mechanics)

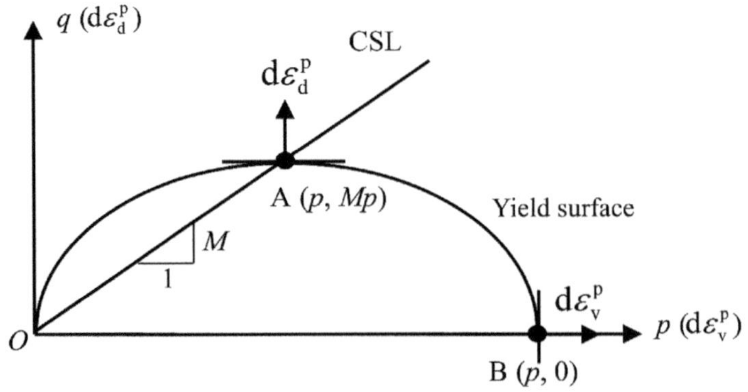

$$dW_P = p\sqrt{(d\varepsilon_v^p)^2 + (Md\varepsilon_d^p)^2} \quad (3.9)$$

By combining Eqs. (3.6) and (3.9), the shear dilation equation of the Modified Cam-clay model can be derived:

$$\frac{d\varepsilon_v^p}{d\varepsilon_d^p} = \frac{M^2 - \eta^2}{2\eta} \quad (3.10)$$

3.2.4 Yield Function

The yield criterion, flow rule, and hardening law of materials are the three fundamental elements of plasticity models. The general expression for the yield surface of work-hardening materials is:

$$f(\sigma, \mathrm{H}) = 0 \quad (3.11)$$

in which, H is the hardening parameter (internal variable), which determines the evolution law of subsequent yield surfaces, namely the hardening law of the material. The Modified Cam-clay model adopts the orthogonal flow rule, where the direction of plastic strain flow is orthogonal to the plastic potential surface. Assuming that under stress increments dp and dq, the increments of plastic strains $d\varepsilon_v^p$ and $d\varepsilon_d^p$ are generated, respectively. They satisfy the following relationship:

$$dp d\varepsilon_v^p + dq d\varepsilon_d^p = 0 \quad (3.12)$$

By combining Eqs. (3.10) and (3.12), the following differential equation can be obtained:

$$\frac{dp}{dq} + \frac{M^2p^2 - q^2}{2pq} = 0 \quad (3.13)$$

Solving this differential equation and considering the boundary conditions $p = p_x$ and $q = 0$, there is:

$$q^2 + M^2p^2 - M^2p_xp = 0 \quad (3.14)$$

where p_x represents the intersection point of the yield surface with the p-axis. Equation (3.14) represents the plastic potential function derived based on the shear dilation equation and the orthogonal flow rule.

The Modified Cam-clay model adopts an associated flow rule, where the yield function is the same as the plastic potential function. The shape of the yield surface in the Modified Cam-clay model is elliptical, as shown in Fig. 3.3. In the yield function, the hardening parameter p_x represents the mean stress during isotropic compression.

According to Eq. (3.4), for normally consolidated clay, the plastic strain corresponds uniquely to the mean stress, therefore, the plastic strain can also serve as the hardening parameter. Solving Eqs. (3.14) and (3.4) simultaneously and further simplifying, it can be obtained that:

$$g = f = \ln\frac{p}{p_0} + \ln\left(1 + \frac{q^2}{M^2p^2}\right) - \frac{1}{c_p}\varepsilon_v^p = 0 \quad (3.15)$$

in which, p_0 is the intersection point of the initial yield surface with the p-axis, which can be determined based on the initial stress state. Equation (3.15) represents the yield function with plastic volumetric strain ε_v^p as the hardening parameter, and there is $H = \varepsilon_v^p / c_p$.

The Cam-clay model and the Modified Cam-clay model have significant scientific and practical value. Their theoretical frameworks are comprehensive, with clear physical interpretations in the derivation process. Based on critical state theory, they unify strength and deformation aspects of soils, requiring fewer parameters that can be determined through conventional triaxial tests. These models can effectively predict the stress–strain behavior of saturated normally consolidated remolded clay, making them among the most widely used constitutive models in geotechnical engineering. As noted by Shen (2000), the introduction of Roscoe's renowned Cam-clay model in 1963 marked a departure from classical theories of linear elastic porous media and rigid-plastic materials, leading to a deepening understanding of the nonlinear deformation characteristics of soils. Thus, the Cam-clay model can be considered the beginning of modern soil mechanics.

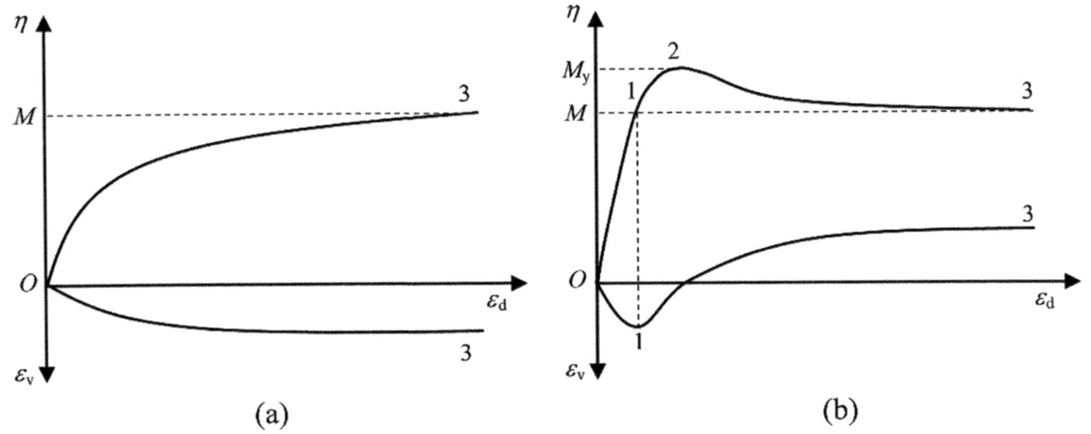

Fig. 3.4 Stress–strain relationships in triaxial tests: **a** normally consolidated clay, and **b** over-consolidated clay (reprinted from Chen et al. 2020, with permission of Chinese journal of theoretical and applied mechanics)

3.3 Unified Hardening Model

The UH model represents a prominent advancement in extending the applicability of the Modified Cam-clay model (Yao et al. 2008, 2009; Yao 2015). Crucially, the UH model encompasses both hardening and softening, together with shear contraction and shear dilation, and thereby generates a unified description of soil behaviour during loading. As a result, the UH model expands the application scope of the Modified Cam-clay model to over-consolidated soils.

3.3.1 Mechanical Properties of Over-Consolidated Soils

The typical stress–strain relationships of normally consolidated and over-consolidated soils in triaxial drained compression tests are shown in Fig. 3.4. Normally consolidated soils undergo hardening and shear contraction during shearing until reaching the critical state of plastic flow (point 3). In contrast, although ultimately reaching the critical state, over-consolidated soils undergo hardening (O $\rightarrow$ 2 stage) followed by softening (2 $\rightarrow$ 3 stage) during shearing, and undergo shear contraction (O $\rightarrow$ 1 stage) followed by shear dilation (1 $\rightarrow$ 3 stage). The Modified Cam-clay model, based on experimental results of normally consolidated soils, struggles to describe the softening and shear dilation characteristics of over-consolidated soils. The following section analyzes the mechanical properties of over-consolidated soils and their description methods in the UH model.

(1) Compressive Hardening

As shown in Fig. 3.5, normally consolidated soils exhibit swelling during isotropic unloading, leading to over-consolidation, while over-consolidated soils undergo plastic deformation during reloading. The reloading line approximates

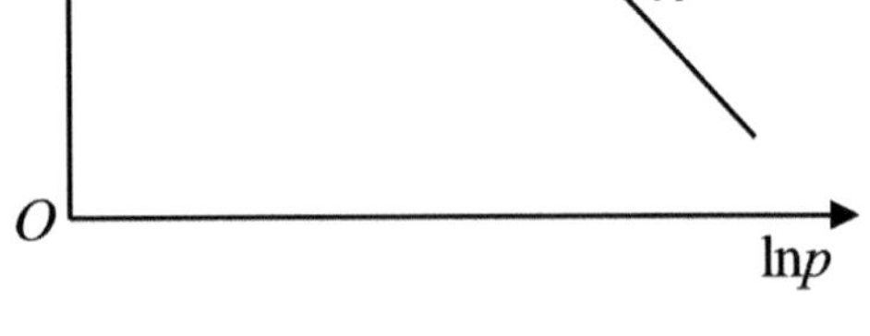

Fig. 3.5 Schematic diagram of the reloading line of over-consolidated clay (reprinted from Yao 2015, with permission of Chinese journal of geotechnical engineering)

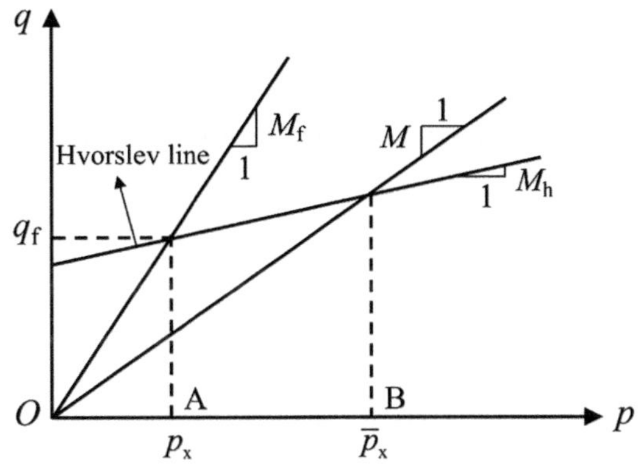

Fig. 3.6 Schematic diagram of reloading line of over-consolidated line (reprinted from Chen et al. 2020, with permission of Chinese journal of theoretical and applied mechanics)

a straight line in the $e - \ln p$ coordinates, indicating the compressive hardening behavior of over-consolidated soils during recompression. By employing a derivation method similar to that used for normally consolidated soils, the plastic volumetric strain of over-consolidated soils during isotropic compression can be obtained.

$$d\varepsilon_v^p = \frac{\bar{\lambda} - \kappa}{1 + e_0}\frac{dp}{p} \quad (3.16)$$

(2) Friction

In Fig. 3.6, point A represents the over-consolidated state with a mean stress of p_x, and its corresponding pre-consolidation pressure point B with a mean stress of $\bar{p}_x$. The peak strength of over-consolidated soil is higher than that of normally consolidated soil, hence the peak strength envelope in the $p - q$ coordinates (i.e., the Hvorslev line) lies above the CSL. The Hvorslev line indicates that the peak strength of over-consolidated soil is proportional to the mean stress, reflecting the frictional behavior of over-consolidated soil. As shown in Fig. 3.6, the current Hvorslev line is determined by the intersection of the CSL and the perpendicular line passing through point B, and consequently, the peak strength q_f corresponding to point A can be determined by the intersection of the Hvorslev line and the perpendicular line passing through point A. Based on geometric relations, the expression for M_f is given by:

$$M_f = \left(\frac{1}{R_0} - 1\right)(M - M_h) + M \quad (3.17)$$

where M is the critical state stress ratio, M_f is the peak stress ratio corresponding to point A, and M_h is the slope of the Hvorslev line.

(3) Dilatancy

The model still employs the shear-dilation equation of the Modified Cam-clay model (Eq. 3.10), and the corresponding plastic potential equation is:

$$g = q^2 + M^2p^2 - M^2p_xp \quad (3.18)$$

3.3.2 Yield Surface

Unlike normally consolidated soil, over-consolidated soil exhibits a phenomenon of shear contraction followed by shear dilation during the hardening stage. In other words, the plastic strain ε_v^p experiences non-monotonic changes as the mean stress p_x increases. Therefore, p_x and ε_v^p are not uniquely corresponding, and ε_v^p cannot be continued to be used as the hardening parameter. Assuming the hardening parameter of the UH model is denoted as H, compared with Eq. (3.15), the expression of the yield surface for over-consolidated soil is given by:

$$f = \ln\frac{p}{p_0} + \ln\left(1 + \frac{q^2}{M^2p^2}\right) - H = 0 \quad (3.19)$$

The degree of over-consolidation of over-consolidated soil decreases continuously with shear progress, eventually reaching a critical state, where the strength decreases to the critical state stress ratio M and degrades into normally consolidated soil. To describe the state changes of over-consolidated soil during shear, Yao et al. (2009) introduced the concepts of reference yield

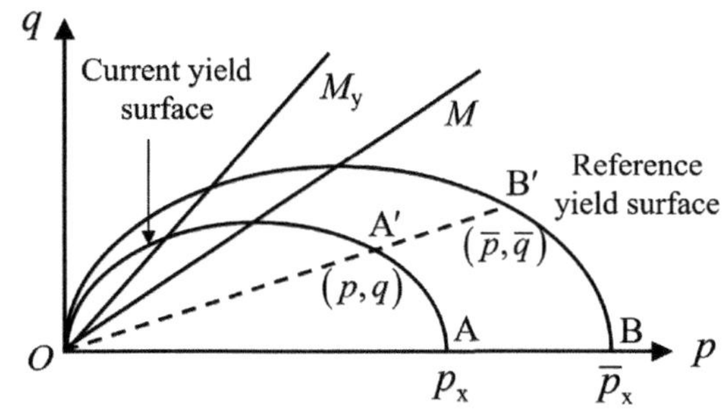

Fig. 3.7 Current and reference yield surfaces in the UH model (reprinted from Yao et al. 2009, with permission of Emerald publishing limited)

surface, over-consolidation parameter, and current strength. The reference yield surface is defined as the Modified Cam-clay model's yield surface of normally consolidated soil with plastic strain equal to that of the current yield surface of over-consolidated soil. Therefore, the over-consolidated state point A corresponds to the normally consolidated state point B, which lies on the reference yield surface. Assuming A′ (p, q) is a point on the current yield surface, its corresponding stress point on the reference yield surface with the same stress ratio is denote as B′ ($\overline{p}$, $\overline{q}$). Thus, the expression of the reference yield surface is obtained as shown in Fig. 3.7:

$$f = \ln\frac{\overline{p}}{\overline{p}_0} + \ln\left(1 + \frac{\overline{q}^2}{M^2\overline{p}^2}\right) - \frac{1}{c_p}\varepsilon_v^p = 0 \quad (3.20)$$

in which, $\overline{p}_0$ is the intersection point of the initial reference yield surface and the p-axis. The definition of the over-consolidation parameter R is:

$$R = \frac{p}{\overline{p}} \quad (3.21)$$

Solving Eq. (3.20) for $\overline{p}$, substituting it back into Eq. (3.21), and considering $q/p = \overline{q}/\overline{p}$, it can be obtained that:

$$R = \frac{p}{\overline{p}_0}\left(1 + \frac{q^2}{M^2p^2}\right)\exp\left(-\frac{\varepsilon_v^p}{c_p}\right) \quad (3.22)$$

R characterizes the degree of proximity between the current yield surface and the reference yield surface, namely the degree of over-consolidation. The larger the R value, the lower the degree of overconsolidation. When $R = 1$, the over-consolidated soil degrades to normally consolidated soil.

The parameter R_0 in Eq. (3.17) represents the over-consolidation parameter at point A in the isotropic compression state before the soil starts shearing. Additionally, the peak strength corresponding to the current state of the over-consolidated soil is defined as the current strength M_Y. Referring to Eq. (3.17), there is:

$$M_Y = \left(\frac{1}{R} - 1\right)(M - M_h) + M \quad (3.23)$$

To ensure that when R is very small, the value of M_Y is not greater than the slope of the zero-tension line, which is 3, Yao et al. (2012) proposed a modified formula for M_Y by fitting a parabola to the zero-tension line and the Hvorslev line:

$$M_Y = 6\left[\sqrt{\frac{k}{R}\left(1 + \frac{k}{R}\right)} - \frac{k}{R}\right] \quad (3.24)$$

and there is:

$$k = \frac{M^2}{12(3 - M)} \quad (3.25)$$

3.3.3 Hardening Rule

To describe the evolution of the yield surface of over-consolidated soil, Yao et al. (2008) proposed a unified hardening parameter based on the loading characteristics and results of isotropic shear and isotropic loading tests:

$$H = \frac{1}{c_p}\int \frac{M_Y^4 - \eta^4}{M^4 - \eta^4} d\varepsilon_v^p \quad (3.26)$$

For normally consolidated soil, $M = M_Y$, and the unified hardening parameter can degenerate into the hardening parameter of normally consolidated soil, and the UH model can degenerate into the Modified Cam-clay model.

The rationality of the unified hardening parameter is discussed in light of the stress–strain characteristics of triaxial shear tests on over-consolidated soil shown in Fig. 4b:

(1) Stage O → 1: $0<\eta<M$. This stage is characterized by shear contraction and hardening. Point 1 ($\eta = M$) marks the boundary between shear contraction and shear dilation, also known as the critical point. According to the plastic potential function and the orthogonality flow rule as shown in Eq. (3.18), when the stress ratio is below M, the soil undergoes shear contraction ($d\varepsilon_v^p > 0$). From Eq. (3.26), it follows that $dH > 0$ Therefore, the unified hardening parameter can describe the hardening behavior in this stage effectively.

(2) Stage 1 → 2: $M<\eta<M_{Y-P}$ (where M_{Y-P} represents the current strength of the soil at point 2, which is also the peak strength during the shear process). This stage is characterized by shear dilation and hardening. Point 2 ($\eta = M_Y$) marks the boundary between hardening and softening. When the stress ratio exceeds M, the soil undergoes shear dilation ($d\varepsilon_v^p<0$). From Eq. (3.26), it follows that $dH > 0$. Therefore, the unified hardening parameter can describe the hardening behavior and the transition from shear contraction to shear dilation in this stage effectively.

(3) Stage 2 → 3: $\eta \geq M_Y \geq M$. This stage is characterized by shear dilation and softening. During this stage, the current strength decreases, and the stress ratio also decreases but remains slightly higher than the current strength until reaching the critical state, where both the stress ratio and the current strength decrease to M ($\eta = M_Y = M$, point 3). Similar to the previous stage, the soil undergoes shear dilation ($d\varepsilon_v^p<0$). According to Eq. (3.26), $dH<0$. Therefore, the unified hardening parameter can facilitate the transition from hardening to softening in the stress–strain relationship of over-consolidated soil.

3.4 Constitutive Model for Breakable Soils

3.4.1 Influence of Particle Size Distribution on Stress–Strain Relationship of Soil

In Sect. 3.2.2, the evolution law of particle breakage in soil was established. Based on this, establishing a constitutive model for soil with particle breakage requires exploring the influence of particle breakage on the stress–strain relationship of soils.

The PSD change caused by particle breakage significantly affects the mechanical properties of soil, leading to extensive experimental and numerical studies by many scholars. Liu (2014) conducted triaxial tests on samples of Hostun sand and glass beads with different PSDs and similar average d_{50} particle size (i.e., the particle size corresponding to 50% cumulative particle masses). The distribution was controlled by the non-uniformity coefficient c_u (where $c_u = d_{60}/d_{10}$). A higher c_u value indicates a wider distribution range of particle sizes, with higher proportions of particles both larger and smaller than the average particle size d_{50}. Triaxial tests were conducted with confining pressures below 1,000 kPa, which resulted in minimal further particle breakage. The results of isotropic compression tests on Hostun sand and glass beads are shown in Fig. 3.8. It can be observed that with increasing distribution width (increasing c_u), the NCL shifts downward in the $e - \ln p$ plane, while the slope remains almost constant. At the same stress level, samples with wider distributions have smaller void ratios, indicating that a wider distribution requires a lower void ratio to maintain structural stability at the same stress level.

Yan and Dong (2011) used the discrete element method (DEM) to simulate triaxial drained compression tests of particle materials with the same initial void ratio but different PSDs. The distribution control was consistent with the previous description, where c_u was varied while d_{50} remained constant. As shown in Fig. 3.9, when c_u is 1.0, the particle material exhibits characteristics

Fig. 3.8 Isotropic compression tests on soils with different particle size distributions: **a** Hostun sand, and **b** glass beads (Liu 2014)

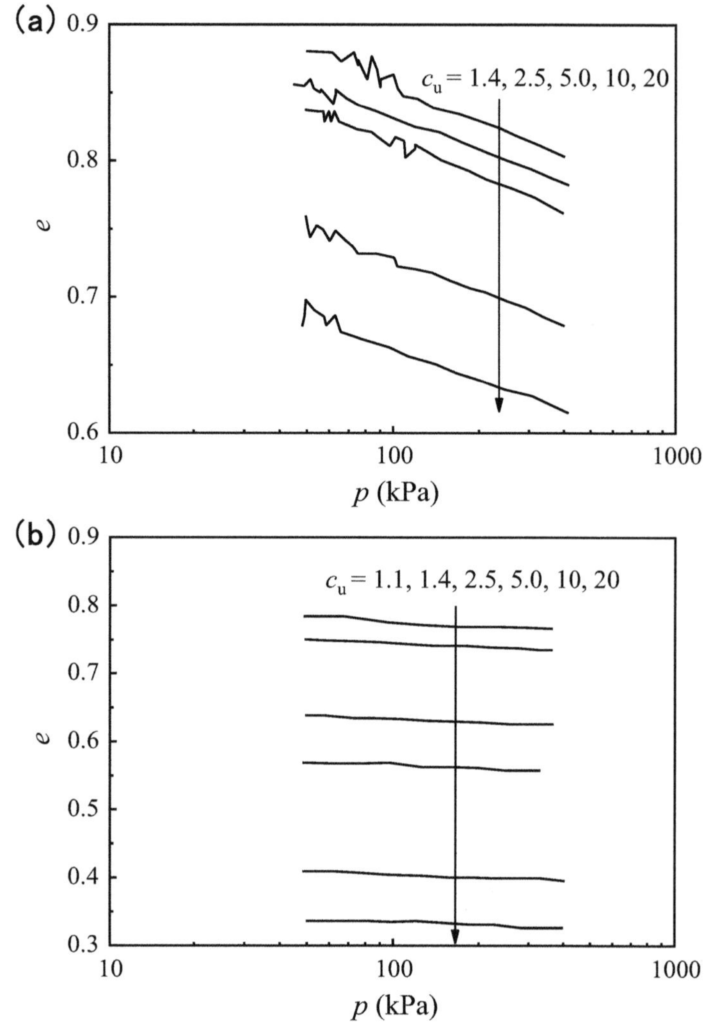

of dense sand, with peak strength and softening behavior, accompanied by significant shear dilation. With an increase in distribution width, the stress–strain relationship of the particle material gradually approaches that of medium-dense sand and loose sand. When c_u increases to 2.17, the material fully exhibits characteristics of loose sand, with no peak strength and sustained shear dilation. However, particle materials with different distributions ultimately reach the same critical state, where the critical state stress ratio M is unaffected by distribution changes (under the same confining pressure for different specimens). Similar patterns are observed in DEM simulations conducted by other researchers (Liu 2014; Li et al. 2015; Jiang et al. 2018; Yang and Luo 2018). In summary, changes in particle distribution significantly influence the stress–strain relationship of particle materials during shear. Moreover, it has been shown that with a constant void ratio, as the distribution widens, the stress–strain relationship of a specimen gradually transitions from that of dense sand to that of loose sand.

3.4.2 Yield Surface and Loading/Unloading Criteria

Experiments and DEM simulation results indicate that an increase in c_u (widening of the distribution and increased content of fine particles)

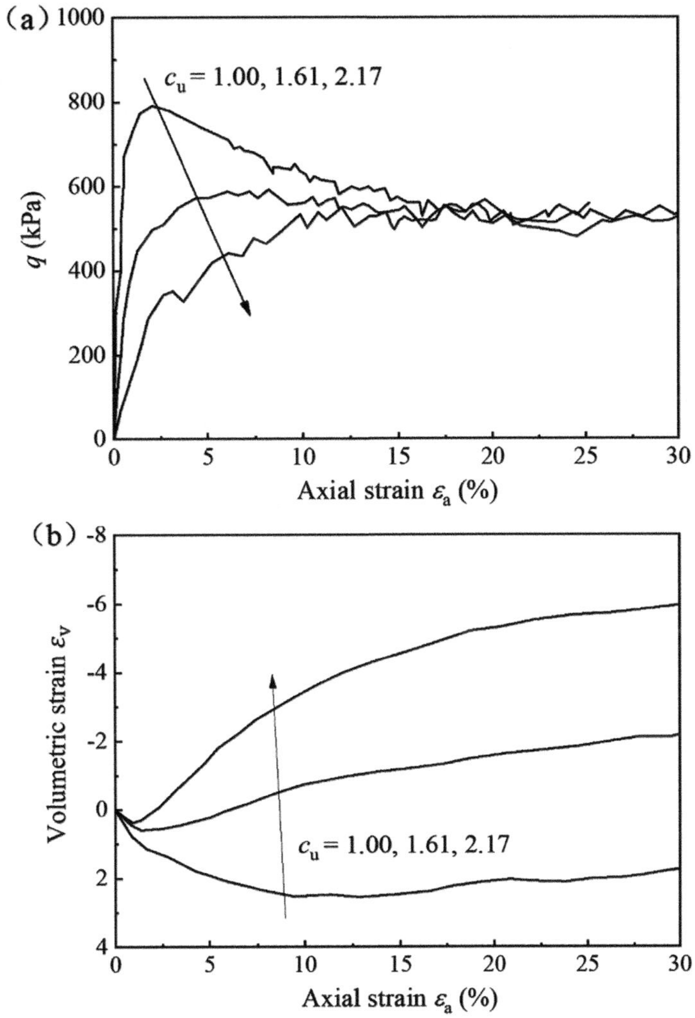

Fig. 3.9 DEM modelling of triaxial drained compression tests on soils with different particle size distributions: **a** deviatoric stress and axial strain, and **b** volumetric strain and axial strain (reprinted from Yan and Dong 2011, with permission of American society of civil engineers)

leads to a decrease in the peak strength of granular materials and a transition from shear dilation to shear contraction. This transition exceeds the applicability range of the Modified Cam-clay model, while it can be described by the UH model. A constitutive model for soils with particle breakage is established based on the UH model in the following.

The change in particle size distribution resulting from particle breakage induces a translation of the NCL in the $e - \ln p$ plane, which in turn can lead to variations in pre-consolidation pressure. As depicted in Fig. 3.10, under isotropic compression conditions, for a soil state located at point A with a non-uniformity coefficient of c_{u1}, the corresponding normally consolidated state is represented by point B, with a pre-consolidation pressure of $\overline{p}_{x1}$. However, after particle breakage, as the non-uniformity coefficient increases to c_{u2}, due to the translation of the NCL, the normally consolidated state corresponding to point A shifts from point B to point B′. Consequently, the pre-consolidation pressure decreases from $\overline{p}_{x1}$ to $\overline{p}_{x2}$. The reduction rate of pre-consolidation pressure caused by particle breakage is described by the parameter k_d $(0 < k_d \leq 1)$:

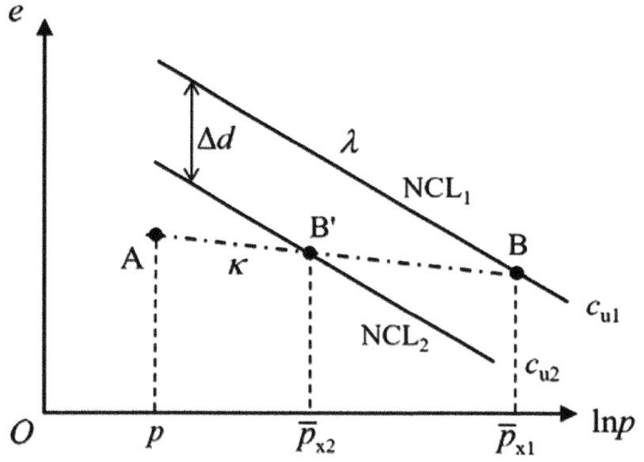

Fig. 3.10 Schematic diagram of the change in initial consolidation pressure due to NCL translation (Ma 2022; Ma and Chen 2024)

$$\overline{p}_{x2} = k_d \overline{p}_{x1} \tag{3.27}$$

Considering the particle breakage, the reference yield surface of the soil at point A is given by:

$$f = \overline{q}^2 + M^2\overline{p}^2 - M^2\overline{p}_{x2}\overline{p} \tag{3.28}$$

in which,

$$\overline{p}_{x2} = k_d\overline{p}_{x1} = k_d\overline{p}_{x0}\exp\left(\frac{1}{c_p}\varepsilon_v^p\right) \tag{3.29}$$

Combining Eqs. (3.28) and (3.29), it can be obtained that:

$$f = \ln\frac{\overline{p}}{\overline{p}_0} + \ln\left(1 + \frac{\overline{q}^2}{M^2\overline{p}^2}\right) + \ln k_d - \frac{1}{c_p}\varepsilon_v^p = 0 \tag{3.30}$$

Replacing the Modified Cam-clay model hardening parameters in the reference yield surface with the unified hardening parameters proposed by the UH model, the yield surface of the constitutive model for soils with particle breakage:

$$f = \ln\frac{p}{p_0} + \ln\left(1 + \frac{q^2}{M^2p^2}\right) + \ln k_d - \frac{1}{c_p}\int\frac{M_Y^4 - \eta^4}{M^4 - \eta^4}d\varepsilon_v^p = 0 \tag{3.31}$$

in which, the hardening parameter is the plastic strain, and M is unaffected by particle breakage as mentioned above. In this model, the degree of soil hardening is affected by both the stress state and the degree of particle breakage, i.e., the value of k_d, which is related to the displacement Δd of NCL, as follows:

$$\Delta d = -(\lambda - \kappa)\ln k_d \tag{3.32}$$

Under the framework of Modified Cam-clay model (UH model is an extension of Modified Cam-clay model), the NCL and CSL of soils are always parallel lines in the $e - lnp$ plane, and the distance between them is $(\lambda - \kappa)\ln 2$. That is to say, the particle breakage could induce the same displacement Δd of NCL and CSL. By analysing the test results in Fig. 3.8, the displacement of NCL has a hyperbolic relationship with the non-uniformity coefficient c_u. In addition, Hu et al. (2011) analyzed the test data in Yamamuro and Lade (1996) and found that a hyperbolic function can be used to describe the relation between the breakage index B_r and the displacement of CSL. Therefore, in this model, the displacement Δd of NCL is related to the soil particle breakage index B_r by a hyperbolic function, there is:

$$\Delta d = (N_0 - N_u)\frac{B_r}{\rho + B_r} \tag{3.33}$$

where N_0 and N_u are the intercepts of the NCL corresponding to the initial grading before particle breakage and the ultimate grading after complete particle breakage, respectively, and ρ is a model constant. By solving Eqs. (3.39) and (3.40) simultaneously, it can be obtained that:

$$k_d = exp\left[-\frac{B_r(N_0 - N_u)}{(\lambda - \kappa)(\rho - B_r)}\right] \tag{3.34}$$

Introducing the parameter k_d, the current yield surface of the particle breakage soil is a spatial surface in the $p - q - k_d$ coordinates. As shown in Fig. 3.11, the size of the yield surface section depends on the value of k_d. When the plastic strain is equal (i.e., on the same yield surface), the section of the yield surface contracts as k_d

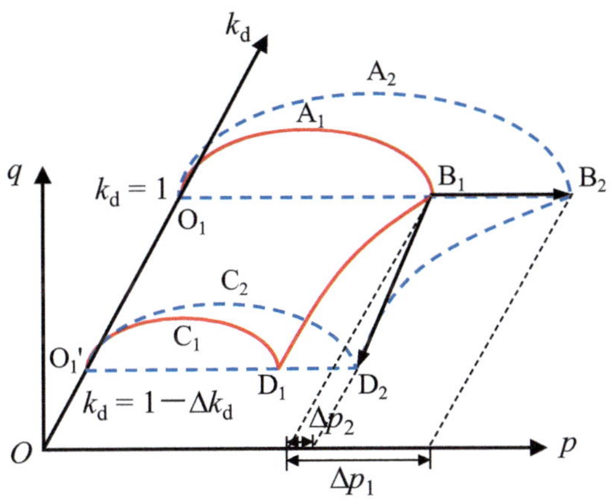

Fig. 3.11 Three-dimensional yield surface of the constitutive model for soils with particle breakage

decreases. For the soil where particles have not yet broken under isotropic compression conditions, the stress point is located at point B_1 on the yield surface $O_1A_1B_1 - O_1'C_1D_1$. When the mean stress increases by Δp_1 during loading and no breakage occurs, the yield surface expands to $O_1A_2B_2 - O_1'C_2D_2$. However, if particle breakage occurs during loading, k_d decreases from 1 to $1 - \Delta k_d$, and only an incremental stress Δp_2 is required to expand the yield surface to $O_1A_2B_2 - O_1'C_2D_2$, producing the same plastic deformation. In other words, the decrease in k_d causes the soil to undergo larger plastic deformation under the same stress increment, thereby allowing the model to describe the deformation due to the weakening of soil skeleton, which is caused by the particle breakage.

The continuity condition of the yield surface is given by:

$$df = \frac{\partial f}{\partial \boldsymbol{\sigma}} : d\boldsymbol{\sigma} + \frac{\partial f}{\partial k_d} dk_d - \frac{1}{c_p} \frac{M_Y^4 - \eta^4}{M^4 - \eta^4} d\varepsilon_v^p = 0 \tag{3.35}$$

By employing the associated flow rule, the increment of plastic volumetric strain is given by:

$$d\varepsilon_v^p = d\lambda \frac{\partial f}{\partial p} \tag{3.36}$$

Combining Eqs. (3.35) and (3.36), it can be obtained that:

$$d\lambda = \frac{\frac{\partial f}{\partial \boldsymbol{\sigma}} : d\boldsymbol{\sigma} + \frac{\partial f}{\partial k_d} dk_d}{\frac{1}{c_p} \frac{M_Y^4 - \eta^4}{M^4 - \eta^4} \frac{\partial g}{\partial p}} \tag{3.37}$$

When $d\lambda > 0$, new plastic strain increments occur, indicating that the soil is under loading conditions. Conversely, when $d\lambda < 0$ (or $d\lambda = 0$), the soil is under unloading (or neutral loading) conditions.

3.4.3 Elastoplastic Constitutive Relationship

The stress–strain relationship of the elastoplastic model is given by:

$$d\boldsymbol{\sigma} = \boldsymbol{D}^e : (d\varepsilon - d\varepsilon^p) \tag{3.38}$$

where $\boldsymbol{D}^e$ is the elastic stiffness tensor, which is expressed in matrix form, according to Voigt's rule, as follows:

$$[\boldsymbol{D}^e] = \frac{E}{(1+\mu)(1-2\mu)} \begin{bmatrix} (1-\mu) & \nu & \nu & 0 & 0 & 0 \\ \nu & (1-\mu) & \nu & 0 & 0 & 0 \\ \nu & \nu & (1-\mu) & 0 & 0 & 0 \\ 0 & 0 & 0 & \frac{1-2\mu}{2} & 0 & 0 \\ 0 & 0 & 0 & 0 & \frac{1-2\mu}{2} & 0 \\ 0 & 0 & 0 & 0 & 0 & \frac{1-2\mu}{2} \end{bmatrix} \tag{3.39}$$

where μ is the Poisson's ratio, and E is the Young's modulus, which can be calculated by the following equation:

$$E = \frac{(1-2\mu)(1+e_0)}{\kappa} \sigma_{ii} \tag{3.40}$$

Therefore, the elastoplastic constitutive relationship of the model can be derived as follows:

$$d\boldsymbol{\sigma} + \frac{\boldsymbol{D} : \frac{\partial g}{\partial \boldsymbol{\sigma}} \frac{\partial f}{\partial k_{\mathrm{d}}}}{-\frac{\partial f}{\partial H} \frac{\partial H}{\partial \varepsilon_{\mathrm{v}}^{\mathrm{p}}} \frac{\partial g}{\partial p} + \frac{\partial f}{\partial \boldsymbol{\sigma}} : \boldsymbol{D} : \frac{\partial g}{\partial \boldsymbol{\sigma}}} \frac{\partial k_{\mathrm{d}}}{\partial B_{\mathrm{r}}} dB_{\mathrm{r}} = \left(\boldsymbol{D} - \frac{\boldsymbol{D} : \frac{\partial f}{\partial \boldsymbol{\sigma}} \otimes \frac{\partial g}{\partial \boldsymbol{\sigma}} : \boldsymbol{D}}{-\frac{\partial f}{\partial H} \frac{\partial H}{\partial \varepsilon_{\mathrm{v}}^{\mathrm{p}}} \frac{\partial g}{\partial p} + \frac{\partial f}{\partial \boldsymbol{\sigma}} : \boldsymbol{D} : \frac{\partial g}{\partial \boldsymbol{\sigma}}} \right) : d\boldsymbol{\varepsilon} \tag{3.41}$$

Compared to the constitutive relationship of conventional elastoplastic models, the left side of the equation above contains an additional term related to the increment of

k_{d}. It is evident that introducing the parameter k_{d} corresponds to introducing an additional stress tensor associated with particle breakage in the elasto-plastic constitutive relationship. The deformation of the soil under the influence of this additional stress tensor represents the deformation induced by particle breakage.

3.4.4 Model Evaluation

(1) Parameter Study

The model, based on the UH model, introduces three additional model constants related to particle breakage: N_{u}, ρ, and χ'. Their influence on the computational results was investigated through calculations. The values of the UH model constants are listed in Table 3.1, while the values of the constants related to particle breakage are marked on the corresponding figures.

The constants N_{u} and ρ describe the relationship between the particle breakage index B_{r} and the rate of decrease in pre-consolidation pressure k_{d}, reflecting the degree of influence of particle breakage on the mechanical properties of the soil. k_{d} is directly proportional to N_{u} and ρ, meaning that for equal levels of particle breakage indicated by B_{r}, higher values of N_{u} and ρ correspond to higher k_{d}, indicating less influence of particle breakage on the stress–strain relationship of the soil. Therefore, in Figs. 3.12 and 3.13, as N_{u} and ρ increase, the stress–strain relationship of the soil approaches that of soil without particle breakage.

The constant $\chi\prime$ describes the evolution of particle breakage in the loading process, where higher values of $\chi\prime$ indicate greater resistance to particle breakage. Therefore, in Fig. 3.14, as $\chi\prime$ increases, the stress–strain relationship of the soil approaches that of soil without particle breakage.

(2) Model Performance

To validate the model's reliability, it was employed to predict the stress–strain relationships of different breakable soils in triaxial tests, using the parameters specified in Table 3.2.

The results of isotropic compression and triaxial drained compression tests of Calcareous sand, along with the model predictions, are shown in Fig. 3.15. As the particle breakage progresses, the slope of the isotropic compression line for Calcareous sand shows a trend of initially increasing and then decreasing, which the model can describe reasonably well. Additionally, the stress–strain relationship of Calcareous sand in triaxial drained compression tests was predicted rather accurately by the model. The development of breakage index in the tests are shown in Fig. 15d. For the tests with confining pressure of 100 kPa, the breakage of particles was relatively weak (B_{r} was less than 0.025), correspondingly, the sand specimens exhibited a peak strength and apparently shear dilatancy in the tests in Figs. 15b and c. With the increase in confining pressure, the degree of particle breakage increased. More large particles broken into small particles under higher stress level, which enhanced the shear contraction of specimens. This is an important reason for the

Table 3.1 Model parameters used in the parameter study

UH model					Soil breakage		
λ	κ	ν	M	N_0	N_{u}	ρ	$\chi\prime$
0.097	0.01	0.3	1.75	1.8	—	—	—

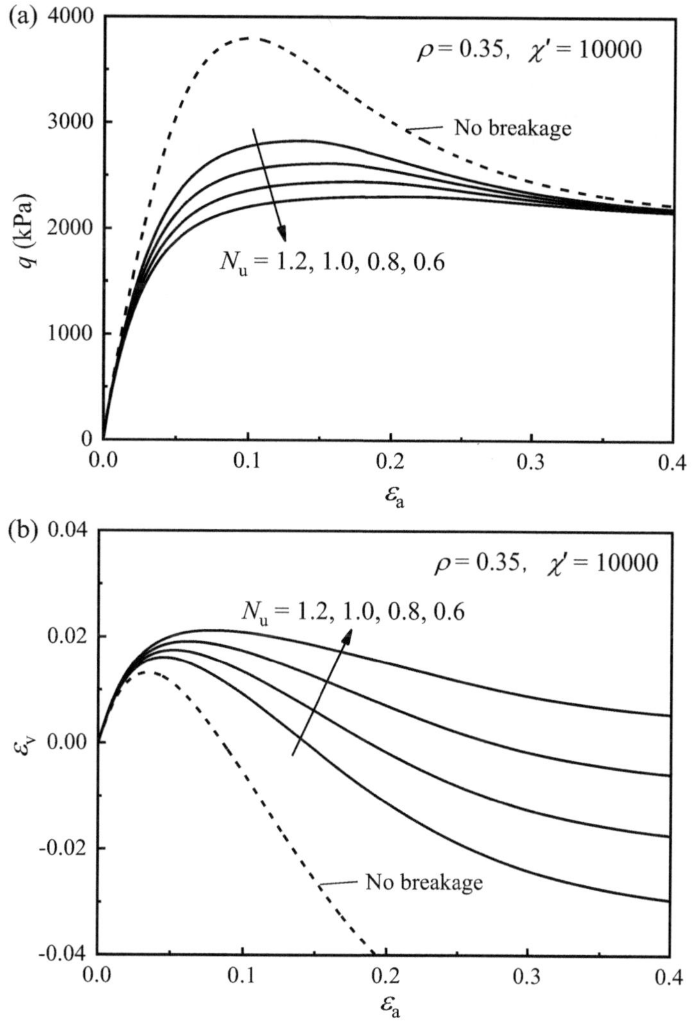

Fig. 3.12 Influences of constant N_u on the calculated results: **a** deviatoric stress and axial strain, and **b** volumetric strain and axial strain (Ma 2022; Ma and Chen 2024)

transmission of shear dilatancy to shear contraction with the increase in confining pressure in Fig. 15c.

The results of triaxial drained compression tests on Silica sand compared with model predictions are presented in Fig. 3.16. As can be seen, there is good agreement between the experimental results and the predicted stress–strain curves. Additionally, the predicted breakage index by the model was also comparable with the test data. Similar with the Calcareous sand, with the increase in particle breakage degree, the shear dilatancy gradually changed to shear contraction.

The particle breakage evolution of different soils as shown in Fig. 3.12 has been predicted by this model. As shown in Fig. 3.17, the predicted results of the model are in good agreement with the experimental results, indicating that the model can effectively predict the evolution of particle breakage. Overall, the proposed constitutive model for breakage soils can reasonably describe the breakage evolution of soils and its influences on the stress–strain of soils.

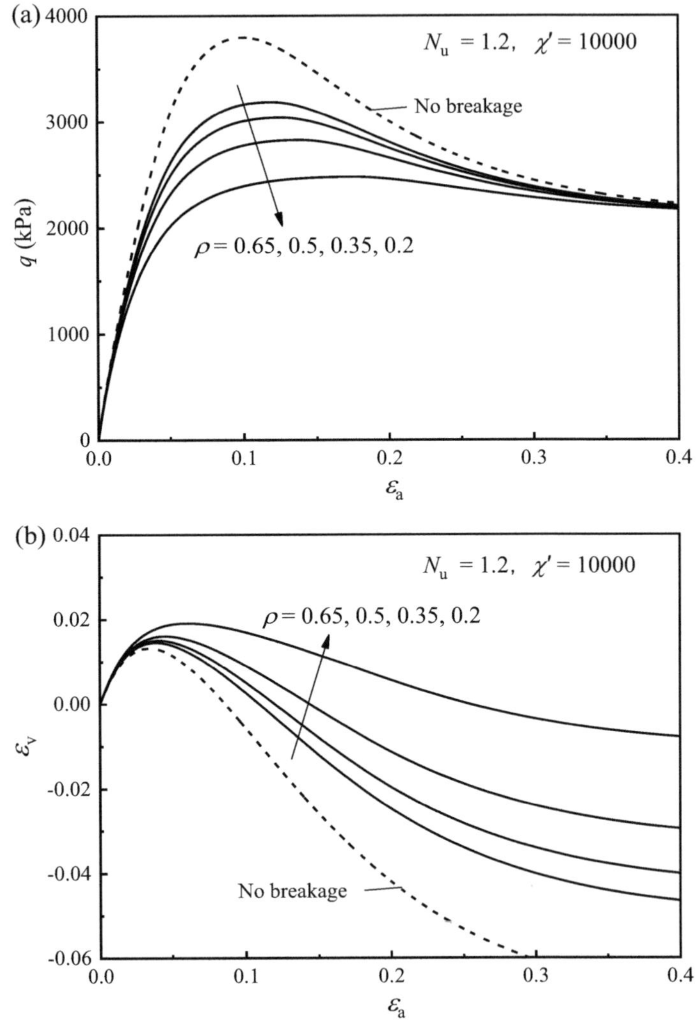

Fig. 3.13 Influences of constant ρ on the calculated results: **a** deviatoric stress and axial strain, and **b** volumetric strain and axial strain (Ma 2022; Ma and Chen 2024)

3.5 Constitutive Model for Natural Gas Hydrates

The mechanical properties of gas hydrate-bearing sediments (GHBS) are significantly influenced by the hydrate content. Various scholars have proposed constitutive relationships that consider the reinforcing effects of hydrates on sediments, based on different theoretical frameworks such as linear elasticity (Sun et al. 2019), nonlinear elasticity (Miyazaki et al. 2012), elastoplasticity (Moridis et al. 2019), and critical state theory (Shin and Santamarina 2019). Given that model parameters need to be easily obtained through basic unit cell tests and readily incorporated into numerical simulations, the constitutive relationship based on modified elasticity (Santamarina and Ruppel 2010) is the most widely applied.

Existing studies have shown that hydrate saturation has a minimal effect on Poisson's ratio but a significant impact on the elastic modulus, which can be expressed as (Santamarina and Ruppel 2010):

$$E = E_0\left(\frac{\sigma_c}{\sigma_{c0}}\right)^{a_H} + c_H E_H (s_H)^{b_H} \tag{3.42}$$

where E is the elastic modulus, σ_c and σ_{c0} are the confining pressure and initial confining pressure,

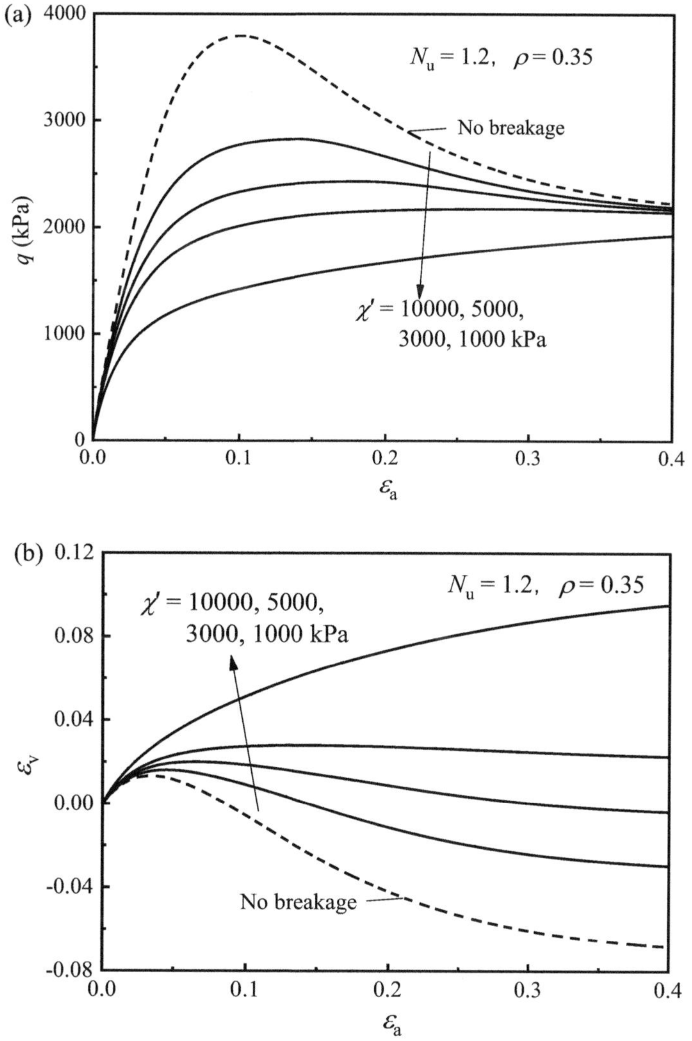

Fig. 3.14 Influences of constant χ' on the calculated results: **a** deviatoric stress and axial strain, and **b** volumetric strain and axial strain (Ma 2022; Ma and Chen 2024)

Table 3.2 Parameters of constitutive model for soils with particle breakage

Sand	λ	κ	v	M	N_0	N_u	ρ	χ'(kPa)
Calcareous sand Daouadji et al. (2001)	0.097	0.01	0.25	1.75	1.89	1.2	0.35	10,000
Silica sand Yu et al. (2017)	0.076	0.009	0.25	1.37	1.45	1.25	0.20	30,000
Carbonate sand Wang et al. (2020)	0.135	0.01	0.35	1.600	1.37	1.30	0.55	5000
Coral sand Yu et al. (2018)	0.075	0.01	0.2	1.5	1.3	1.2	0.2	50,000

respectively, E_0 and E_H are the elastic modulus of sediment without hydrates and with hydrates, respectively, a_H is the sensitivity parameter of the elastic modulus of sediment without hydrates to confining pressure σ_c, b_H is a dimensionless parameter related to hydrates, and c_H is a parameter related to pore characteristics.

By neglecting the effect of stress state changes on the elastic modulus of sediments and setting b_H = 1, Eq. (3.42) can be further simplified to:

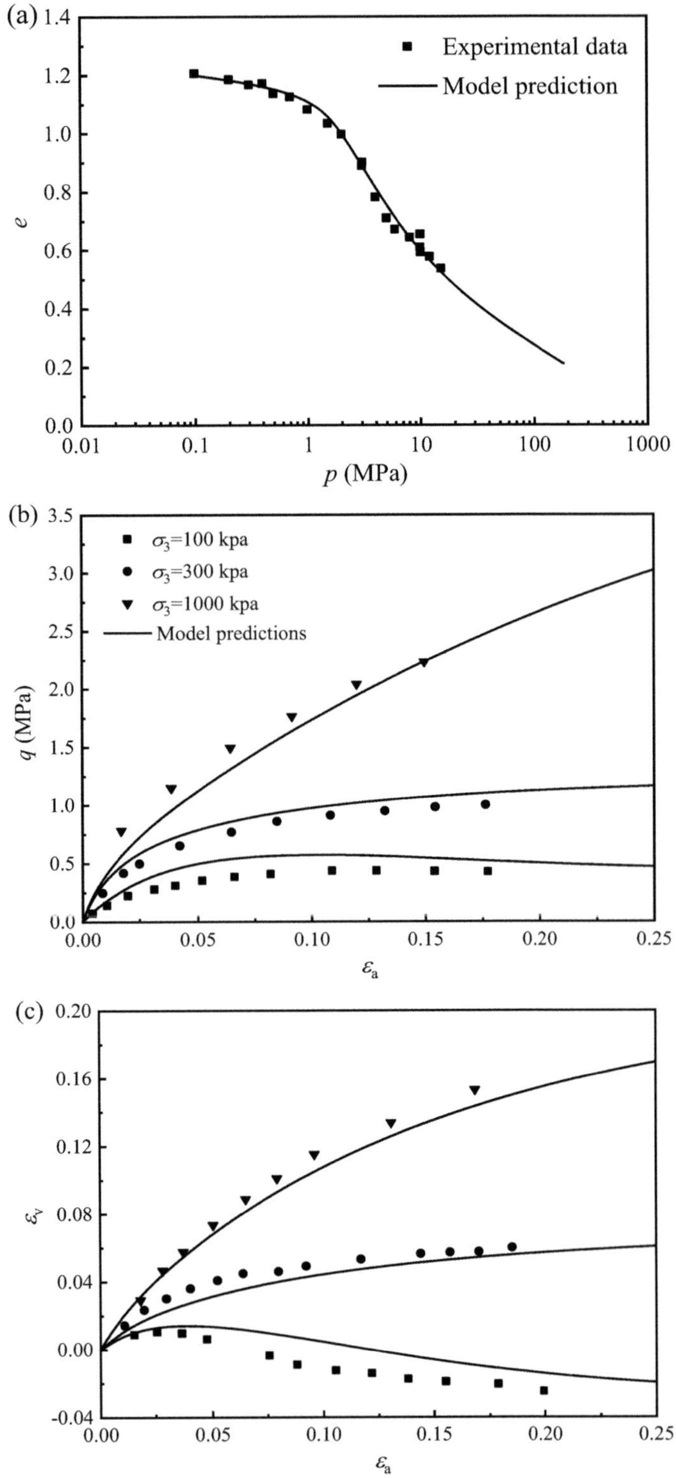

Fig. 3.15 Model predictions for the triaxial tests of Calcareous sand: **a** void ratio and mean stress in isotropic compression tests, **b** deviatoric stress and axial strain in triaxial drained compression tests, **c** volumetric strain and axial strain in triaxial drained compression tests, and **d** breakage index in triaxial drained compression tests. (Ma 2022; Ma and Chen 2024)

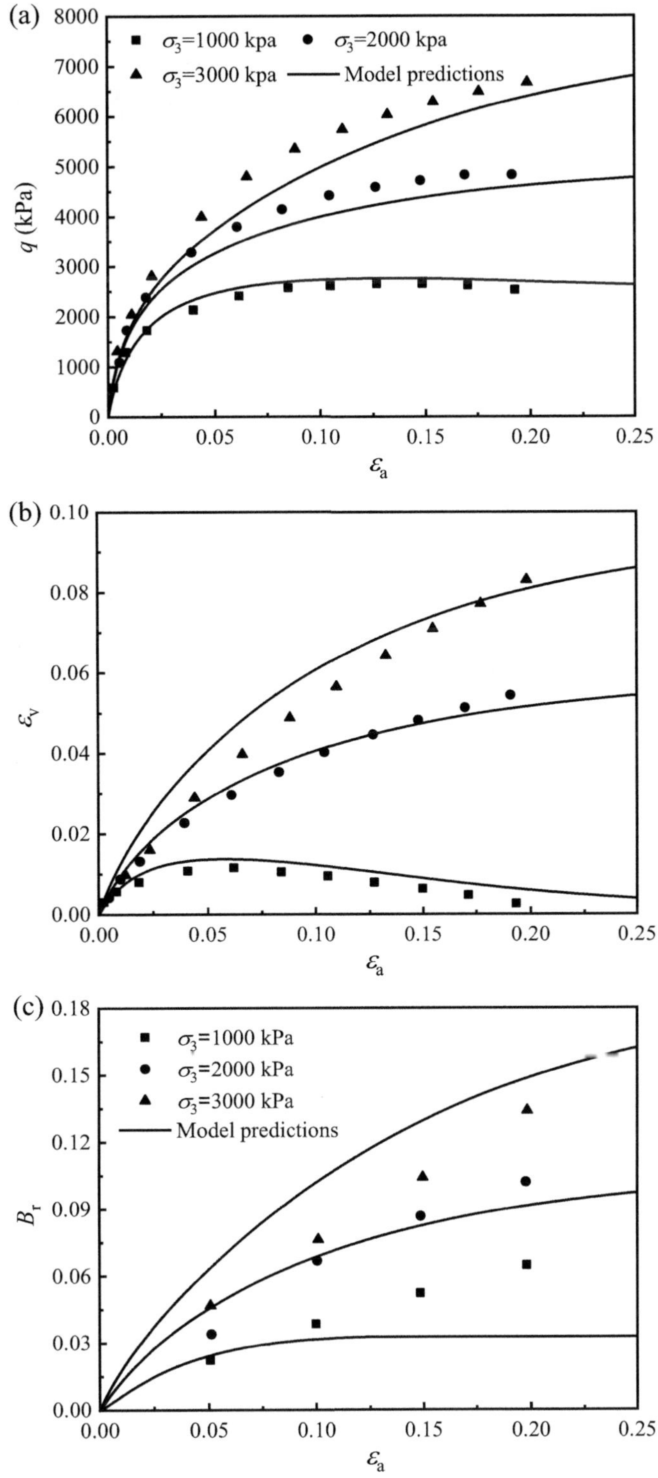

Fig. 3.16 Model predictions for the triaxial tests of Silica sand: **a** deviatoric stress and axial strain, **b** volumetric strain and axial strain, and **c** breakage index in triaxial drained compression tests. (Ma 2022; Ma and Chen 2024)

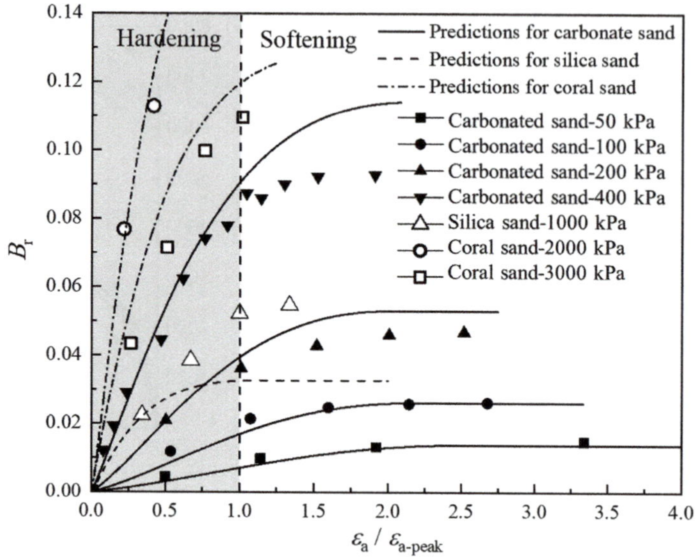

Fig. 3.17 Model predictions for the breakage evolutions of different soils

$$E = E_0 + c_H E_H s_H^t = E_0 + c_H E_H s_H^{t_I}(1 - D_H) \tag{3.43}$$

where t is the time during the decomposition process, t_I is the initial moment, and s_H^t and $s_H^{t_I}$ are the hydrate saturations at time t and t_I, respectively. Equation (3.43) indicates that the elastic modulus of hydrate-bearing sediments is closely related to the degree of decomposition of hydrate.

The modified elastic model is clear in its physical meaning, has few parameters, and is easy to use in large-scale numerical models. However, some parameters, such as E_0 and E_H, are difficult to determine through conventional tests. The purpose here is to obtain these parameters through conventional compression tests, triaxial compression tests, etc.

3.5.1 Volumetric Strain and Modified Elastic Model

Exploitation of natural gas hydrates disrupts the equilibrium state of solid-phase hydrates, causing them to decompose into water and gas. This reduction in the solid-phase component of GHBS increases their pore volume and weakens their intergranular cementation. This results in the release of water and gas, leading to a significant change in pore pressure and thus sediment softening and volume contraction.

Based on indoor triaxial tests and considering the stress and deformation characteristics of GHBS during depressurization decomposition, the volumetric strain for GHBS that accounts for solid phase decomposition can be obtained according to previous study (Yuan et al. 2022). As illustrated in Fig. 3.18, assuming GHBS is

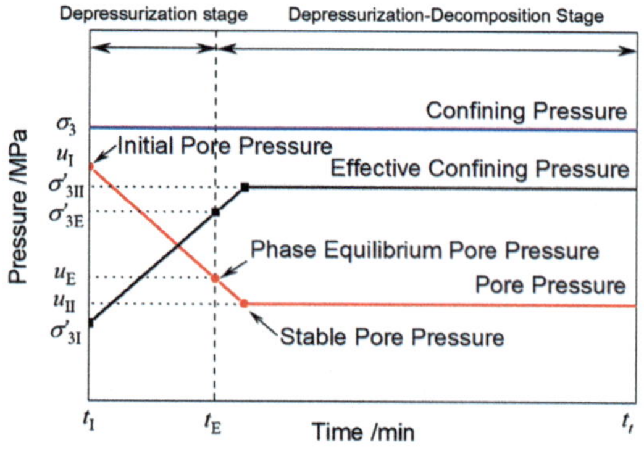

Fig. 3.18 Change in stress states with time in GHBS (reprinted from Yuan et al. 2022, with permission of Chinese journal of geotechnical engineering)

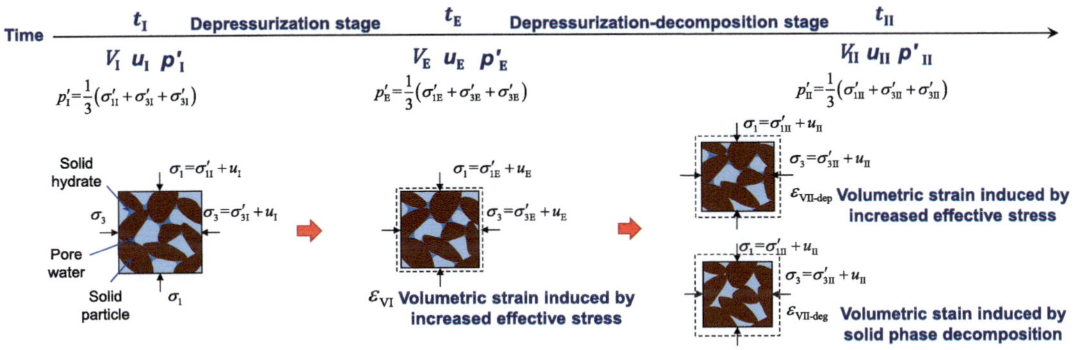

Fig. 3.19 Volumetric strain paths of depressurization-decomposition process in GHBS (modified from Yuan et al. 2022)

initially saturated with pore pressure u_I and confining and axial stresses σ_3 and σ_1, respectively. By reducing pore pressure through fluid drainage, an equilibrium state is reached at time t_E. The stress state of the unit cell is established in two stages, namely depressurisation and depressurisation–decomposition, as defined below, with t_E as the dividing point between these stages.

(I) Depressurization stage. By reducing the pore pressure at a constant rate over the time period $(t_E - t_I)$, the pore pressure decreases from the initial pore pressure u_I to the phase equilibrium pore pressure u_E. During this stage, the hydrates do not decompose, and the expulsion of pore fluid increases the effective confining pressure from σ'_{3I} to σ'_{3E}, causing the volume of the element to shrink from V_I to V_E along the depressurization path shown in Fig. 3.19.

(II) Depressurization-decomposition stage. As hydrates undergo depressurization and decomposition, the effective stress of the unit cell continuously changes, leading to a complex deformation process, and there are various possible stress and decomposition paths for the unit cell. Assuming that the hydrate decomposition rate is independent of the stress level, the strain at any given time during this stage depends only on the decomposition time and the final effective stress state, meaning that the deformation is independent of the stress path. As illustrated by the depressurization-decomposition path in Fig. 3.19, although there are multiple paths between t_I and t_{II}, the final deformations are consistent. Taking the path $V_E \rightarrow V'_{II} \rightarrow V_{II}$ as an example, hydrate decomposition is first considered, followed by the depressurization process. The equilibrium pore pressure u_E decreases to the stable pore pressure u_{II}, while the effective confining pressure increases from σ'_{3E} to σ'_{3II}. Hydrate decomposition generates water and gas, which are expelled during the depressurization process. The volumetric strain during this stage consists of two parts: (1) the volumetric strain caused by depressurization, which is due to the expulsion of pore fluid along the path $V_I \rightarrow V'_{II}$, and (2) the volumetric strain resulting from hydrate decomposition under a certain effective stress along the path $V'_I \rightarrow V_{II}$.

(1) Volumetric Strain During the Depressurization Stage

The volume change path of this process is illustrated by $V_I \rightarrow V_E$ in Fig. 3.19, involves volume contraction of the unit cell due to the drainage of pore fluid during the depressurization process, while hydrate decomposition does not occur during this stage.

Depressurization tests are typically conducted under triaxial stress conditions, considering the correlation between volumetric strain and the mean effective principal stress p'. When the depressurization rate is constant over the time period $(t_E - t_I)$, the mean effective stress p' is a linear function of time t, changing from p'_I to p'_E.

As the compression index χ_I remains unchanged, the total volumetric strain can be obtained. Neglecting the creep effects of GHBS, the volumetric strain ε_{VI} at any time t can be linearly interpolated from total volumetric strain during this stage ($t_I \le t < t_E$), expressed as:

$$\varepsilon_{VI} = \frac{t - t_I}{t_E - t_I}\chi_I \ln\left(\frac{p'_E}{p'_I}\right) \tag{3.44}$$

where t is the time during the depressurization process; ε_{VI} is the volumetric strain caused by depressurization at time$t$; t_E is the time required for the unit cell to decrease from the initial pore pressure to the equilibrium pore pressure; $p'_I = \frac{\sigma'_{1I} + 2\sigma'_{3I}}{3}$ is the initial mean effective principal stress, and $p'_E = \frac{\sigma'_{1E} + 2\sigma'_{3E}}{3}$ is the mean effective principal stress at equilibrium state; e_I and e_E are the void ratios under p'_I andp'_E, respectively; $\chi_I(s_H) = \frac{\lambda_I}{(1+e_I)}$ is the modified slope considering the initial porosity; λ_I is the slope of the $e - \ln p'$ line, which is related to the initial saturation of hydratess_H.

(2) Volumetric Strain During the Depressurization-Decomposition Stage

As shown in Fig. 3.19, the volumetric strain during the stage II of depressurization-decomposition ($V_E \rightarrow V_{II}$) can be divided into two parts: the volumetric strain due to depressurization ($V_E \rightarrow V'_{II}$) and the volumetric strain due to hydrate decomposition ($V'_{II} \rightarrow V_{II}$). After the pore pressure drops to the equilibrium pressure, keep the pore pressure reduction rate unchanged and continue to drop to the predetermined pore pressure u_{II}. During this stage, pore fluid is drained from the unit, and the effective confining stress acts on the softened soil skeleton, resulting in volumetric strain. Additionally, the volumetric strain behavior of hydrate decomposition during $V_E \rightarrow V_{II}$ is obtained by the degree of decomposition D_H determined by the Kim-Bishnoi kinetic model.

Based on the total strain theory, the volumetric strain during the depressurization-decomposition stage ($t_E \le t \le t_{II}$) can be expressed as:

$$\begin{aligned}
\varepsilon_{VII} &= \frac{V_E - V_{II}}{V_E} = 1 - \frac{V'_{II}}{V_E}\frac{V_{II}}{V'_{II}} \\
&= 1 - \left(1 - \frac{\Delta V_{II-dep}}{V_E}\right)\left(1 - \frac{\Delta V_{II-deg}}{V'_{II}}\right) \\
&= 1 - \left(1 - \varepsilon_{VII-dep}\right)\left(1 - \varepsilon_{VII-deg}\right) \\
&= 1 - \left[1 - \chi_t(t)\ln\left(\frac{p'_{II}}{p'_E}\right)\right](1 - \varepsilon_\infty D_H) \\
&\approx \chi_t(t)\ln\left(\frac{p'_{II}}{p'_E}\right) + \varepsilon_\infty D_H
\end{aligned} \tag{3.45}$$

in which, ε_{VII} is the volumetric strain of the unit cell at time t during $V_E \rightarrow V_{II}$. ε_∞ is the volumetric strain induced by complete decomposition of hydrates in GHBS, which is related to the initial saturation of hydrates s_H, and effective confining pressures σ'_{3E}, and σ'_{3II}; $p'_{II} = \frac{\sigma'_{1II} + 2\sigma'_{3II}}{3}$ is the mean effective principal stress of the unit cell when it reaches the stable pore pressure; and $\chi_t(t)$ is the modified slope considering the initial porosity, reflecting the compression characteristics of GHBS, which is related to the hydrate mass in the unit cell at time t.

To analyze the relationship between the volumetric strain from Yuan's study (2022) and that from the modified elastic model, the incremental forms of the volumetric strain can be respectively obtained based on Eqs. (3.43), (3.44), and (3.45):

$$\begin{aligned}
d(\mathrm{tr}(\boldsymbol{\varepsilon})) &= \frac{\chi_t}{\mathrm{tr}(\boldsymbol{\sigma}')}d\left(\mathrm{tr}\left(\boldsymbol{\sigma}'\right)\right) \\
&+ \left[\varepsilon_\infty + (\chi_\infty - \chi_I)\ln\frac{\mathrm{tr}(\boldsymbol{\sigma}')}{\mathrm{tr}(\boldsymbol{\sigma}'_E)}\right]dD_H
\end{aligned} \tag{3.46}$$

$$\begin{aligned}
d(\mathrm{tr}(\boldsymbol{\varepsilon})) &= \frac{1 - 2\mu}{E}d\left(\mathrm{tr}\left(\boldsymbol{\sigma}'\right)\right) \\
&+ \frac{1 - 2\mu}{E}\mathrm{tr}\left(\boldsymbol{\sigma}'\right)\frac{c_H E_H s_H^{t_I}}{E}dD_H
\end{aligned} \tag{3.47}$$

in which, $\mathrm{tr}(\boldsymbol{\varepsilon})$ is the trace of the strain tensor $\boldsymbol{\varepsilon}$, $\mathrm{tr}(\boldsymbol{\sigma}')$ is the trace of the effective stress tensor $\boldsymbol{\sigma}'$, $\mathrm{tr}(\boldsymbol{\sigma}'_E)$ is the trace of the effective stress tensor $\boldsymbol{\sigma}'_E$

at the phase equilibrium moment, μ is the Poisson's ratio, ε_∞ is the volumetric strain when the hydrate is fully decomposed under a certain deviatoric stress, χ_{I} and χ_∞ are the slopes of the $e - \ln p'$ line for GHBS at the initial moment and when fully decomposed, respectively. χ_t is related to χ_{I} and χ_∞, and there is $\chi_t(t) = \chi_{\mathrm{I}} + (\chi_\infty - \chi_{\mathrm{I}})D_{\mathrm{H}}$ (Yuan et al. 2022).

By comparing Eqs. (3.46) and (3.47), it can be obtained that:

$$E = \frac{(1-2\mu)\mathrm{tr}(\boldsymbol{\sigma}')}{\chi_t} = \frac{(1-2\mu)\mathrm{tr}(\boldsymbol{\sigma}')}{\chi_{\mathrm{I}} + (\chi_\infty - \chi_{\mathrm{I}})D_{\mathrm{H}}} \approx \frac{(1-2\mu)\mathrm{tr}(\boldsymbol{\sigma}')}{\chi_{\mathrm{I}}} \left(1 - \left(\frac{\chi_\infty}{\chi_{\mathrm{I}}} - 1\right)D_{\mathrm{H}} + \left(\left(\frac{\chi_\infty}{\chi_{\mathrm{I}}} - 1\right)D_{\mathrm{H}}\right)^2\right) \quad (3.48)$$

Through comparison between Eqs. (3.43) and (3.48), there are:

$$E_0 = \frac{(1-2\mu)\mathrm{tr}(\boldsymbol{\sigma}')}{\chi_{\mathrm{I}}}\left(3 - 3\frac{\chi_\infty}{\chi_{\mathrm{I}}} + \left(\frac{\chi_\infty}{\chi_{\mathrm{I}}}\right)^2\right) \quad (3.49)$$

$$c_{\mathrm{H}}E_{\mathrm{H}}s_{\mathrm{H}}^{t_1} = \frac{(1-2\mu)\mathrm{tr}(\boldsymbol{\sigma}')}{\chi_{\mathrm{I}}}\left(\frac{\chi_\infty}{\chi_{\mathrm{I}}} - 1\right)\left(2 - \frac{\chi_\infty}{\chi_{\mathrm{I}}} - \left(\frac{\chi_\infty}{\chi_{\mathrm{I}}} - 1\right)D_{\mathrm{H}}\right) \quad (3.50)$$

Based on previous derivations and comparisons with triaxial tests for the volumetric strain, the physical meanings and precise expressions of the parameters in the modified elastic model are determined, providing guidance for using this mode in large-scale hydrate exploitation numerical simulations.

3.5.2 Model Validation and Parameter Analysis

Through the above calculations, the volumetric strain of GHBS at any time t during the depressurization-decomposition process can be obtained based on Eq. (3.43), and the degree of decomposition D_{H} can be determined according to Eq. (3.22). The volumetric strain values calculated using the modified elastic model is compared and validated against the depressurization-decomposition tests of GHBS conducted by Hyodo et al. (2013) and Choi et al. (2020). The latter two groups of researchers both used Toyoura sand as the sediment skeleton and prepared samples of GHBS in laboratories for depressurisation–decomposition tests. During the tests, the temperature was kept constant, and the total stress remained unchanged (12 MPa and 6.21 MPa, respectively), with hydrate decomposition achieved by reducing the pore pressure. The volumetric strain–time and decomposition degree-time curves of K_0-consolidated samples undergoing depressurization decomposition are compared, as shown in Figs. 3.20 and 3.21.

As shown in Figs. 3.20 and 3.21, the volumetric strain gradually increases and stabilizes with the decrease in pore pressure and the decomposition of hydrates. The model can reflect two characteristics of hydrate depressurization decomposition: ① Before the pore pressure drops to equilibrium pressure, the sample volume decreases as the pore pressure decreases. This is because the volume change of the sample before hydrate decomposition is due to the expulsion of pore water. When the depressurization rate is high, the volume strain decreases nearly linearly. ② When the pore pressure drops below the equilibrium pressure, the hydrate cementation gradually decreases with the progress of decomposition until it disappears. The water and gas produced by decomposition are expelled from the sample, and the soil particles are arranged more tightly under effective confining pressure, resulting in an increase in volume deformation with hydrate decomposition. The model can effectively characterize the sudden change in volumetric strain rate after hydrate decomposition begins and can reflect the contribution of hydrate decomposition to volumetric strain.

Figure 3.22 illustrates the influence of depressurization rate r_p on the volumetric strain

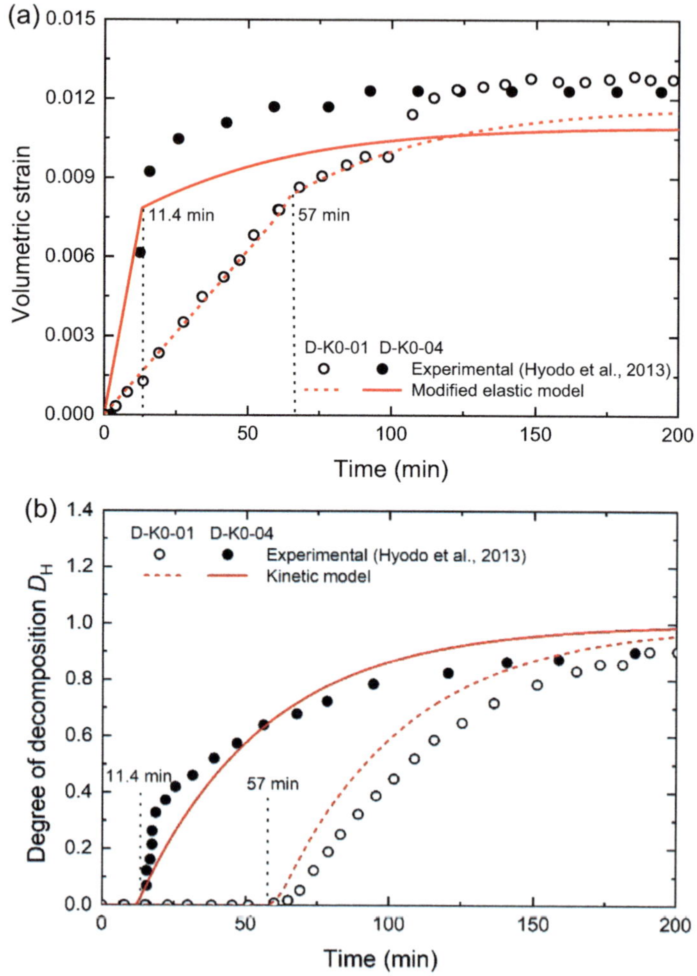

Fig. 3.20 Comparison of results of Hyodo's experiments and proposed models: **a** volumetric strain versus time, and **b** degree of decomposition versus time

of GHBS during the depressurization-decomposition process of hydrates. Before the pore pressure reaches the equilibrium pressure, a lower depressurization rate results in a slower fluid drainage rate and consequently a slower increase in volumetric strain during the depressurization stage. This suggests that a lower depressurization rate is conducive to stabilizing the GHBS in the early stages of exploitation, albeit with an increase in the time required to reduce the initial pore pressure to the hydrate equilibrium pore pressure. Conversely, a higher depressurization rate accelerates fluid drainage from the pores, leading to a faster growth in volumetric strain of GHBS and hence rapid deformation. Additionally, hydrate decomposition began after a shorter depressurisation period and thus was complete within a shorter total time.

Figure 3.23 illustrates the relationship between volumetric strain and degree of decomposition during the depressurization-decomposition process of hydrates. During the depressurization stage, when hydrate decomposition has not yet begun, the volumetric strain evolves. As it proceeds to the decomposition stage, depressurization and decomposition simultaneously cause volume strain, and then the decompression pressure remains stable, and the volume strain increases linearly with the decomposition of hydrates. The variation in hydrate mass does not depend on the

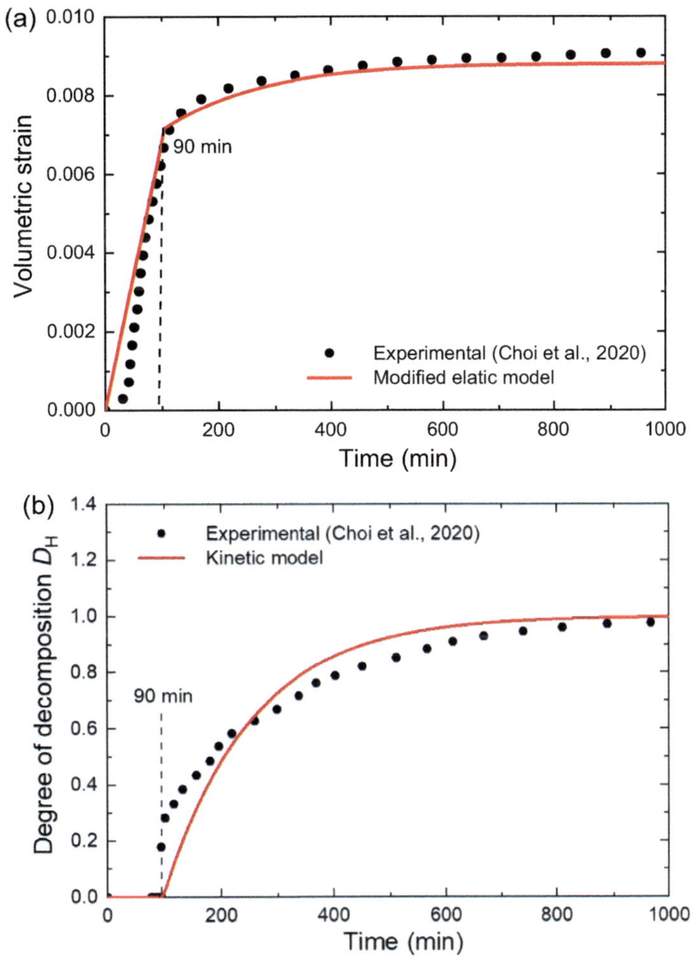

Fig. 3.21 Comparison of results of Choi's experiments and proposed models: **a** volumetric strain versus time, and **b** degree of decomposition versus time

depressurization rate. Selecting an appropriate depressurization rate in engineering applications can balance sediment stability and extraction efficiency.

Figure 3.24 compares the effect of pore pressure reduction $\Delta u = u_{\mathrm{I}} - u_{\mathrm{II}}$ on the volumetric strain of GHBS during hydrate decomposition. As the stable pore pressure u_{II} decreased, i.e., the pore pressure continued to decrease, there was an increase in effective stress. Consequently, there was an increase in the volume of pore fluids that drained, resulting in an increase in the density of the samples and an increase in the stable volumetric strain. The gas production increases with the decrease in stable pore pressure, leading to improved efficiency in hydrate exploitation. In practice, the lower pore pressure in the sediment after exploitation compared to the surrounding environment creates a pressure difference, leading to seawater intrusion and an increase in pore pressure, while the effective stress decreases. This may cause sediment rebound. The lower the stable pore pressure, the greater the increase in effective stress during pore water recovery, and the more significant the potential rebound of the sediment.

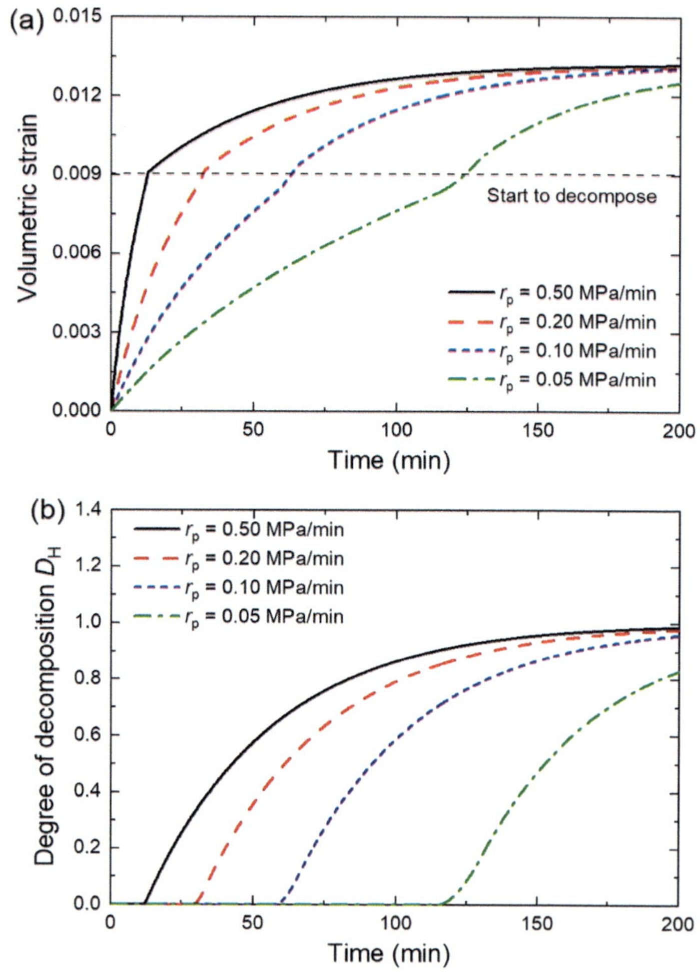

Fig. 3.22 Influence of depressurization rate r_p on volumetric strain of GHBS: **a** volumetric strain versus time, and **b** degree of decomposition versus time ((**a**) is reprinted from Yuan et al. 2022, with permission of Chinese journal of geotechnical engineering)

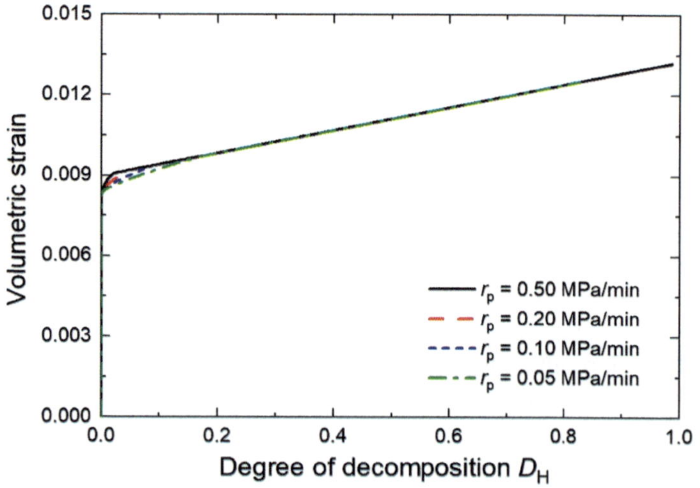

Fig. 3.23 Variation of volumetric stain of GHBS with degree of hydrate decomposition

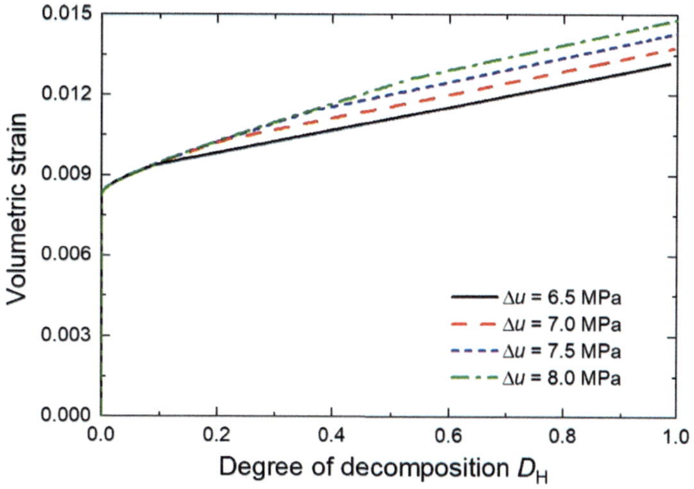

Fig. 3.24 Influence of pore pressure reduction Δu on volumetric strain of GHBS (reprinted from Yuan et al. 2022, with permission of Chinese journal of geotechnical engineering)

To safely and effectively explore the natural methane hydrate, it is crucial to examine the mechanical behavior of methane hydrate-bearing sediments (MHBS). Natural methane hydrate unevenly distributes in pores or bonds with soil particles in MHBS, changing the mechanical behavior of MHBS including stiffness, shear strength and dilatancy. An anisotropic critical state model for MHBS considering hydrate pore-filling and cementing effects was developed to describe this characteristic (Zhu et al. 2024). Based on the unified critical state model for both clay and sand, an equivalent hydrate ratio is defined to address pore-filling effect. Cohesive strength and its hardening law are introduced to characterize hydrate cementation. To describe the anisotropic behavior, inherent anisotropy of soil particles and hydrates are modeled separately, and rotation hardening is introduced to describe the stress-induced anisotropy. Detailed predictions indicate that hydrate pore-filling affects the hydrate-dependent stiffness and dilatancy of MHBS, which become more pronounced with increasing hydrate saturation. Cementing effect increases initial stiffness and peak strength of MHBS. Pronounced influence of inherent anisotropic parameters on pre-peak stress-strain relation of MHBS is noted, and increasing hydrate saturation enhances the effect of hydrate anisotropy. These predictions contribute to a better understanding of the relation between hydrate morphologies and MHBS mechanical properties.

3.6 Constitutive Model for Municipal Solid Waste

3.6.1 Influences of Biodegradation on the Compressibility of Municipal Solid Waste

Chen et al. (2010) conducted multiple-loading one-dimensional compression tests on MSW samples at different ages. The components, bulk density, and initial void ratio of waste samples at different ages were kept the same. Each load level was maintained for 3 days, and Na_2CO_3 solution was injected back into the samples during the compression tests to inhibit biochemical degradation. Therefore, it can be assumed that almost all deformation generated during the tests was primary compression.

Figure 3.25 shows the results of one-dimensional compression tests on waste samples with ages of 3 days and 235 days. It can be seen that with increasing age, the compression lines of MSW moved downwards but the slope increased. These results indicate that after biochemical degradation, the instantaneous compressibility of MSW aged 235 days was less than that of MSW aged 3 days (fresh MSW). The

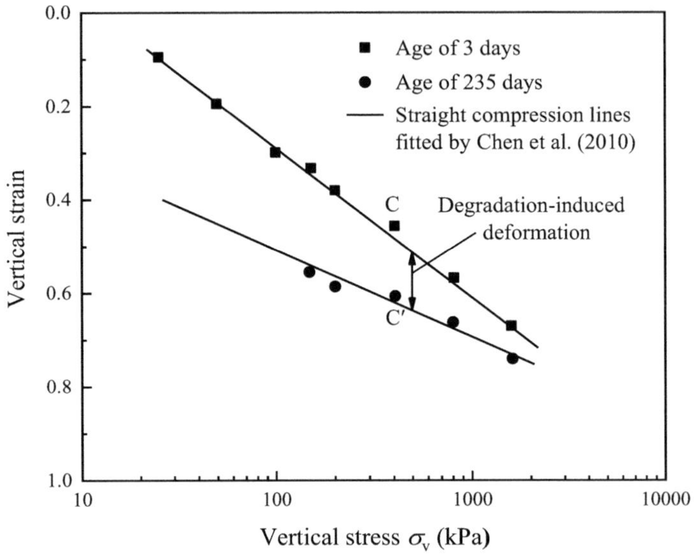

Fig. 3.25 Results of one-dimensional compression tests of MSWs with different ages (Ma 2022; Ma et al. 2024)

vertical distance between compression lines of waste at different ages can be considered as the compression deformation caused by biochemical degradation, which obviously decreases with increasing stress level. This is similar to the phenomenon of strain hardening exhibited by soil under isotropic compression conditions, where both the increase in load and the compression deformation caused by biochemical degradation are inversely proportional to the stress level.

3.6.2 Compression Line Equations of Degraded Municipal Solid Waste

Ma et al. (2024) established a novel compression curve equation of MSW to consider the influences of degradation on the compression characteristics of MSW. Three basic assumptions were made: (i) the compression curve of fresh MSW is a straight line in the *e*–ln*p* plane, which is the same as the compression curves of conventional soils, (ii) the resilience curves of fresh and degraded MSW are both straight lines in the *e*–ln*p* plane and their gradients are independent on the degradation, and (iii) the degradation-induced deformation of MSW is totally plastic.

The plastic strain of fresh solid waste under isotropic compression conditions is given by:

$$\varepsilon_v^p = \frac{\lambda - \kappa}{1 + e_0} \ln \frac{p}{p_0} \tag{3.51}$$

In Fig. 3.25, the compression deformation of degraded solid waste is inversely proportional to the stress level. To describe this experimental phenomenon, a parameter p_b ($p_b \geq 0$) is introduced, where its physical meaning is the additional mean stress required to produce deformation equal to the compression deformation caused by degradation under isotropic compression conditions. With the introduction of p_b, the plastic strain of degraded solid waste under isotropic compression conditions becomes:

$$\varepsilon_v^p = \frac{\lambda - \kappa}{1 + e_0} \ln \frac{p + p_b}{p_0 + p_{b0}} \tag{3.52}$$

where p_{b0} is the initial value of p_b.

The parameter p_b is related to the degree of degradation of MSW. For fresh MSW, $p_b = 0$, and Eq. (3.52) degenerates to Eq. (3.51); for degraded MSW, $p_b > 0$. In the model, it is assumed that the parameter p_b is calculated according to the following equation:

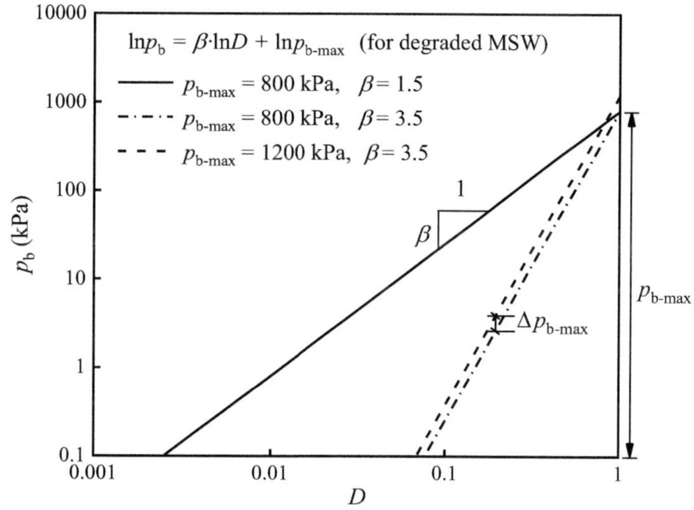

Fig. 3.26 The change rule of variable p_b with parameters p_{b-max} and β of degraded MSW in the dual logarithmic coordinates (Ma 2022; Ma et al. 2024)

$$p_b = p_{b-max} D^{\beta} \tag{3.53}$$

in which, p_{b-max} and β are positive constants, and D represents the degradation degree of MSW, calculated according to the following equation:

$$D = 1 - \frac{m_{sd}^{t}}{m_{sd}^{ini}} \tag{3.54}$$

where m_{sd}^{t} and m_{sd}^{ini} represent the mass of degradable solid phase in the waste at time t and initial time, respectively. For fresh waste, $m_{sd}^{t} = m_{sd}^{ini}$, and $D = 0$; for completed degraded MSW, $m_{sd}^{t} = 0$, and $D = 1$.

Calculating the parameter p_b requires two model constants, p_{b-max} and β. For completely degraded MSW, where $D = 1$, $p_b = p_{b-max}$, indicating that p_{b-max} represents the maximum value p_b can attain during the biochemical degradation process. According to Eqs. (3.52) and (3.53), the maximum deformation induced by degradation depends on the value of p_{b-max}. Therefore, p_{b-max} should be closely related to the total loss of solid phase in the waste skeleton. When other mechanical properties are relatively consistent, the higher the proportion of degradable solid phase in the waste skeleton, the greater the value of p_{b-max}. Additionally, the introduction of the constant β describes the possible nonlinear relationship between p_b and D. As shown in Fig. 3.26, for degraded MSW ($D > 0$), the $p_b - D$ relationship is linear in double-logarithmic coordinates, where β represents the slope of the line controlling the growth rate of $\ln p_b$ with $\ln D$; $\ln p_{b-max}$ represents the intercept of the line, indicating the value of $\ln p_b$ when $\ln D = 0$.

Parameters $p_{b\text{-max}}$ and β can be determined by the long-term degradation compression test of MSW. Based on the definition of variable p_b (Eq. (3.52)), there is a correspondence between the value of p_b and the plastic deformation of MSW. Thus, if the deformations of MSW of two different degradation degrees (excluding the deformation of fresh MSW) were known, the two sets of values of (p_b, D) could be obtained; then, $p_{b\text{-max}}$ and β could be determined based on the straight-line relationship between $\ln p_b$ and $\ln D$ shown in Fig. 3.26.

The differential form of Eq. (3.52) is:

$$d\varepsilon_v^p = \frac{\lambda - \kappa}{1 + e_0} \frac{1}{p + p_b} (dp + dp_b) \tag{3.55}$$

According to Eq. (3.55), when the waste is instantaneously compressed, that is, $dp_b = 0$ and $dp > 0$, the plastic strain modulus decreases with the increase of p, which is consistent with experimental observations. Therefore, introducing the parameter p_b into the compression line equation can describe the influence of

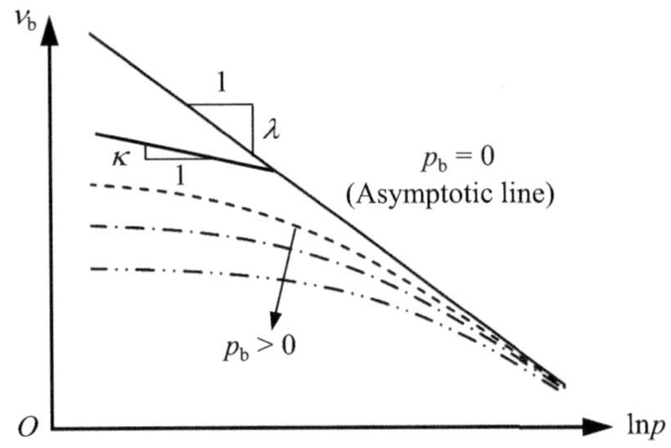

Fig. 3.27 Diagram of compression lines of degraded MSW (Ma 2022; Ma et al. 2024)

degradation on the instantaneous compressibility of the waste.

Based on Eq. (3.52), the compression line equation for degraded waste is given by

$$v_b = v_0 - \left[(\lambda - \kappa)\ln\frac{p + p_b}{p_0 + p_{b0}} + \kappa\ln\frac{p}{p_0}\right] \tag{3.56}$$

in which, v_b represents the specific volume of MSW, differing from that of conventional soils due to the loss of solid components, given by $v_b = V/V_{s0}$, where V is the total volume of the waste and V_{s0} is the initial volume of MSW; v_0 represents the value of v_b at the initial time, where $v_b = V_0/V_{s0}$. The compression lines for degraded MSW using this equation are shown in Fig. 3.27. With the introduction of parameter p_b, the compression lines for degraded MSW form a set of curves, with the compression line for fresh MSW being their asymptote. As the stress level increases, the distance between the compression lines for MSW with different degradation degrees decreases, consistent with experimental observations.

3.6.3 Yield Surface and Loading/Unloading Criteria

The experiments indicate that the MSW specimens consistently exhibit a compression trend during the shearing process, as observed in Karimpour-Fard et al. (2011), Lian (2012) and Ramaiah and Ramana (2017), without transitioning from shear contraction to shear dilation. Therefore, the constitutive model for MSW can be established based on the Modified Cam-clay model. Another three assumptions were made to obtain the yield surface of the model: (iv) the plastic volume and shear deformations of MSW, including the deformations induced by stress and degradation, satisfy the shear-dilatancy equation of the Modified Cam-clay model and the associated flow rule, (v) the gradient of critical state line M is not influenced by the degradation of MSW (McDougall 2007; Babu et al. 2010; Feng et al. 2016a), and (vi) the Poisson's ratio of MSW is independent on the degradation.

Based on the above assumptions, the yield surface of the MSW model is:

$$f = q^2 + M^2p^2 - Mpp_{xb} = 0 \tag{3.57}$$

where p_{xb} is the intersection point of the yield surface with the p-axis. The subscript of "b" indicates that it is influenced by degradation. According to Eq. (3.52), p_{xb} can be obtained:

$$p_{xb} = (p_0 + p_{b0})\exp\left(\frac{1 + e_0}{\lambda - \kappa}\varepsilon_v^p\right) - p_b \tag{3.58}$$

By combining Eqs. (3.57) and (3.58), another form of the yield surface equation can be obtained:

$$f = \ln\left[p\left(1 + \frac{q^2}{M^2p^2}\right) + p_b\right] - \ln(p_0 + p_{b0}) - \frac{1}{c_p}\varepsilon_v^p = 0 \tag{3.59}$$

The variable p_b, which is initially proposed as the 1-D additional mean stress under the isotropic condition in Eq. (3.52), was introduced in the yield surface through the shear dilatancy equation. Similar to the MCC model, the plastic volume strain can be regarded as the hardening variable in Eq. (3.59). Obviously, not only the stress state but also the degradation (i.e., the value of p_b) can influence the hardening degree of MSW, reflecting the influences of degradation on the stress–strain behaviors of MSW.

After introducing the parameter p_b, the model's yield surface becomes a spatial surface in the $p-q-p_b$ coordinates. As shown in Fig. 3.28, the size of the yield surface section depends on the value of p_{xb}, which is related to p_b. According to Eq. (3.58), p_{xb} and p_b are linearly related within the same yield surface, so the intersection of the yield surface with $p-p_b$ forms a straight line, with a slope of 1. From Fig. 3.28, there are two pathways for the waste hardening (expansion of the yield surface): Firstly, increasing stress. For fresh MSW under isotropic compression conditions ($p_b=0$), the stress point is at B_1 on the yield surface $OA_1B_1-O_1C_1D_1$. When the mean stress increases from p_{x1} to $p_{x1}+\Delta p_x$, the yield surface expands to $OA_2B_2-O_1C_2D_2$, indicating the waste hardening. Secondly, degradation (increasing p_b). When the mean stress of the waste remains constant at p_{x1}, p_b increases from 0 to Δp_b due to degradation (i.e., the state point moves from B_1 to B_1'), causing the yield surface to expand from $OA_1B_1-O_1C_1D_1$ to $OA_2B_2-O_1C_2D_2$. The subsequent yield surfaces for these two hardening processes are the same, with equal increments of plastic strain. Therefore, Δp_b is equivalent to Δp_x in terms of producing plastic deformation, and they are numerically equal.

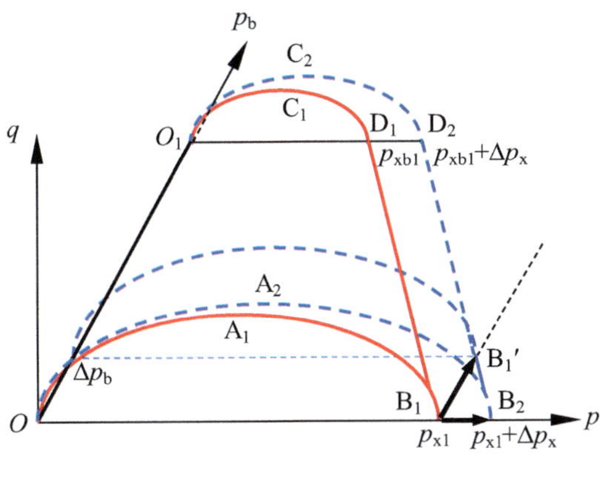

Fig. 3.28 Three-dimensional yield surface of the constitutive model for MSW (Ma 2022; Ma et al. 2024)

The continuity condition for the yield function is given by:

$$df = \frac{\partial f}{\partial \boldsymbol{\sigma}} : d\boldsymbol{\sigma} + \frac{\partial f}{\partial p_b} dp_b + \frac{\partial f}{\partial \varepsilon_v^p} d\varepsilon_v^p = 0 \quad (3.60)$$

Using the associated flow rule, the increment of plastic strain is given by:

$$d\varepsilon_v^p = d\lambda \frac{\partial f}{\partial p} \quad (3.61)$$

in which, $d\lambda$ is the plastic scalar multiplier. Combining Eqs. (3.60) and (3.61), it can be obtained that:

$$d\lambda = \frac{\frac{\partial f}{\partial \boldsymbol{\sigma}} : d\boldsymbol{\sigma} + \frac{\partial f}{\partial p_b} dp_b}{-\frac{\partial f}{\partial \varepsilon_v^p}\frac{\partial g}{\partial p}} \quad (3.62)$$

In this model, loading results in plastic deformation, otherwise it's unloading or neutral loading. Hence, according to Eq. (3.62), the loading–unloading criterion is:

$$\begin{cases} \frac{\partial f}{\partial \boldsymbol{\sigma}} : d\boldsymbol{\sigma} + \frac{\partial f}{\partial p_b} dp_b > 0, loading \\ \frac{\partial f}{\partial \boldsymbol{\sigma}} : d\boldsymbol{\sigma} + \frac{\partial f}{\partial p_b} dp_b = 0, netural\ loading \\ \frac{\partial f}{\partial \boldsymbol{\sigma}} : d\boldsymbol{\sigma} + \frac{\partial f}{\partial p_b} dp_b < 0, unloading \end{cases} \quad (3.63)$$

Taking isotropy as an example, the loading–unloading paths of the model are shown in Fig. 3.29. As mentioned earlier, the intersection of the yield surface with the $p-p_b$ plane is a straight line with a gradient of 1. Therefore, when the angle α between the path and the horizontal direction is in the range of 135–180°, the path points towards the inner side of the yield surface projection, indicating an unloading path, such as path 01. When the angle α is within the range of 0–135°, the path points to the outer side of the projected yield surface, indicating a loading path (α=135° represents neutral loading), such as paths 02 and 03. Based on the stress variation, the loading region can be further divided into the stress reduction region II and the stress increase region III. In region I, although the waste skeleton weakens due to degradation,

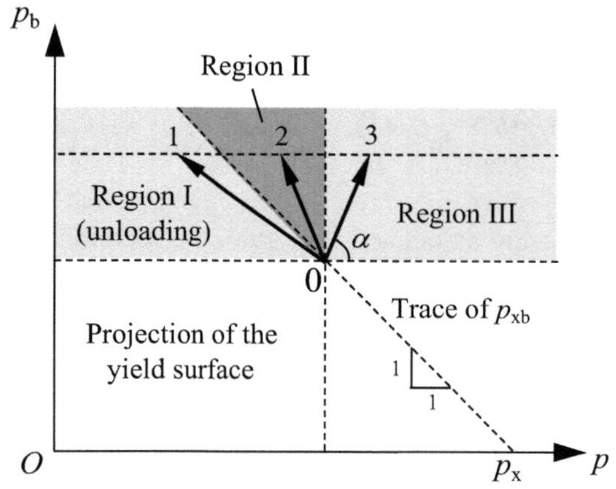

Fig. 3.29 Diagram of loading and unloading paths during isotropic compression of MSW (Ma 2022; Ma et al. 2024)

the magnitude of stress reduction is large, and the balance between stress level and skeleton stiffness is maintained, preventing plastic deformation of the waste. However, in region II, the stress reduction magnitude is small, and the balance between stress level and skeleton stiffness cannot be maintained, resulting in plastic deformation of the waste.

3.6.4 Elastoplastic Constitutive Relationship

The incremental form of the elastoplastic constitutive relation for this model can be derived by combining Eqs. (3.60) to (3.62) and Eq. (3.38):

$$d\boldsymbol{\sigma} + \frac{\boldsymbol{D}^{\mathrm{e}} : \frac{\partial g}{\partial \boldsymbol{\sigma}} \frac{\partial f}{\partial p_{\mathrm{b}}}}{-\frac{\partial f}{\partial \varepsilon_{\mathrm{v}}^{\mathrm{p}}} \frac{\partial g}{\partial p} + \frac{\partial f}{\partial \boldsymbol{\sigma}} : \boldsymbol{D}^{\mathrm{e}} : \frac{\partial g}{\partial \boldsymbol{\sigma}}} \frac{\partial p_{\mathrm{b}}}{\partial D} dD = \left(\boldsymbol{D}^{\mathrm{e}} - \frac{\boldsymbol{D}^{\mathrm{e}} : \frac{\partial f}{\partial \boldsymbol{\sigma}} \otimes \frac{\partial g}{\partial \boldsymbol{\sigma}} : \boldsymbol{D}^{\mathrm{e}}}{-\frac{\partial f}{\partial \varepsilon_{\mathrm{v}}^{\mathrm{p}}} \frac{\partial g}{\partial p} + \frac{\partial f}{\partial \boldsymbol{\sigma}} : \boldsymbol{D}^{\mathrm{e}} : \frac{\partial g}{\partial \boldsymbol{\sigma}}} \right) : d\boldsymbol{\varepsilon} \tag{3.64}$$

Compared to the constitutive relation of conventional elastoplastic models, the left-hand side of the equation includes an additional term related to the increment of p_b, which represents an additional stress tensor equivalent to the deformation effect of degradation. It is important to note that in Eq. (3.52), the parameter p_b is introduced as an additional stress under isotropic compression conditions. Through the above calculations, the one-dimensional additional stress (scalar) is extended to a three-dimensional stress tensor.

3.6.5 Model Evaluation

(1) Long-Term Deformation of MSW

At the landfill scale, the degradation level of the waste can be calculated according to Eq. (3.54). However, at the representative element scale, due to the relatively small volume and simple composition of MSW specimens, the degradation level D is calculated according to the first-order kinetics equation:

$$D = 1 - exp^{-kt} \tag{3.65}$$

in which, t is the degradation time (days), and k is the degradation rate (days^{-1}).

The influence of the constants $p_{b-\max}$ and β on the model's calculation results is discussed through an example first. Using the model to calculate the volumetric deformation caused by degradation, the mean stress p of the waste is 200 kPa, and the initial void ratio e_0 is 2.5. The parameters used for the calculation are shown in Table 3.3. The calculation results are shown in Fig. 3.30. The maximum volumetric strain caused by degradation increases with the increase of the constant $p_{b-\max}$. When $p_{b-\max}$ is the same, the curve with β=1.5 grows faster in the initial stage, and then its growth rate is gradually surpassed by the curve with β=3.5. Eventually, both curves reach the same maximum deformation.

This model is used to predict the results of the long-term one-dimensional compression tests conducted by Chen et al. (2010), with an initial stress ratio of 4.0, and an overlying load of 150 kPa during degradation. The parameter values used for the calculation are referenced from Chen et al. (2010) and Gao and Kavazanjian (2022), as shown in Table 3.4. From Fig. 31a, it can be seen that the model can predict the

Table 3.3 Model parameters used in the case study

λ	κ	ν	M	k(day^{-1})
0.2	0.02	0.3	1.5	0.011

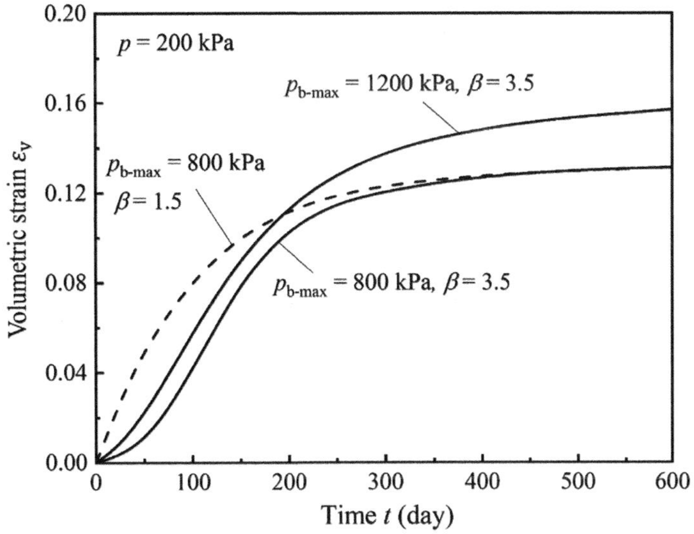

Fig. 3.30 Influences of constants p_{b-max} and β on the predicted volumetric strain caused by degradation (Ma 2022; Ma et al. 2024)

Table 3.4 Model parameters used for predicting the one-dimensional compression tests of MSW (Ma 2022; Ma et al. 2024)

λ	κ	ν	M	k(day^{-1})	p_{b-max}	β
0.675	0.08	0.3	1.5	0.00658	1000	2.9

compression lines of waste at 3 days and 235 days reasonably well. In Fig. 31b, due to the closed state of the vent valve during the Day 20–50 period of the test, the waste rebounded under the pressure of gas. Excluding the deformation data during this period, the model's prediction of the long-term vertical strain of the waste is very close to the experimental data.

(2) Triaxial Test of MSW

The model is adopted to predict the stress–strain relationship of fresh MSW and 4-year MSW in triaxial drained compression tests. The experimental data are from Machado et al. (2008), and the calculation parameters are shown in Table 3.5. Due to the lack of experimental data for 4-year MSW, it is difficult to accurately determine the values of k, p_{b-max}, and β. Therefore, the value of p_b for the 4-year MSW is obtained by fitting the experimental data. The initial void ratio of the MSW specimen is 3.2, and the confining pressure for the triaxial test is 100 and 200 kPa.

The test data and the model predictions for the triaxial compression tests are shown in Fig. 3.32. The stress–strain behaviors of MSW at high strain level are not the focus of consideration in the engineering practice; thus, the maximum axial strain in the comparation was selected to 15% (Zhan et al. 2008). When the axial strains of MSW specimens are lower than 15%, the model predictions are comparable with the test data. An interesting phenomenon shown by the test data is that the curve of degraded MSW is above the curve of fresh MSW in the q–ε_a plane. In other

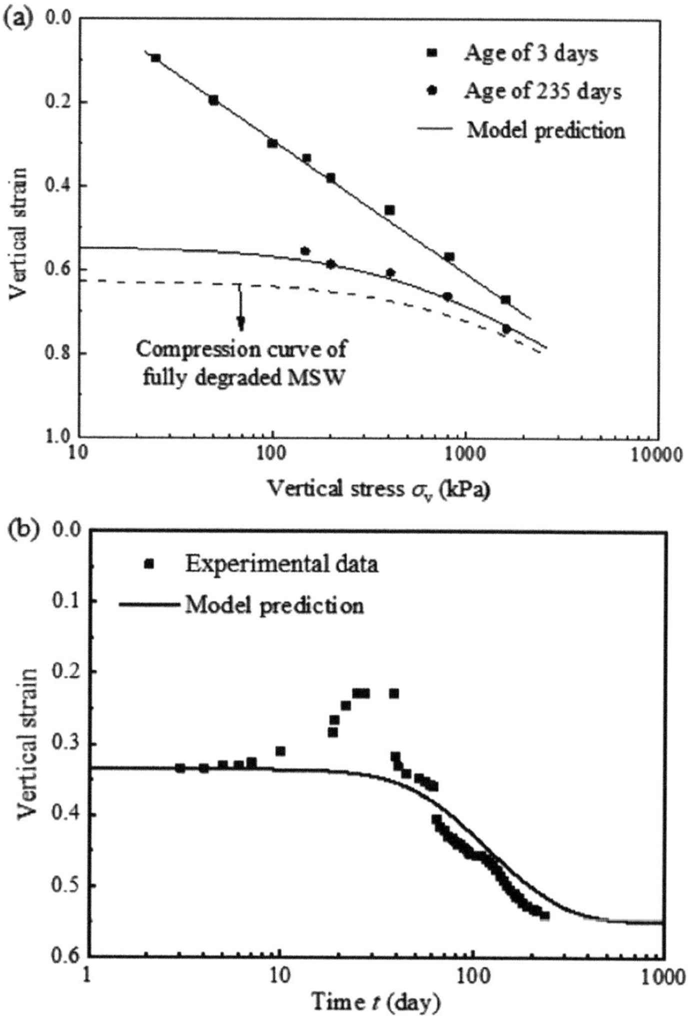

Fig. 3.31 Results of one-dimensional compression tests and model predictions: **a** compression lines of MSWs with various ages, and **b** long-term vertical strain of MSW (Ma 2022; Ma et al. 2024)

Table 3.5 Model parameters used for predicting the triaxial tests of MSW

Age	λ	κ	v	M	p_0
Fresh	0.25	0.02	0.25	1.2	0
4 years	0.25	0.02	0.25	1.2	700

words, at the same axial strain, the deviatoric stress q of degraded MSW is larger than that of fresh MSW. Similar phenomenon was also observed from the tests conducted by Bareither et al. (2012) and Feng et al. (2016b). As shown in Fig. 32a, through introducing the variable p_b, of which the physical significance is an additional mean stress, the constitutive model for MSW can reflect the above influence of degradation on the shear stress–strain behaviors of MSW in the triaxial tests.

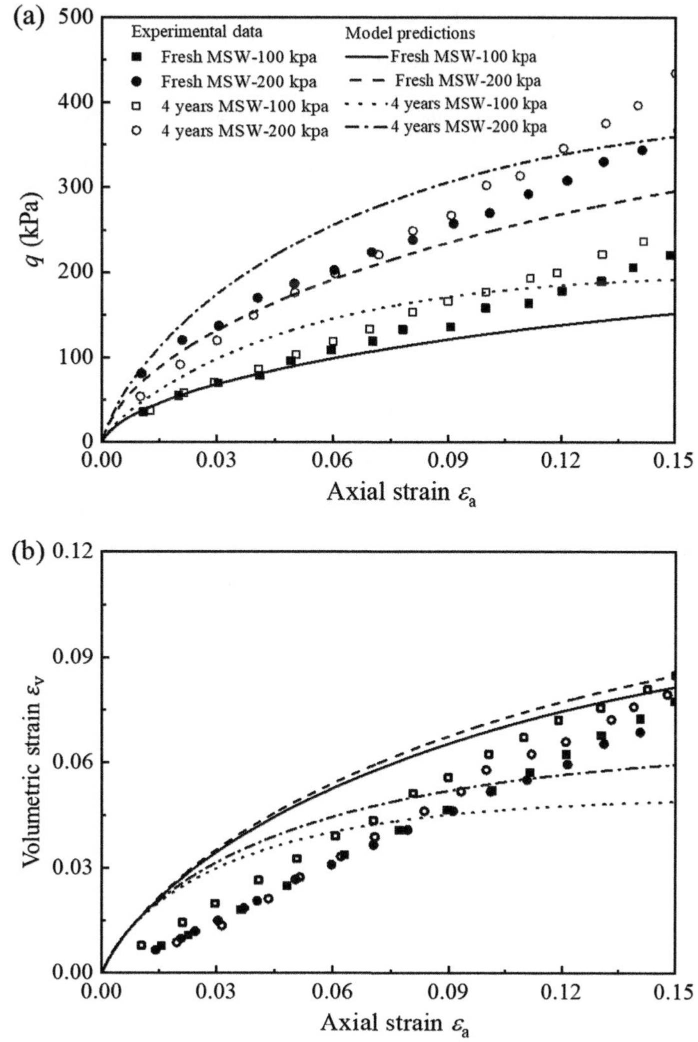

Fig. 3.32 Results of triaxial tests of MSW and model predictions: **a** deviatoric stress and axial strain, and **b** volumetric strain and axial strain (Ma 2022; Ma et al. 2024)

References

Babu GLS, Reddy KR, Chouksey SK (2010) Constitutive model for municipal solid waste incorporating mechanical creep and biodegradation-induced compression. Waste Manag (Elmsford) 30(1):11–22

Bareither CA, Benson CH, Edil TB (2012) Effects of waste composition and decomposition on the shear strength of municipal solid waste. J Geotech Geoenviron Eng 138(10):1161–1174

Chen YM, Ke H, Fredlund DG, Zhan LT, Xie Y (2010) Secondary compression of municipal solid wastes and a compression model for predicting settlement of municipal solid waste landfills. J Geotech Geoenviron Eng 136(5):706–717

Chen YM, Ma PC, Tang Y (2020) Constitutive models and hypergravity physical simulation of soils. Chin J Theoret Appl Mech 52(4):901–915 (in Chinese)

Choi JH, Lin JS, Dai S, Lei L, Seol Y (2020) Triaxial compression of hydrate-bearing sediments undergoing hydrate dissociation by depressurization. Geomech Energy Environ 23:100187

Daouadji A, Hicher P, Rahma A (2001) An elastoplastic model for granular materials taking into account grain breakage. Eur J Mech Solids 20(1):113–137

Feng SJ, Cao BY, Bai ZB, Yin ZY (2016a) Constitutive model for municipal solid waste considering the effect of biodegradation. Géotechnique Lett 6(4):244–249

Feng SJ, Gao KW, Chen YX, Li Y, Zhang LM, Chen HX (2016b) Geotechnical properties of municipal solid waste at Laogang Landfill, China. Waste Manage 63:354–365

Gao W, Chen YM, Zhan LT, Bian XC (2015) Engineering properties for high kitchen waste content municipal solid waste. J Rock Mech Geotech Eng 7 (6):646–658
Gao W, Kavazanjian E Jr (2022) A constitutive model for municipal solid waste considering mechanical creep and biodegradation-induced compression. Acta Geotech 17(1):37–63
Hu W, Yin ZY, Dano C, Hicher P (2011) A constitutive model for granular materials considering grain breakage. Sci China Technol Sci 54(8):2188–2196
Hyodo M, Yoneda J, Yoshimoto N, Nakata Y (2013) Mechanical and dissociation properties of methane hydrate-bearing sand in deep seabed. Soils Found 53 (2):299–314
Jiang MD, Yang ZX, Barreto D, Xie YH (2018) The influence of particle-size distribution on critical state behavior of spherical and non-spherical particle assemblies. Granular Matter 20:1–15
Karimpour-Fard M, Machado SL, Shariatmadari N, Noorzad A (2011) A laboratory study on the MSW mechanical behavior in triaxial apparatus. Waste Manage 31(8):1807–1819
Ke H, Liu JL, Chen YM, Zhan LT, Xu XB (2010) Biodegradation-compression tests on municipal solid waste subjected to different vertical pressures. Chin J Geotech Eng 32(10):1610–1615 (in Chinese)
Li G, Liu YJ, Dano C, Hicher PY (2015) Grading-dependent behavior of granular materials: from discrete to continuous modeling. J Eng Mech 141 (6):04014172
Li JC (2018) Studies on static and seismic stability of landfills with high water level by centrifugal model tests. Ph.D. thesis, Zhejiang University, Hangzhou, China (in Chinese)
Li XS, Dafalias YF (2000) Dilatancy for Cohesionless soils. Géotechnique 50(4):449–460
Lian BQ (2012) Study on large strain triaxial tests of municipal solid waste. Master thesis, Zhejiang University, Hangzhou, China (in Chinese)
Liu YJ (2014) Modeling the influence of the particle size distribution on the critical state mechanical behavior of granular material. Ph.D. thesis, Shanghai University, Shanghai, China (in Chinese)
Ma PC (2022) Study on constitutive models for soils with degradable solid skeleton and their applications. Ph.D. thesis, Zhejiang University, Hangzhou, China (in Chinese)
Ma PC, Chen YM (2024) A skeleton weakening constitutive model for breakable particle soils. Computers and Geotechnics (under review)
Ma PC, Chen YM, Xu WJ, Hu J (2024) A novel constitutive model for degradable soil: municipal solid waste as a case study. Computers and Geotechnics (under review)
Machado SL, Vilar OM, Carvalho MDF (2008) Constitutive model for long term municipal solid waste mechanical behavior. Comput Geotech 35(5):775–790
McDougall JA (2007) Hydro-bio-mechanical model for settlement and other behaviour in landfilled waste. Comput Geotech 34(4):229–246
Miyazaki K, Tenma N, Aoki K, Yamaguchi T (2012) A nonlinear elastic model for triaxial compressive properties of artificial methane-hydrate-bearing sediment samples. Energies 5(10):4057–4075
Moridis GJ, Queiruga AF, Reagan MT (2019) Simulation of gas production from multilayered hydrate-bearing media with fully coupled flow, thermal, chemical and geomechanical processes using TOUGH+ Millstone. Part 1: Numerical modeling of hydrates. Transp Porous Media 128:405–430
Ramaiah BJ, Ramana GV (2017) Study of stress-strain and volume change behavior of emplaced municipal solid waste using large-scale triaxial testing. Waste Manage 63:366–379
Reddy K, Hettiarachchi H, Gangathulasi J, Bogner J, Lagier T (2009) Geotechnical properties of synthetic municipal solid waste. Int J Geotech Eng 3(3):429–438
Roscoe KH, Burland JB (1968) On the generalized stress-strain behaviour of wet clay. Cambridge University Press, Engineering plasticity, Cambridge, pp 535–609
Roscoe KH, Schofield AN, Thurairajah A (1963) Yielding of clays in states wetter than critical. Géotechnique 13(3):211–240
Rowe PW (1962) The stress-dilatancy relation for static equilibrium of an assembly of particles in contact. Proc Royal Soc London. Series A. Mathem Phys Sci 269(1339):500–527
Santamarina JC, Ruppel C (2010) The impact of hydrate saturation on th mechanical, electrical, and thermal properties of hydrate-bearing sand, silts, and clay. In Geophysical Characterization of Gas Hydrates, Society of Exploration Geophysicists, pp 373–384
Schofield AN, Wroth P (1968) Critical state soil mechanics. McGraw-hill, London
Shen ZJ (2000) Theoretical Soil Mechanics. China Water & Power Press (in Chinese)
Shin H, Santamarina JC (2019) An implicit joint-continuum model for the hydro-mechanical analysis of fractured rock masses. Int J Rock Mech Min Sci 119:140–148
Singh MK, Sharma JS, Fleming IR (2009) Shear strength testing of intact and recompacted samples of municipal solid waste. Can Geotech J 46(10):1133–1145
Sun X, Li YH, Liu Y, Song YC (2019) The effects of compressibility of natural gas hydrate-bearing sediments on gas production using depressurization. Energy 185:837–846

Wang ZN, Wang G, Ye QG (2020) A constitutive model for crushable sands involving compression and shear induced particle breakage. Comput Geotech 126:103757

Xie Y, Chen YM, Tang XW, Ke H, Zhan LT (2005) Development and application of biodegradation-compression test apparatus for municipal solid waste. Chin J Geotech Eng 27(5):571–578 (in Chinese)

Yamamuro JA, Lade PV (1996) Drained sand behavior in axisymmetric tests at high pressures. J Geotech Geoenviron Eng 122(2):109–119

Yan WM, Dong J (2011) Effect of particle grading on the response of an idealized granular assemblage. Int J Geomech 11(4):276–285

Yang J, Luo XD (2018) The critical state friction angle of granular materials: does it depend on grading? Acta Geotech 13:535–547

Yao YP (2015) Advanced UH models for soils. Chin J Geotech Eng 37(2):193–217 (in Chinese)

Yao YP, Sun DA, Matsuoka H (2008) A unified constitutive model for both clay and sand with hardening parameter independent on stress path. Comput Geotech 35(2):210–222

Yao YP, Hou W, Zhou AN (2009) UH model: three-dimensional unified hardening model for overconsolidated clays. Géotechnique 59(5):451–469

Yao YP, Zhang BY, Zhu JG (2012) Behaviors, constitutive models and numerical simulation of soils. Chin Civil Eng J 45(3):127–150 (in Chinese)

Yuan SM, Wang LJ, Zhu B, Chen YM (2022) Volumetric stain analysis model for gas hydrate-bearing sediment considering effects of hydrate dissociation. Chin J Geotechn Eng 44(6):1044–1052 (in Chinese)

Yu FW (2017) Characteristics of particle breakage of sand in triaxial shear. Powder Technol 320:656–667

Yu FW (2018) Particle breakage in triaxial shear of a coral sand. Soils Found 58(4):866–880

Zhan TL, Chen YM, Ling WA (2008) Shear strength characterization of municipal solid waste at the Suzhou landfill, China. Eng Geol 97(3–4):97–111

Zhu B, Yuan SM, Wang LJ, Liu YJ, Chen YM (2024) A critical state constitutive model for methane hydrate-bearing sediments considering hydrate pore-filling and cementing effects. Int J Numer Anal Methods Geomech 0:1–17

Liquid–Gas Seepage Constitutive Models for Skeleton-Degradable Soils

4

Abstract

This chapter elucidates the properties of liquid–gas flows and develops a dual-porosity flow constitutive model for skeleton-degradable soils. The degradation of the soil skeleton significantly alters soil pore characteristics such as porosity, pore size, and tortuosity, affecting liquid and gas seepage properties. Considering MSW as an example, this chapter discusses its intrinsic permeability, water retention curve, and relative permeability. The effects of compression, degradation, and anisotropy on the flow characteristics of MSW are examined. To account for the heterogeneity of porous media such as MSW, a dual-porosity flow model is developed for developing new water retention curve and to analyze two-phase flow. Finally, the evolution of pore water and pore gas pressures in HFWC and LFWC MSW landfills is analyzed and compared, revealing faster saturation increments and higher pore gas pressures in HFWC landfills compared to those in LFWC MSW landfills. This comprehensive summary offers valuable insights into the dynamic behaviors of liquid and gas flows in skeleton-degradable soils.

Keywords

Skeleton-degradable soil · Liquid–gas flow · Municipal solid waste · Intrinsic permeability · Anisotropy · Compression · Degradation · Water retention curve · Relative permeability coefficient · Dual-porosity flow model · Leachate level · Landfill

4.1 General Introduction

Soil is a fragmented porous medium with interconnected pores. The movement of liquid and gas substances through soil predominantly occurs through convection driven by pressure gradients. The velocity of this convection in saturated or completely dry states is primarily determined by the intrinsic permeability of the soil. However, under unsaturated conditions, the convection velocity is also influenced by the water retention curve (WRC) and relative permeability. Notably, the degradation of the soil skeleton alters its pore characteristics such as porosity, pore size, and tortuosity, leading to changes in liquid and gas seepage properties.

This chapter takes MSW as an example and elaborates on its intrinsic permeability, WRC, and relative permeability. In addition, this

Y. Chen, *Soil Degradation-Consolidation Theory and Its Applications*,
https://doi.org/10.1007/978-981-97-7985-7_4

chapter analyses the effects of compression and degradation on the flow characteristics of MSW, and introduces a dual-porosity flow model for heterogeneous porous media such as MSW. Based on this model, two-phase flow governing equations considering the mutual inhibition of liquid and gas migration are derived. Finally, the evolutions of pore water pressure and pore gas pressure in landfills are analyzed.

4.2 Intrinsic Permeability

The intrinsic permeability of soil, reflecting the ease of fluid passage, is solely determined by the pore structure of the soil and is unaffected by the properties of the fluid. This parameter is typically determined by testing the liquid-phase and gas-phase permeability coefficients of soil under fully saturated and completely dry conditions, respectively. The specific relations used to derive the intrinsic permeability using these coefficients are presented in Eq. (4.1) and Eq. (4.2). Notably, the intrinsic permeability obtained from liquid-phase permeability tests is generally lower than that obtained from gas-phase permeability tests. This discrepancy is attributed to the Klinkenberg effect (Klinkenberg 1941), which refers to the slippage of gas molecules on particle surfaces.

$$k_s = \frac{k_i \rho_L g}{\mu_L} \tag{4.1}$$

$$k_d = \frac{k_i \rho_G g}{\mu_G} \tag{4.2}$$

where k_i denotes the intrinsic permeability (m^2); k_s represents the liquid-phase permeability coefficient under fully saturated conditions (m/s); k_d denotes the gas-phase permeability coefficient under completely dry conditions (m/s); ρ_L and ρ_G refer to the densities of the liquid and gas phases, respectively (kg/m^3); μ_L and μ_G indicate the viscosities of the liquid and gas phases, respectively (kg/m/s); and g represents the acceleration due to gravity (m/s^2).

4.2.1 Effects of Compression and Degradation

The intrinsic permeability of MSW is predominantly influenced by both compression and degradation. Several studies have documented these influences. For instance, through large-scale Pitsea compression cell tests, Powrie and Beaven (1999) demonstrated that the saturated permeability coefficient (k_s) of waste decreased from approximately 1×10^{-4} m/s to 1×10^{-7} m/s as the vertical effective stress increased from 20 kPa to 500 kPa. Additionally, through accelerated degradation and permeability tests, Hossain et al. (2009) demonstrated that k_s decreased gradually from 1×10^{-4} m/s to 1×10^{-5} m/s as degradation progressed. This decrease was attributed to the reduced particle sizes of degradable solid-phase particles and pore filling by the resulting fine particles. Throughout the degradation process of waste samples, the fine particle mass fraction (residue on a 200-mesh sieve) increased from 10% to 39%.

Compared with waste at shallow landfill depths, waste at deeper landfill depths experiences greater compression and longer degradation periods, resulting in a gradual decrease in permeability coefficient with depth. Zhan et al. (2014) conducted pumping tests at the Qizishan Landfill in Suzhou and subsequently performed laboratory tests on the samples collected from different depths. They observed that as the burial depth increased from 2.5 m to 17.5 m, the saturated permeability coefficient decreased from 5.5×10^{-6} m/s to 1.4×10^{-7} m/s. To quantitatively characterize the variation in the intrinsic permeability of MSW with compression and degradation, Ke et al. (2017) prepared artificial waste samples with different degrees of degradation and performed saturated permeability coefficient tests under varying confining pressures using a triaxial permeability apparatus. The results, depicted in Fig. 4.1, revealed that as the porosity decreased from 0.75 to 0.60, the intrinsic permeability coefficients of samples with the

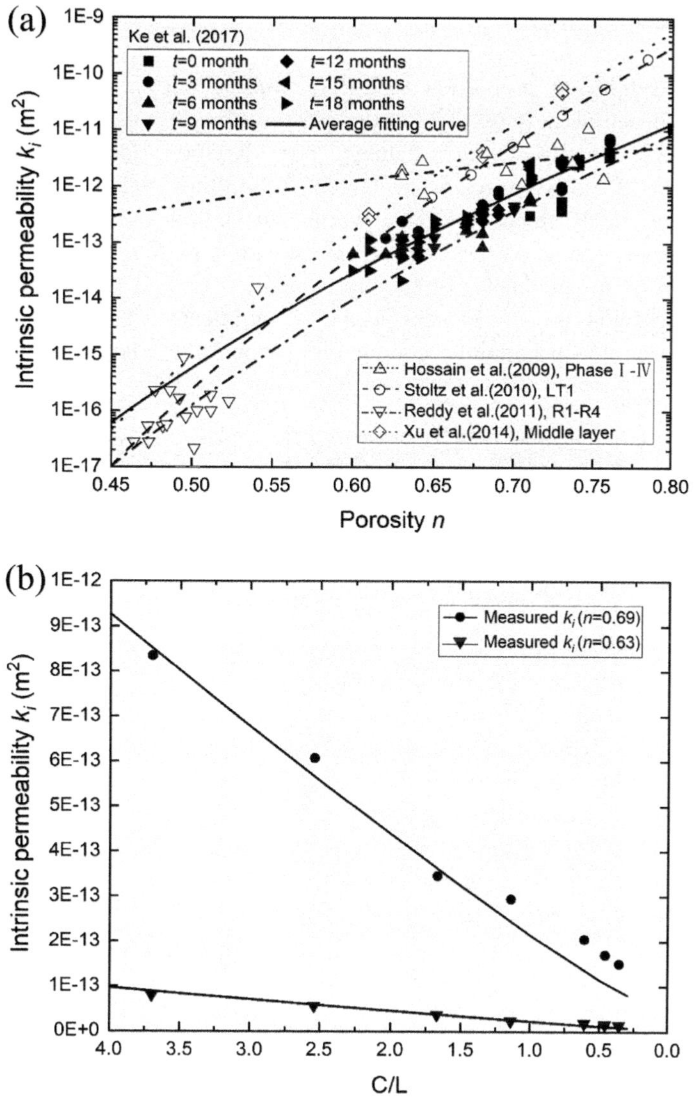

Fig. 4.1 Variation in the intrinsic permeability (k_i) of MSW with compression and degradation: **a** variation of k_i with n for specimens with same degradation age, and **b** variation of k_i with C/L for specimens with the same porosity (reprinted from Ke et al. 2017, with permission from Elsevier)

same degradation age decreased by approximately 2–3 orders of magnitude. Furthermore, as the ratio of cellulose to lignin content (C/L) decreased from 4.7 (fresh waste) to 0.4 (completely degraded waste), the intrinsic permeability coefficients of samples with the same porosity decreased by approximately one order of magnitude. Based on these experiments, a predictive model capturing the variation of intrinsic permeability with both compression and degradation is proposed, as presented in Eq. (4.3). Within this framework, degradation effects are reflected through the stabilization index (C/L), while compression effects are reflected through the porosity (n).

$$k_i = \left[(A_0 - A_\infty)\left(\frac{\text{C/L} - \text{C/L}_\infty}{\text{C/L}_0 - \text{C/L}_\infty}\right) + A_\infty\right] n^B \frac{n^3}{(1-n)^2} \tag{4.3}$$

where C/L and C/L_∞ represent the content ratios of cellulose to lignin in fresh MSW and completed degraded MSW, respectively; and A_0, A_∞, and B denote model parameters (m^2).

4.2.2 Anisotropy

A significant proportion of MSW consists of flattened plastics, textiles, and paper, resulting in MSW initially having a flattened pore distribution. As MSW undergoes layered landfilling, substantial settlement occurs within the landfill body, causing the pore arrangement within the MSW to gradually orient horizontally. This horizontal pore alignment leads to significant anisotropy in hydraulic conductivity, where the horizontal hydraulic conductivity exceeds the vertical hydraulic conductivity.

Notably, the ratio of horizontal hydraulic conductivity to vertical hydraulic conductivity is termed as the anisotropy of seepage (A). Ke et al. (2022) utilised computed tomography (CT) to visualise the pore structure of MSW and used a self-developed bidirectional permeameter to test the anisotropy of MSW under different pressures. The captured CT images, depicted in Fig. 4.2, demonstrated that as the load increased, the two-

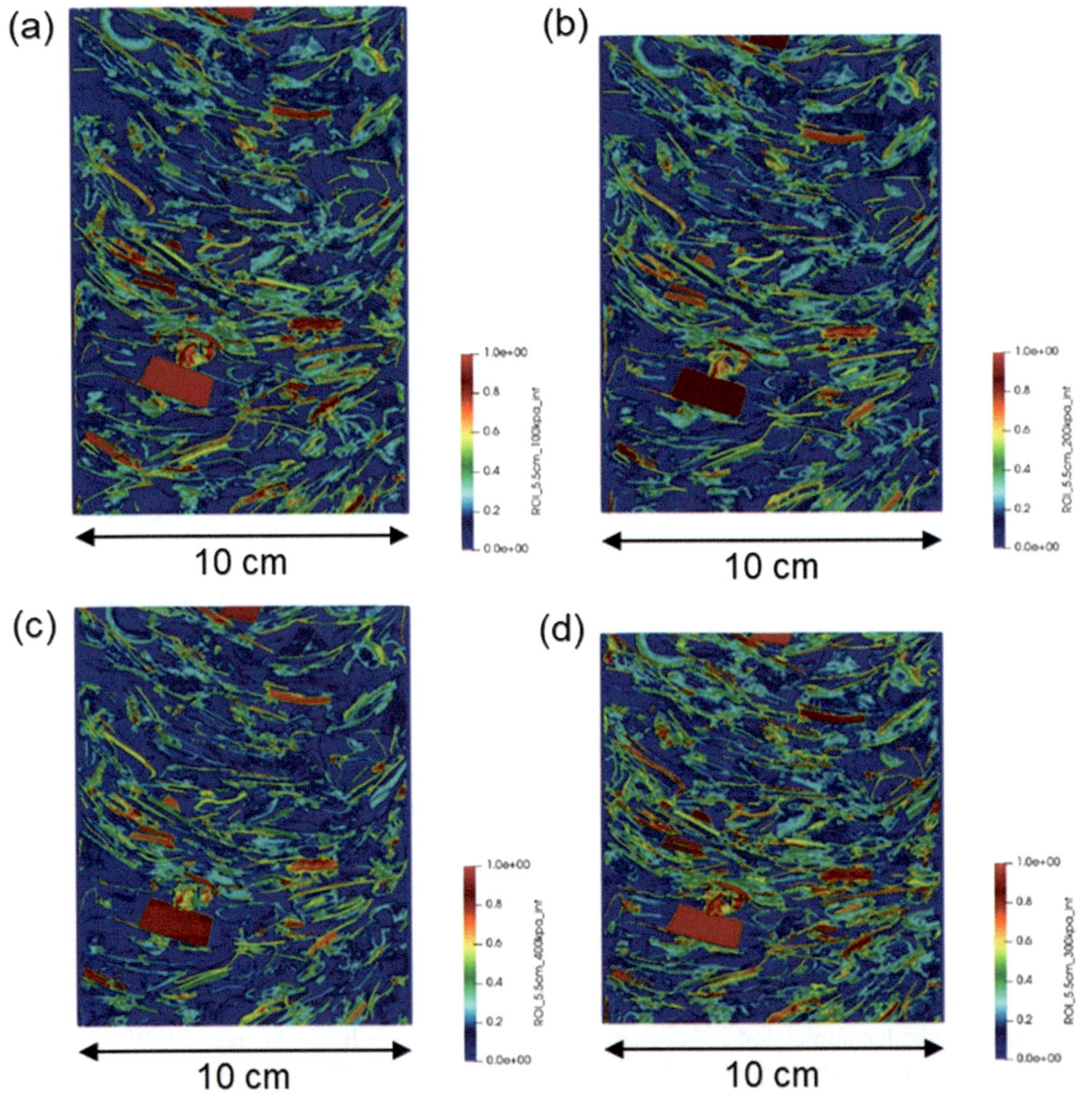

Fig. 4.2 CT images of MSW samples subjected to different vertical stresses: **a** 100 kPa, **b** 200 kPa, **c** 300 kPa, and **d** 400 kPa (reprinted from Ke et al. 2022, with permission from Elsevier)

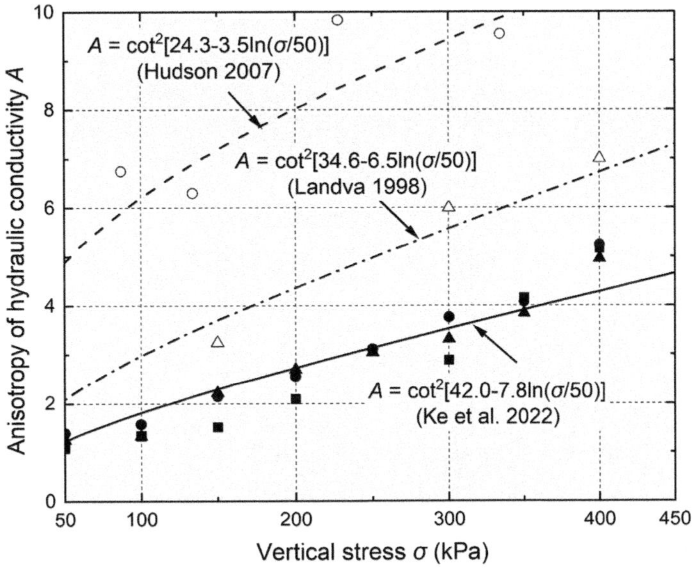

Fig. 4.3 Variation in the anisotropy of MSW with vertical stress (reprinted from Ke et al. 2022, with permission from Elsevier)

dimensional arrangement of particles tended to align horizontally, causing the pore channels to flatten accordingly. Meanwhile, their anisotropy test results, illustrated in Fig. 4.3, revealed that as the vertical stress increased from 100 to 400 kPa, the anisotropy of HFWC MSW from China increased from 2 to 5, while that of LFWC MSW from the UK increased from 6 to 10. Notably, the oriented two-dimensional arrangement of particles is primarily responsible for waste anisotropy, resulting in a greater anisotropy value of seepage at an increased load.

Ke et al. (2022) examined the compression-induced changes in pore arrangement angles and integrated them with the Poiseuille equation for cylindrical pipe flows. Ultimately, they formulated a model describing the variation in anisotropy with overlying stress, as presented in Eq. (4.4). Figure 4.3 compares the calculation results of this model with actual measurement results.

$$A = \frac{k_h}{k_v} = cot^2\theta = \cot^2\left[\theta_{50} - \alpha ln\left(\frac{\sigma}{50}\right)\right] \quad (4.4)$$

where A denotes the anisotropy value of seepage, θ represents the pore arrangement angle (°), θ_{50} denotes the initial pore arrangement angle under a vertical pressure of 50 kPa (°), α signifies the pore angle compression coefficient (°/kPa), and σ denotes the vertical pressure (kPa). The value ranges of θ_{50} and α are 20–50° and 3–8°/kPa for MSW, respectively.

4.3 Water Retention Curve

In landfill environments, gases produced by waste prevent the landfill body from reaching full saturation. Even below the leachate level, waste remains unsaturated owing to the accumulation of LFG. The WRC, which reflects the relationship between soil water content and matric suction, is crucial for describing the unsaturated hydraulic characteristics of MSW.

4.3.1 Test Setup

The WRC for MSW is typically obtained using two methods (Fredlund and Rahardjo 1993). One of these methods involves axis translation techniques utilizing a pressure plate apparatus or Tempe apparatus for testing. The samples used are typically small, with diameters usually less than 10 cm. Gas pressure control is achieved using pressure gauges and pressure-regulating valves, with an accuracy of approximately

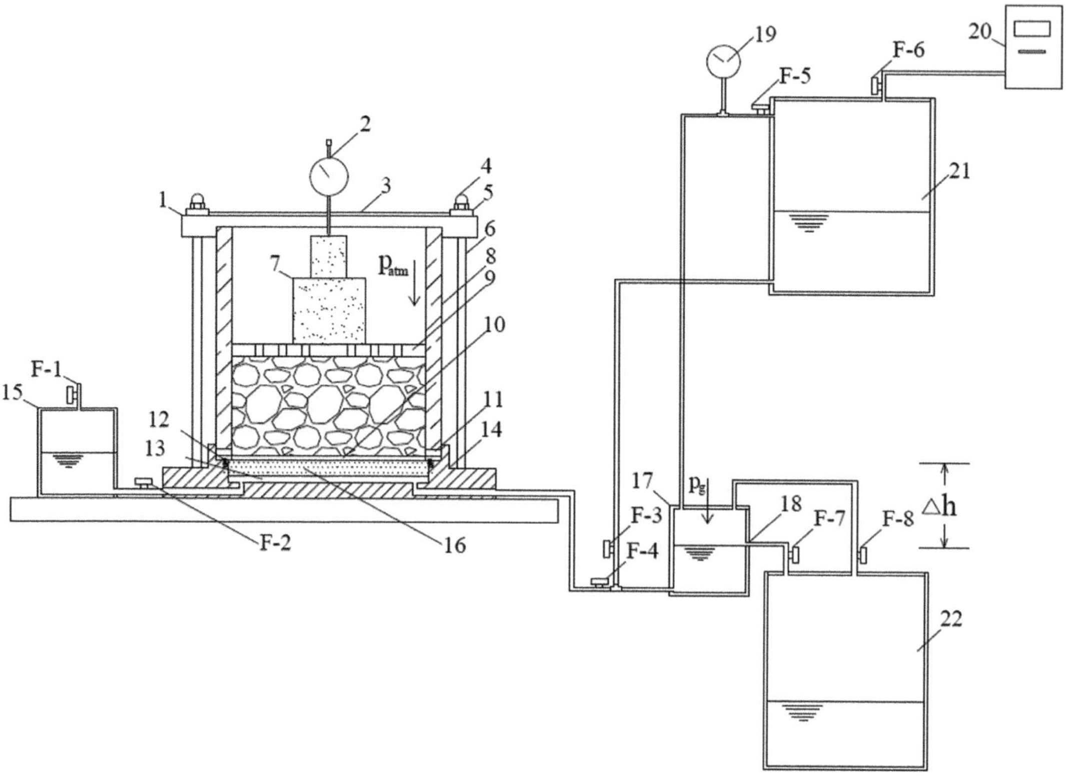

1 – Flange, 2 – Dial indicator, 3 – Membrane, 4 – Cap, 5 – Gasket, 6 – Bolt, 7 – Concrete block, 8 – Sample bucket, 9 – Perforated plate, 10 – Filter paper, 11 – Rubber ring, 12 – Structural adhesive, 13 – Annular groove, 14 – O-ring, 15 – Flushing bucket, 16 – High inlet value ceramic plate, 17 – Hanging water bucket, 18 – Overflow outlet of the hanging water bucket, 19 – Negative pressure gauge, 20 – Pump, 21 – Water replenishment tank, 22 – Water storage tank, F-1–F-6 – Valves

Fig. 4.4 Large-scale WRC testing device of unsaturated MSW (Yang 2016)

10 kPa. However, the larger particle sizes of MSW require representative samples with larger dimensions. Global studies have demonstrated that the water content of waste changes rapidly within the low matric suction range below 10 kPa, complicating accurate measurements using the pressure plate or Tempe apparatus. The second method involves the hanging column technique, where a 10 cm change in the water head difference (Δh) corresponds to a 1 kPa change in matric suction. Hence, this technique requires a larger operating area for measurements in the high matric suction range. Notably, for MSW, the residual volumetric water content on the WRC corresponds to a matric suction value exceeding 50 kPa, requiring a laboratory height of over 5 m. This specification is impractical for conventional laboratories and inconvenient for practical operations. To overcome the shortcomings of existing testing equipment, Yang (2016) designed a large-scale WRC testing device of unsaturated MSW (Patent number: CN201510289047.9), with large dimensions, high precision, and a wide measurement range. The schematic of the device is depicted in Fig. 4.4.

The primary components of the device include the following:

(1) Sample chamber

The sample chamber comprises a sample bucket and base. The sample bucket, made of organic glass, has an inner diameter of 27 cm and a height of 30 cm. The base, also made of organic glass, features a circular groove on top to facilitate liquid flow within the sample chamber. A high-air-entry-value ceramic plate is positioned within the central groove of the base, sealed with structural adhesive along its edges. On the surface of the ceramic plate, layers comprising the filter paper, silicone gaskets, and the sample bucket are placed sequentially from bottom to top. Eight evenly distributed screws secure flanges onto the base of the sample bucket. Caps are tightened on top of these screws. The high-air-entry-value ceramic plate, manufactured by the Soilmoisture company (U.S.), features a diameter of 27.6 cm, a thickness of 1.0 cm, and an air entry value of 1 bar.

(2) Matric suction control system

The matric suction control system primarily comprises a hanging water bucket, vacuum pump, and negative pressure gauge. Matric suction levels of samples are regulated by adjusting the height of the water column and applying a negative pressure using a vacuum pump.

(3) Water storage tank, water replenishment tank, and flushing bucket

During experiments, water storage and replenishment tanks maintain a constant water volume in the hanging water bucket. Changes in the water volume within these tanks are recorded to determine variations in sample water content. The flushing bucket removes bubbles from the pipeline between the bucket and the hanging water bucket, ensuring uninterrupted water flow through the pipeline.

The matric suction control method of the abovementioned device involves two methods: the hanging water column method and the negative pressure extraction method. In the first method, the fundamental control principle involves adjusting the matric suction of the sample by manipulating the water head. This is achieved by lowering the water level in the hanging water bucket, thus creating a height difference in the water head between the hanging water bucket and ceramic plate surface. This creates a negative pore water pressure on the ceramic plate surface while maintaining zero air pressure on the water surface, effectively controlling the matric suction of the sample. However, the range of matric suction controlled explicitly by the water head method is restricted by the height of the hanging water column, typically below 10 kPa. To extend this control range, a vacuum pump is utilized to reduce the pressure within the hanging water column to increase the negative pore water pressure acting on the ceramic plate surface and thus enhance the matric suction of the sample.

In summary, first, at matric suctions of less than 10 kPa, precise control of the matric suction of a sample is achieved by adjusting the water head. Second, at matric suctions of 10–90 kPa, further control of the matric suction of a sample is achieved by using the vacuum pump to extract negative pressure from the hanging water column. Figure 4.5 shows the WRCs of MSW measured by the large-scale testing device. It can be seen that there are abundant data points throughout the entire range of matric suction.

4.3.2 Effects of Compression and Degradation

Experimental studies worldwide have demonstrated that the WRC of MSW is influenced by both degradation and compaction. For instance, Jang et al. (2002) obtained the water characteristic curves of MSW using an improved Tempe apparatus and examined the effect of compaction on these curves. Their experiments revealed that as waste compaction increased, the water characteristic curves became shallower, saturated water content decreased, and residual water content increased. Furthermore, Kazimoglu et al. (2006) employed an enhanced volume pressure plate apparatus to examine synthetic MSW samples, discovering a strong correlation between the experimentally obtained water characteristic curves and those simulated by the van Genuchten model. Wei (2007) adopted the Tempe and pressure plate apparatuses to derive the water characteristic curves of synthetic MSW

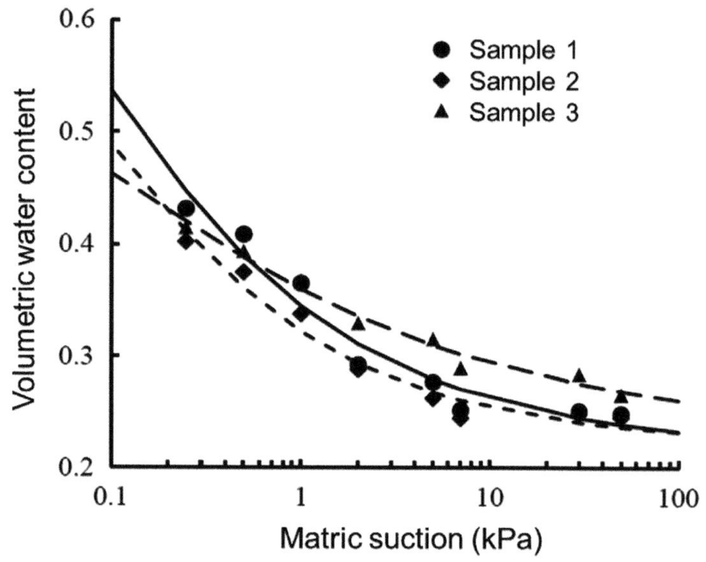

Fig. 4.5 WRCs of MSW measured by the large-scale testing device (Yang 2016)

samples, further investigating the influence of waste composition and porosity ratio on these curves. Their findings revealed that for waste with the same porosity ratio, higher organic matter content led to higher water content at the same matric suction. Meanwhile, for waste with the same composition, a higher porosity ratio resulted in higher saturated water content. Wu et al. (2012) recorded the water characteristic curves of borehole samples at varying landfill depths using a pressure plate apparatus, discovering that with increasing landfill depth (i.e., increasing sample age), fine pores, the air entry value, and residual water content all increased, whereas the saturated water content decreased.

Xu et al. (2014) examined the particle size distribution and WRCs of borehole samples at various depths of the Suzhou Qizishan Landfill. Their findings, depicted in Fig. 4.6, revealed that as the landfill depth and sample age increased, waste compaction and small particle content also increased. This resulted in a decrease in the saturated volumetric water content, an increase in the residual volumetric water content, and an increase in the water retention capacity. Typically, the WRCs of waste are represented by the van Genuchten model, as indicated in Eq. (4.5). Table 4.1 lists the model parameter values derived from representative experiments.

$$\theta_e = \frac{\theta - \theta_r}{\theta_s - \theta_r} = \left(1 + |\alpha_v \psi|^{n_v}\right)^{-m_v} \tag{4.5}$$

where θ_e denotes the effective volumetric water content; ψ represents matric suction (kPa); θ, θ_r, and θ_s indicate the volumetric water, residual volumetric water, and saturated volumetric water contents, respectively; and α_v, n_v, and m_v denote parameters of the van Genuchten model ($m_v = 1 - 1/n_v$).

Figure 4.7 illustrates the influence mechanisms of compression and degradation on the water retention characteristics of MSW. In particular, compression reduces pore volume, ultimately decreasing the saturated volumetric water content while increasing the residual volumetric water content. Consequently, the rate of decrease in the volumetric water content of MSW slows as matric suction increases. However, the effect of degradation on the WRC is the opposite to the effect of compression on the WRC. During biochemical degradation, solid mass loss reduces the particle size of MSW and increases its pore volume. Consequently, biochemical degradation increases the saturated volumetric water content while decreasing the residual volumetric water content of MSW, resulting in a faster decrease in MSW volumetric water content with increasing matric suction.

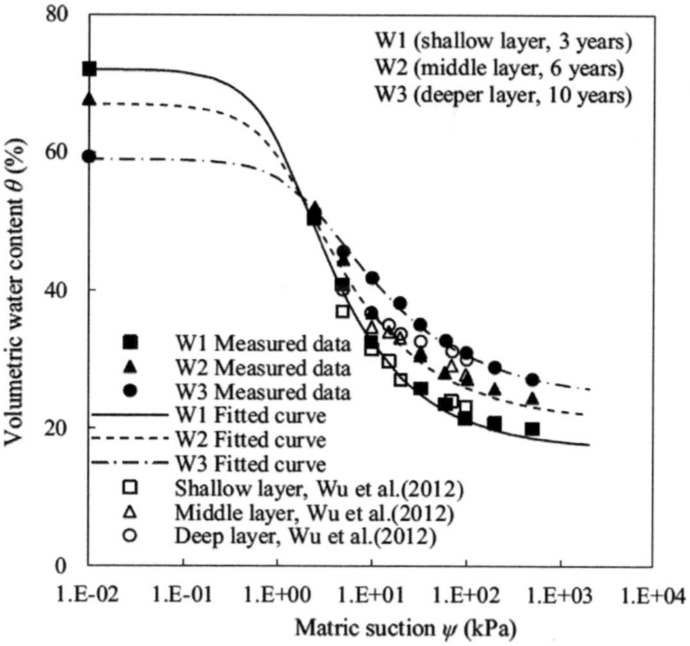

Fig. 4.6 WRCs of different-aged MSWs extracted from varying landfill depths (reprinted from Xu et al. 2014, with permission from Canadian Science Publishing)

Table 4.1 Van Genuchten model parameters for various MSWs

Type		α_v	n_v	m_v	θ_s	θ_r
MSW, China	Shallow	0.88	1.58	0.36	0.72	0.17
	Middle	0.82	1.51	0.33	0.67	0.21
	Deep	0.45	1.44	0.30	0.59	0.24
MSW, U.S	High compaction	3.38	1.89	0.47	0.60	0.21
	Medium compaction	2.92	1.60	0.38	0.53	0.22
	Low compaction	1.18	1.33	0.25	0.41	0.03
MSW, Australia	–	1.4	1.6	0.38	0.58	0.14

4.4 Relative Permeability Coefficients of Liquid and Gas

The unsaturated permeability of a fluid is typically expressed using its relative permeability, which reflects the influence of one fluid on the permeability of another fluid through pores. This interaction is primarily attributed to the forces acting among different fluids and changes in flow channels. If a fluid completely occupies the connected pore volume, its relative permeability is one. Conversely, if only a residual amount of this fluid remains within pores, restricting its movement, its relative permeability is zero.

Based on porous media fluid mechanics, the relationship between unsaturated permeability and relative permeability can be expressed as follows:

$$k_L = k_s k_{rL} = \frac{k_i \rho_L g}{\mu_L} k_{rL} \tag{4.6}$$

$$k_G = k_d k_{rG} = \frac{k_i \rho_G g}{\mu_G} k_{rG} \tag{4.7}$$

where k_L denotes the unsaturated liquid-phase permeability (m/s), k_{rL} represents the relative permeability of the liquid-phase, k_G indicates the unsaturated gas-phase permeability (m/s), and

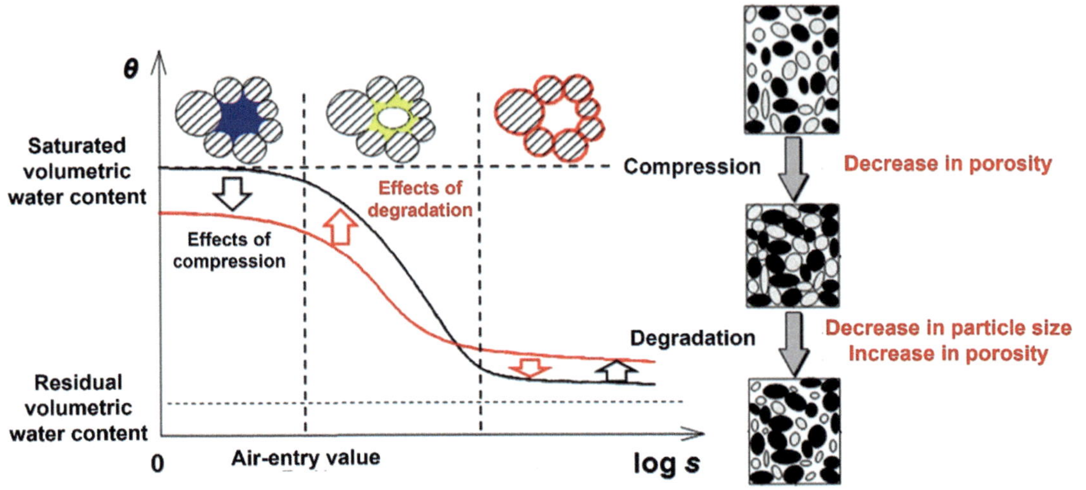

Fig. 4.7 Mechanisms whereby compression and degradation affect the water retention characteristics of MSW (Liu 2016)

k_{rG} signifies the relative permeability of the gas-phase.

In the past few years, numerous tests have been conducted to examine the unsaturated liquid and gas permeability characteristics of MSW. Notably, direct measurements of the unsaturated liquid-phase permeability are often difficult. Zhang (2007) conducted seepage tests on samples extracted from the middle section of the Qizishan Landfill in Suzhou. Monitoring of matric suction during the seepage process enabled determination of the WRC, which was analysed to identify the water content and flow rate of the waste samples. The hydraulic gradient was then computed based on matric suction. Ultimately, the unsaturated liquid-phase permeability of the waste was obtained using Darcy's law. The results indicated that as the liquid-phase saturation of the waste decreased (with increased matric suction), its unsaturated liquid permeability gradually decreased, consistent with the reports of Breitmeyer and Benson (2011).

The unsaturated gas-phase permeability characteristics of MSW have been extensively tested. For instance, Wei (2007) conducted indoor experiments to examine the gas-phase unsaturated permeability characteristics of MSW samples collected from the Qizishan Landfill in Suzhou. Their findings revealed that for a given composition and saturation, the unsaturated gas-phase permeability (k_G) of the waste samples increased exponentially with increasing void ratios. Furthermore, for a given void ratio, the gas permeability at low saturation exceeded that at high saturation, and this discrepancy increased with increasing void ratios. Under controlled void ratios and saturation levels, the gas permeability of the waste samples increased with sample age. Stoltz et al. (2010) also conducted indoor experiments on the unsaturated gas-phase permeability of fresh waste obtained from a landfill site in France. Their findings revealed that k_G increased gradually with increasing gas saturation.

Notably, the liquid- and gas-phase relative permeabilities depend on the effective saturation and are typically described using the van Genuchten model based on the WRC:

$$k_{rL} = \theta_e^{\gamma_v}[1 - (1 - \theta_e^{1/m_v})^{m_v}]^2 \tag{4.8}$$

$$k_{rG} = (1 - \theta_e)^{\gamma_v}(1 - \theta_e^{1/m_v})^{2m_v} \tag{4.9}$$

where γ_v denotes a parameter of the van Genuchten model.

Figure 4.8 compares the results derived from the above equation and those obtained from indoor experiments (Xu et al. 2014). Notably, the

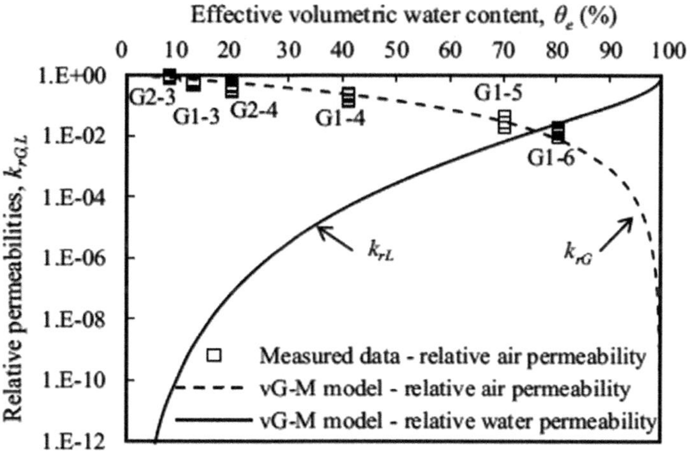

Fig. 4.8 Relationship between the relative permeabilities and effective volumetric water content of MSW (reprinted from Xu et al. 2014, with permission from Canadian Science Publishing)

simulated gas-phase relative permeability results obtained based on the WRC closely align with the experimental results. In particular, as the water content increases, the gas-phase relative permeability gradually decreases, whereas the liquid-phase relative permeability gradually increases. The trend of decreasing gas-phase relative permeability with increasing water content was more pronounced at high saturation than at low saturation.

4.5 Dual-Porosity Flow Model

MSW has a diverse composition, complex particle shapes, and various particle sizes. This means that MSW contains pores of different sizes and therefore exhibits heterogeneous flow dominated by preferential flow through large pores. Consequently, traditional homogeneous flow theories fail to accurately capture the water and gas transport processes within landfills, leading to suboptimal engineering designs for well pumping, gas extraction, gas injection, and leachate recirculation, making it difficult to achieve effective liquid and gas control in landfills.

Warren and Root (1963) proposed the classic dual-porosity model to describe the flow characteristics of heterogeneous media. The key difference between dual-porosity and single-porosity media is that two permeabilities and two porosities exist at any given point in the former. In fracture–pore structures, the permeability of the fracture system significantly exceeds that of the matrix system. In such structures, the fracture system serves as the primary flow pathway, while the matrix system acts as the primary storage space. This results in the formation of two relatively independent hydrodynamic systems, with fluid exchange occurring between them. In actual fracture–pore structures, fractures are mostly random. Flow mechanics studies typically idealize these fractures as a network of vertically intersecting fractures, dividing the formation into rectangular grain-sized blocks.

4.5.1 Dual-Porosity Water Retention Curve

Meng (2020) unified the WRCs of the fracture and matrix domains using a segmented van Genuchten model, as illustrated in Fig. 4.9. Mathematical expressions of the WRCs of the fracture domain and matrix domain are presented in Eqs. (4.10) and (4.11), respectively. Meanwhile, Eq. (4.12) presents the mathematical expression for the integrated dual-porosity WRC. Notably, the segmented van Genuchten model comprehensively captures the dynamic changes in water content in both the fracture and matrix domains across the entire range of matric suction. The boundary for the WRCs of the fracture and

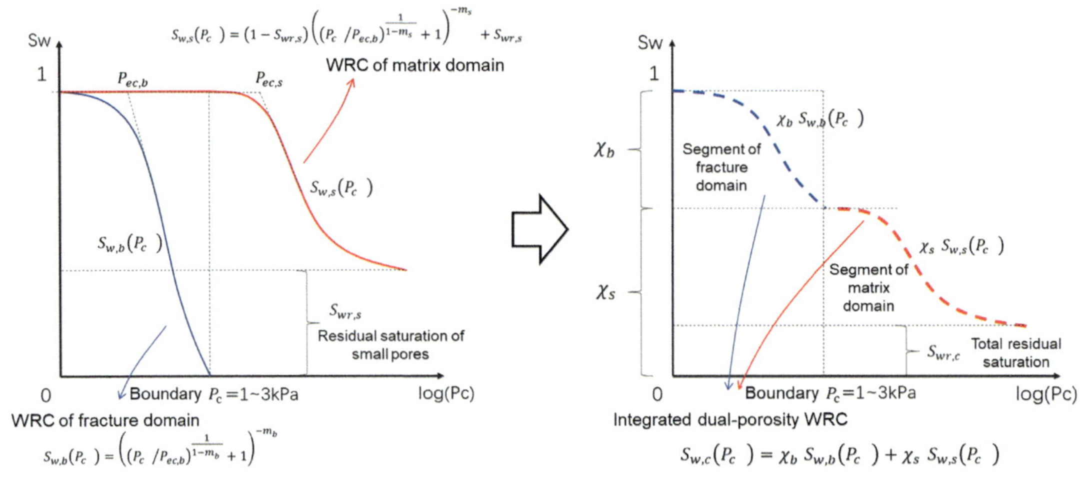

Fig. 4.9 Schematic diagram of dual-porosity WRC (Meng 2020)

matrix domains is typically considered to be a matrix suction of 1–3 kPa.

$$S_{wb} = \left[\left(\frac{P_c}{P_{ecb}}\right)^{\frac{1}{1-m_b}} + 1\right]^{-m_b} \tag{4.10}$$

$$S_{ws} = (1 - S_{wrs})\left[\left(\frac{P_c}{P_{ecs}}\right)^{\frac{1}{1-m_s}} + 1\right]^{-m_s} + S_{wrs} \tag{4.11}$$

$$S_{wc} = \chi_b S_{wb} + \chi_s S_{ws} \tag{4.12}$$

where S_{wb}, S_{ws}, and S_{wc} denote the saturations of the fracture domain, matrix domain, and total system, respectively; χ_b and χ_s represent the volume fractions of the fracture and matrix domains in the total porosity, respectively; P_{ecb} and P_{ecs} signify the entry pressures of the fracture and matrix domains, respectively; m_b and m_s indicate the shape parameters of the fracture and matrix domains, respectively; and S_{wrs} denotes the residual saturation of the matrix domain.

4.5.2 Dual-Porosity Well Flow Model

Ke et al. (2018) and Hu et al. (2020) established a saturated and unsaturated flow migration model of vertical wells in landfills that is based on dual-porosity flow theory. Figure 4.10 illustrates the operation of the model under saturated conditions. When pumping at a constant flow rate (Q) from vertical wells within the landfill, leachate levels within these wells decrease, creating a hydraulic head difference between the wells and their surroundings. Consequently, leachate from both the matrix and fracture domains migrates laterally toward the wells, causing a decline in surrounding leachate levels and establishing a drawdown funnel (AC and BC). During this lateral migration of leachate, leachate mass exchange occurs between the matrix and fracture domains owing to the hydraulic head difference.

According to the principle of water balance within the control volume, the seepage control equations for the fracture and matrix domains are established, as presented in Eqs. (4.13) and (4.14), respectively. In these equations, the first and second terms on the left-hand side represent the horizontal and vertical fluxes flowing into the domain, respectively, while the third term on the left-hand side represents the leachate mass exchange occurring between the fracture and matrix domains. Meanwhile, the term on the right-hand side of these equations represents the net change in leachate mass within the control volume. Furthermore, the expressions for the hydraulic head and hydraulic conductivity, presented in Eqs. (4.15) and (4.16), respectively, across the entire region are obtained using

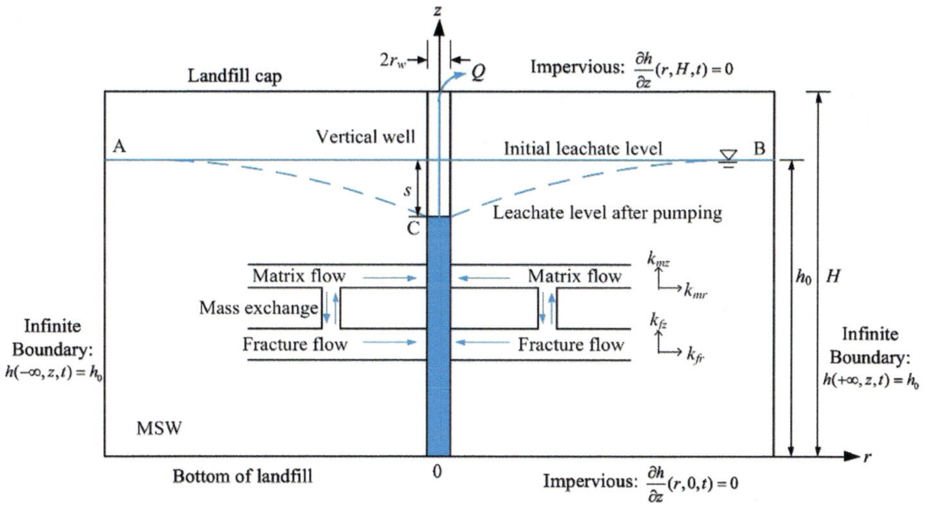

Fig. 4.10 Schematic diagram of the dual-porosity model of leachate flow towards a vertical well (reprinted from Ke et al. 2018, with permission from Elsevier)

weighted averages based on the volume proportions of the fracture and matrix domains.

Fracture domain:

$$k_{fr}\left(\frac{\partial^2 h_f}{\partial r^2}+\frac{1}{r}\frac{\partial h_f}{\partial r}\right)+k_{fz}\frac{\partial^2 h_f}{\partial z^2}+\alpha k_m\left(h_m-h_f\right)=\mu_{sf}\frac{\partial h_f}{\partial t} \tag{4.13}$$

Matrix domain:

$$k_{mr}\left(\frac{\partial^2 h_m}{\partial r^2}+\frac{1}{r}\frac{\partial h_m}{\partial r}\right)+k_{mz}\frac{\partial^2 h_m}{\partial z^2}-\alpha k_m\left(h_m-h_f\right)=\mu_{sm}\frac{\partial h_m}{\partial t} \tag{4.14}$$

Water head relationship:

$$h=w_f h_f+(1-w_f)h_m \tag{4.15}$$

Horizontal hydraulic conductivity relationship:

$$k_r=w_f k_{fr}+(1-w_f)k_{mr} \tag{4.16}$$

In the above equations, h represents the hydraulic head of the entire domain (m); h_f and h_m denote the hydraulic heads of the fracture and matrix domains, respectively (m/s); k_r indicates the horizontal hydraulic conductivity (m/s); k_{fr} and k_{fz} denote the horizontal and vertical hydraulic conductivities of the fracture domain, respectively (m/s); k_{mr} and k_{mz} signify the horizontal and vertical hydraulic conductivities of the matrix domain, respectively (m/s); k_m denotes the hydraulic conductivity at the fracture and matrix domain interface (m/s); μ_{sf} and μ_{sm} represent the specific storage coefficients of the fracture and matrix domains, respectively (m^{-1}); w_f denotes the volume fraction of the fracture domain in the total domain; α represents the coefficient of leachate mass exchange between the two domains (m^{-2}); r denotes the radial distance from the well (m); z represents the vertical distance from the landfill bottom (m); and t indicates the pumping time of the well (s).

The drawdown of a pumping well in a landfill was calculated using the dual-porosity model described above and the classic single-porosity Theis model. The modeling process and parameter values were sourced from the work of Ke et al. (2018), and the obtained results are illustrated in Fig. 4.11. Notably, the drawdown calculated by the Theis model exceeded the

Fig. 4.11 Measured and calculated leachate drawdowns within a landfill (reprinted from Ke et al. 2018, with permission from Elsevier)

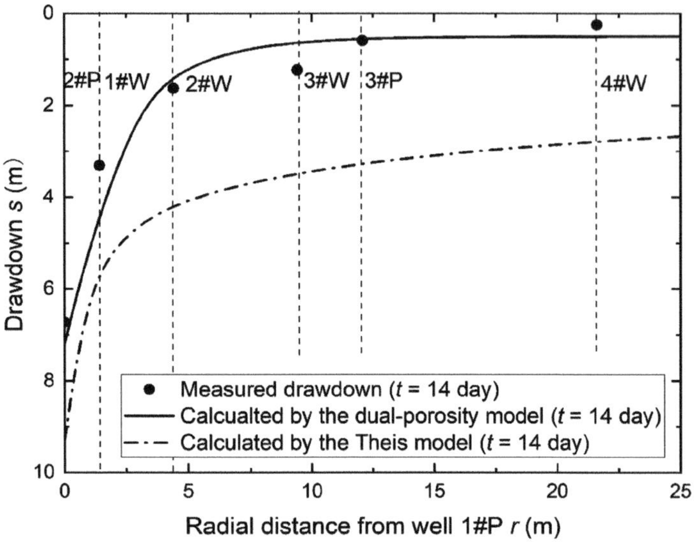

measured drawdown, suggesting drawdown overestimation by the Theis model. Additionally, the drawdown curve calculated using the Theis model had a steeper slope than the measured curve, due to the inability of the Theis model to consider the anisotropy of flow. In contrast, the drawdown computed by the dual-porosity model closely matched the measured drawdown, reflecting the rapid decrease in drawdown with distance. From the measured drawdown data, key parameters suitable for HFWC MSW landfills were determined through fitting: the volume ratio of the fracture ($w_f = 0.45$) and the ratio of hydraulic conductivity between the fracture and matrix domains ($k_{fr}/k_{mr} = 500$).

4.6 Governing Equations for Two-Phase Flow

The mass conservation equations governing two-phase gas–water flows are expressed as

$$\frac{\partial(\rho_w n S_w)}{\partial t}+\frac{\partial M_{wx}}{\partial x}+\frac{\partial M_{wy}}{\partial y} = \rho_w n\frac{\partial S_w}{\partial t}+\frac{\partial M_{wx}}{\partial x}+\frac{\partial M_{wy}}{\partial y} = Q_w \quad (4.17)$$

$$\frac{\partial(\rho_g n S_g)}{\partial t}+\frac{\partial M_{gx}}{\partial x}+\frac{\partial M_{gy}}{\partial y} = nS_g\frac{\partial \rho_g}{\partial t}+\rho_g n\frac{\partial S_g}{\partial t}+\frac{\partial M_{gx}}{\partial x}+\frac{\partial M_{gy}}{\partial y} = Q_g \quad (4.18)$$

where ρ_g and ρ_w denote the densities of gas and water, respectively; n represents the porosity; S_g and S_w signify gas and water saturations, respectively; M_{gx} and M_{gy} indicate the gas fluxes along the horizontal and vertical directions, respectively; M_{wx} and M_{wy} denote the water fluxes along the horizontal and vertical directions, respectively; and Q_w and Q_g represent the water and gas source terms.

The sum of gas-phase and water-phase saturation in porous media equals one:

$$S_g + S_w = 1 \quad (4.19)$$

According to Darcy's law, the water and gas fluxes along the horizontal and vertical directions are expressed as follows:

$$M_{wx} = -\rho_w\frac{k_{ix}k_{rw}}{\mu_w}\frac{\partial P_w}{\partial x}, M_{wy} = -\rho_w\frac{k_{iy}k_{rw}}{\mu_w}\left(\frac{\partial P_w}{\partial y}-\rho_w g\right) \quad (4.20)$$

$$M_{gx} = -\rho_g \frac{k_{ix} k_{rg}}{\mu_g} \frac{\partial P_g}{\partial x}, M_{gy} = -\rho_g \frac{k_{iy} k_{rg}}{\mu_g} \frac{\partial P_g}{\partial y} \tag{4.21}$$

Using the ideal gas equation to describe the relationship between gas pressure and gas density yields the following relation:

$$\rho_g = \frac{(P_g + P_{atm})M}{RT} \tag{4.22}$$

where P_{atm} denotes atmospheric pressure (101.325 kPa); M represents the molar mass of gas; R denotes the ideal gas constant; and T indicates temperature.

Substituting Eqs. (4.19) to (4.22) into Eqs. (4.17) to (4.18) yields the governing equations for two-phase gas–water seepage. By combining Eqs. (4.5) to (4.9), the constitutive relationships for two-phase seepage can be used to model the seepage process.

$$n\frac{\partial S_w}{\partial t} - \frac{k_{iy} k_{rw}}{\mu_w}\left[A\frac{\partial^2 P_w}{\partial x^2} + \frac{\partial}{\partial y}\left(\frac{\partial P_w}{\partial y} - \rho_w g\right)\right] = \frac{Q_w}{\rho_w} \tag{4.23}$$

$$nS_g\frac{\partial P_g}{\partial t} + nP_g\frac{\partial S_g}{\partial t} - \frac{P_g k_{rg} k_{iy}}{\mu_g}\left(A\frac{\partial^2 P_g}{\partial x^2} + \frac{\partial^2 P_g}{\partial y^2}\right) = \frac{Q_g RT}{M} \tag{4.24}$$

4.7 Simulation of Liquid–Gas Pressure Formation in Landfills

Liu (2016) used the two-phase seepage equations defined in Sect. 4.6 to calculate the development of liquid and gas pressure in the 5-m-high landfill unit depicted in Fig. 4.12. Their analysis focused on comparing the reciprocal effects of liquid and gas migration within HFWC and LFWC MSW landfills. The study assumed that MSW was disposed in a single filling round. The upper boundary featured an impermeable cover system for water, allowing LFG to vent freely. Meanwhile, the leachate drainage system of the lower boundary failed, rendering it impervious to both leachate and gas. Table 4.2 summarizes the initial degradable components and LFG generation potentials of HFWC and LFWC MSWs, while Table 4.3 outlines the engineering properties of these MSWs.

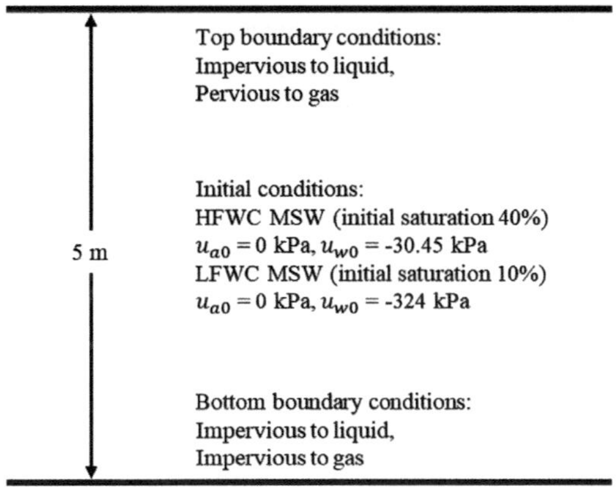

Fig. 4.12 One-dimensional case study of landfill units with different food waste contents (Liu 2016)

Figure 4.13 compares the temporal water saturation changes of landfill units featuring different food waste contents. In the landfill unit containing HFWC, biochemical degradation led to the generation of substantial amounts of leachate within a short period, leading to an increase in the average saturation from 40% to 88%. Owing to the failure of the bottom drainage system, the leachate level in the landfill body increased significantly, and the height of the saturated zone featuring a saturation level of over 90% reached 2.1 m, accounting for 42% of the initial landfill height. Conversely, the landfill unit with LFWC initially had lower water and intracellular water contents, resulting in minimal leachate production through biochemical degradation. Consequently, the saturation level increased only slightly from 10% to 15.5%, and the landfill body remained relatively dry with no notable increase in its leachate level. As depicted in Fig. 4.14, Chen (2010) revealed the presence of elevated primary leachate levels and localized

Table 4.2 Initial degradable components and LFG generation potentials of HFWC and LFWC MSWs (Liu 2016)

MSW	Degradable cellulose content (kg/t, by wet basis)	Intracellular water content (kg/t, by wet basis)	LFG generation potential (m^3/t, by wet basis)
HFWC MSW (China)	147	350	122
LFWC MSW (U.S.)	409	44	339

Table 4.3 Engineering properties of HFWC and LFWC MSWs (Liu 2016)

k_i(m^2)		μ_w(Pa·s)	μ_g(Pa·s)	e_0
9×10^{-12}		1×10^{-3}	1.54×10^{-5}	1.5
α(vG model)		n(vG model)	S_s	S_r
1.4		1.6	0.95	0.2
–	$C_c{}'$	$C_{c\infty}{}'$	$\varepsilon_{S\infty}$	c_s(day^{-1})
HFWC MSW	0.25	0.15	0.33	0.001
LFWC MSW	0.22	0.18	0.20	

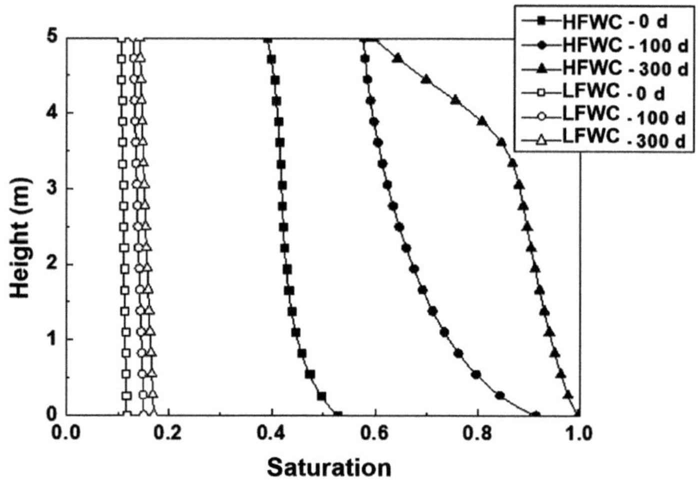

Fig. 4.13 Comparison of the temporal water saturation changes between HFWC landfill and LFWC landfill (Liu 2016)

perched leachate level at the Qizishan Landfill in Suzhou, China, which is consistent with the trends observed in the simulation results displayed in Fig. 4.13.

High saturation levels can significantly impact the gas permeability of MSW, leading to elevated gas pressures within the waste body during periods of gas production. The comparison of pore gas pressure variations in Fig. 4.15 reveals that within the HFWC unit, the bottom pore gas pressure increased to 40.9 kPa, gradually dissipating only after the gas production rate declined (100–300 days after landfilling). Conversely, the LFWC unit experienced minimal changes in gas pressure owing to the low water content of its waste, facilitating rapid drainage and collection of LFG produced by biochemical degradation. However, leachate accumulation within the waste body hindered efficient LFG collection. As illustrated in Fig. 4.16, gas collection at the top of the HFWC unit was lower compared to that from the LFWC unit. The LFG

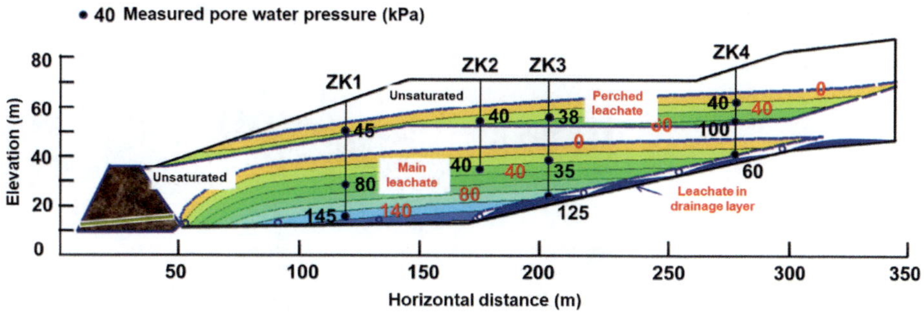

Fig. 4.14 Leachate level distribution of the Qizishan Landfill in Suzhou (Chen 2010)

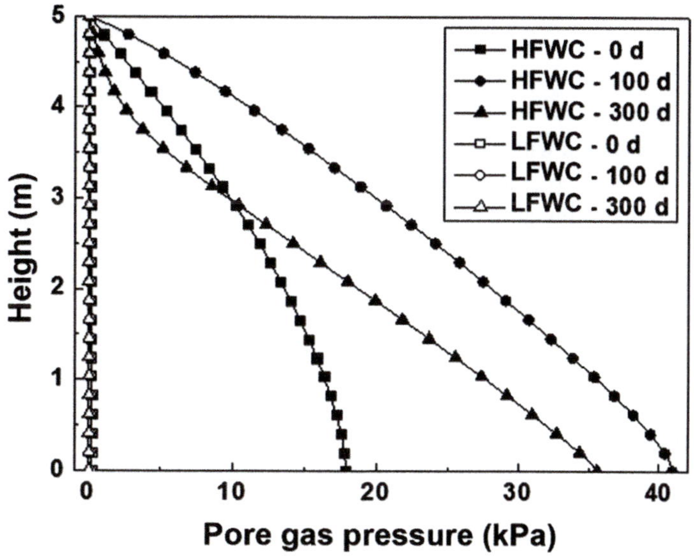

Fig. 4.15 Comparison of pore gas pressure fluctuations between HFWC landfill and LFWC landfill (Liu 2016)

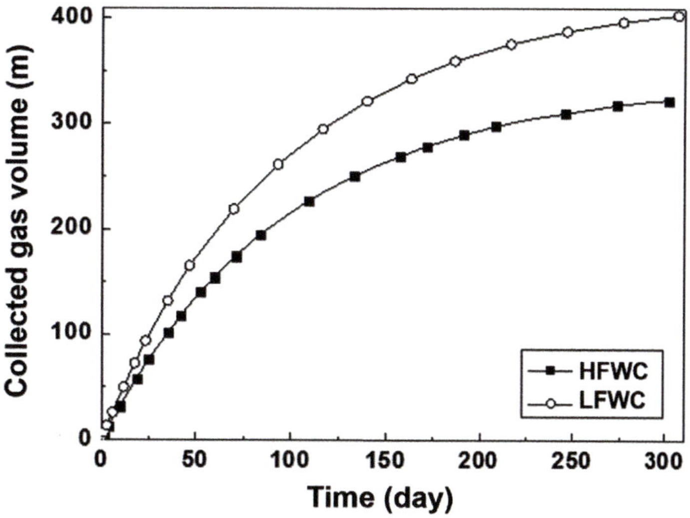

Fig. 4.16 Comparison of LFG collection at the top regions between HFWC landfill and LFWC landfill (Liu 2016)

Fig. 4.17 Leachate spraying caused by LFG accumulation in HFWC landfill (Liu 2016)

accumulated within the waste body also increased pore water pressure. As depicted in Fig. 4.17, during drilling and well installation at a landfill in southern China, elevated pore pressure within the waste body caused leachate to forcefully discharge from boreholes, with leachate spraying more than 5 m above the landfill surface.

References

Breitmeyer RJ, Benson CH (2011) Measurement of unsaturated hydraulic properties of municipal solid waste. In Geo-frontiers 2011: advances in Geotechnical Engineering, pp 1433–1442

Chen YM (2010) Development of leachate mounds and control of leachate-related failures at MSW landfills in humid regions. In the 6th International Congress on Environmental Geotechnics, New Delhi

Fredlund DG, Rahardjo H (1993) Soil mechanics for unsaturated soils. John Wiley & Sons

Hossain MS, Penmethsa KK, Hoyos L (2009) Permeability of municipal solid waste in bioreactor landfill with degradation. Geotech Geol Eng 27(1):43–51

Hu J, Ke H, Lan JW, Chen YM, Meng M (2020) A dual-porosity model for coupled leachate and gas flow to vertical wells in municipal solid waste landfills. Géotechnique 70(5):406–420

Jang YS, Kim YW, Lee SI (2002) Hydraulic properties and leachate level analysis of Kimpo metropolitan landfill, Korea. Waste Manag 22(3):261–267

Kazimoglu YK, McDougall JR, Pyrah IC (2006) Unsaturated hydraulic conductivity of landfilled waste. Unsaturated Soils, pp 1525–1534

Ke H, Hu J, Xu XB, Wang WF, Chen YM, Zhan LT (2017) Evolution of saturated hydraulic conductivity with compression and degradation for municipal solid waste. Waste Manage 65:63–74

Ke H, Hu J, Xu XB, Wu XW, Li YC, Lan JW (2018) Analytical solution of leachate flow to vertical wells in municipal solid waste landfills using a dual-porosity model. Eng Geol 239:27–40

Ke H, Liu Y, Hu J, Qin R, Xu XB, Chen YM (2022) Experimental study on anisotropy of hydraulic conductivity for municipal solid waste. Waste Manage 137:39–49

Klinkenberg LJ (1941) The permeability of porous media to liquids and gases. In Drilling and production practice. Am Petrol Inst 200–213

Liu HL (2016) Research on biodegradation-consolidation-solute migration coupled behaviors of municipal solid waste and landfill stabilization. Ph.D. thesis, Zhejiang University, Hangzhou, China (in Chinese)

Meng M (2020) Study on pore structure and seepage characteristics of MSW based on xCT scanning and digital technology. Master thesis, Zhejiang University, Hangzhou, China (in Chinese)

Powrie W, Beaven RP (1999) Hydraulic properties of household waste and implications for landfills. Proc Inst Civil Eng Geotech Eng 137(4):235–237

Stoltz G, Gourc JP, Oxarango L (2010) Liquid and gas permeabilities of unsaturated municipal solid waste under compression. J Contam Hydrol 118(1–2):27–42

Warren JE, Root PJ (1963) The behavior of naturally fractured reservoirs. Soc Petrol Eng J 3(3):245–255

Wei HY (2007) Experimental and numerical study on gas migration in landfill of municipal solid waste. Ph.D. thesis, Zhejiang University, Hangzhou, China (in Chinese)

Wu HY, Chen T, Wang HT, Lu WJ (2012) Field air permeability and hydraulic conductivity of landfilled municipal solid waste in China. J Environ Manage 98:15–22

Xu XB, Yang QF, Zhan LT, Chen YM, Wang SY (2015) Large-scale measurement device for volume change coefficients of municipal solid waste skeleton and pore water. Patent number: CN201510289047.9, China (in Chinese)

Xu XB, Zhan TT, Chen YM, Beaven RP (2014) Intrinsic and relative permeabilities of shredded municipal solid wastes from the Qizishan landfill, China. Canad Geotech J 51(11):1243–1252

Yang QF (2016) Laboratory research on soil-water characteristics curve of municipal solid waste under bio-mechanical effect. Master thesis, Zhejiang University, Hangzhou, China (in Chinese)

Zhan LT, Xu H, Lan JW, Liu Z, Chen YM (2014) Field and laboratory study on hydraulic characteristics of MSWs. J Zhejiang Univ (Eng Sci) 48(3):478–486 (in Chinese)

Zhang WJ (2007) Experimental and numerical study on water/leachate transport in landfill of municipal solid waste. Ph.D. thesis, Zhejiang University, Hangzhou, China (in Chinese)

Soil Degradation–Consolidation Models

5

Abstract

This chapter introduces the soil degradation–consolidation theory, drawing upon the concepts covered in Chaps. 2, 3, and 4. It offers a comprehensive overview of several degradation–consolidation models and their practical relevance to landfilled MSW and NGHs. The chapter begins by outlining several degradation–consolidation models specifically tailored to skeleton-degradable soils, subsequently deriving governing equations and establishing a finite element numerical solution framework. The chapter then focuses on the degradation and stabilization behaviors of landfilled MSWs with HFWC and LFWC, detailing various degradation and stabilization indicators and exploring anaerobic and aerobic degradation stabilization behaviors. Validations of the established theoretical models and evaluations of degradation stabilization behaviors in HFWC MSW landfills are accomplished through large-scale bioreactor experiments and field surveys. Furthermore, the characteristics of two-dimensional landfill slopes are examined, analyzing the coupling effects of landfill gas (LFG) production and deformation due to the degradation of MSW on the landfill slope stability. Additionally, LFG pressure distributions, horizontal and vertical displacement distributions, and deformation patterns are analyzed. Moreover, NGH degradation and consolidation behaviors are explored by the numerical model and validated by centrifugal hypergravity modeling. Notably, the degradation–consolidation theory describes solid skeleton deformation, liquid and gas flow, solute migration, and heat transport, facilitating its widespread applications to multifield coupling processes within porous media. Leveraging diverse biochemical degradation models, the degradation–consolidation theoretical framework demonstrates direct applicability to various skeleton-degradable soils.

Keywords

Skeleton-degradable soil · Degradation–consolidation model · Municipal solid waste · Natural gas hydrate · Numerical solution · Degradation and stabilization behavior · Stabilization indicator · Large-scale bioreactor experiment · Numerical modeling · Hypergravity experiment

5.1 General Introduction

When MSW is landfilled, the accumulation of leachate, LFG, and heat within landfills can lead to contaminant leakage and landfill slope instability. Hence, multifield coupling models that comprehensively consider various processes such as waste degradation, heat generation, and waste settlement must be adopted during the design and operation of landfills.

Y. Chen, *Soil Degradation-Consolidation Theory and Its Applications*,
https://doi.org/10.1007/978-981-97-7985-7_5

Similar to waste landfilling, the exploitation of NGHs induces solid hydrate dissociation, liquid or gas generation and migration, heat transfer, and soil-skeleton deformation, potentially triggering geotechnical hazards such as borehole inclination and reservoir collapse. Hence, constructing multifield coupling models for NGHs is also crucial. These models must comprehensively consider phenomena such as NGH decomposition, heat generation, and pore pressure changes, thus enabling accurate predictions and mitigation of geoenvironmental risks.

This chapter introduces the degradation–consolidation models for skeleton-degradable soils, applicable to multifield interaction processes in soils experiencing solid-phase mass loss owing to biochemical (such as MSW) or chemical degradation (such as NGHs).

The proposed degradation–consolidation model for skeleton-degradable soils is rooted in the fundamental degradation–consolidation model introduced by Chen et al. (2020) in the context of landfilled MSW. This base framework was developed to consider processes such as anaerobic biochemical reactions, skeleton deformation, liquid–gas flows, and solute migration. Building upon this base model, Li et al. (2021) further considered the influence of temperature on each process, thus establishing a degradation–consolidation model accounting for heat transport processes in anaerobic degradation environments. Recognizing that oxygen (O_2) ventilation could effectively accelerate MSW degradation, Xiao et al. (2022) introduced another degradation–consolidation model capturing the aerobic degradation processes within landfills.

Building upon this degradation–consolidation model for MSW, Ye et al. (2023) further incorporated dynamic hydrate decomposition equations to reflect the conversion of solid-phase hydrates into gas- and liquid-phase hydrates and implemented the mass conservation principles for water and gas phases, thus introducing a degradation–consolidation model for NGHs.

Notably, the fundamental degradation–consolidation model (Chen et al. 2020) and its successors (Li et al. 2021; Xiao et al. 2022; Ye et al. 2023) are all based on the continuum mechanics of porous media and incorporate constitutive relationships for diverse multifield coupling phenomena. These models can be computationally solved using the open-source finite element method code OpenGeoSys (OGS) (Kolditz et al. 2012, 2015; Wang et al. 2009). Furthermore, previous studies have employed these models to analyze the stabilization behaviors of landfilled MSW, deformation development of landfill slopes (Chen et al. 2020; Li et al. 2021; Xiao et al. 2022), and depressurization exploitation of NGHs (Ye et al. 2023). Thus, these models offer valuable insights into the mechanisms and evolution patterns of geoenvironmental disasters in HFWC MSW landfills and hydrate-bearing sediments (HBS).

The degradation–consolidation model for skeleton-degradable soils describes the phenomena of solid skeleton deformation, liquid–gas flows, solute migration, and heat transport based on the principles of momentum, mass, and energy conservation. This framework is universally applicable across the diverse multifield coupling processes of porous media. Hence, by introducing different biochemical degradation models, this theoretical framework can be directly applied to various skeleton-degradable soils.

5.2 Governing Equations of Degradation–Consolidation Models

5.2.1 Degradation–Consolidation Models

(1) Degradation–consolidation model for landfilled MSW

Chen et al. (2020) developed a degradation–consolidation model for landfilled MSW, incorporating factors such as anaerobic biochemical degradation, skeleton deformation, pore water flow, pore gas flow, solute transport, and their temporal interactions, leveraging continuum mechanics principles. This model, illustrated in Fig. 5.1, captures the phenomena of solid-phase mass loss, gas production, and water and solute release caused by anaerobic biochemical degradation. Notably, the MSW solid skeleton deforms owing to solid mass loss, variations in effective stress, gas flow, and solute and water transport. Capillary pressure, gas pressure, displacements, and chemical concentrations are selected as the primary model variables that regulate MSW coupling processes.

In addition to the inherent assumptions of continuum mechanics, the model assumes the following: (1) Solid particles are incompressible. (2) Pore water is continuous and incompressible within void spaces. (3) Pore gases are continuous mixtures within void spaces. (4) Flows of pore water and pore gas are governed by Darcy's law. (5) Solutes do not impact the physical properties of pore water, such as density, viscosity, and permeability. (6) Mechanical dispersion and solid adsorption of solutes are neglected. (7) The coupling processes are isothermal, disregarding heat generation, consumption, and transfer. (8) Phase transitions due to mechanical action are disregarded.

(B) Degradation–consolidation model for landfilled MSW with heat transport effects

The biochemical degradation of MSW typically involves exothermic reactions, wherein processes such as liquid water evaporation and condensation consume and generate heat, thereby influencing temperature changes within the waste body. In addition to the coupling processes considered by the basic degradation–consolidation model depicted in Fig. 5.1 (Chen et al. 2020), the degradation–consolidation frameworks developed by Li et al. (2021) and Xiao et al. (2022) consider heat transport phenomena. These phenomena primarily affect (1) the rate of biochemical reactions, (2) permeability characteristics of liquid–gas flows and solute migration, (3) thermal expansion of the skeleton, and (4) evaporation and condensation of liquid water. Figure 5.2 illustrates the comprehensive framework of a coupled model encompassing biochemical reactions, skeleton deformation, liquid–gas flows, solute migration, and heat transport. This framework is applicable to other skeleton-degradable soils.

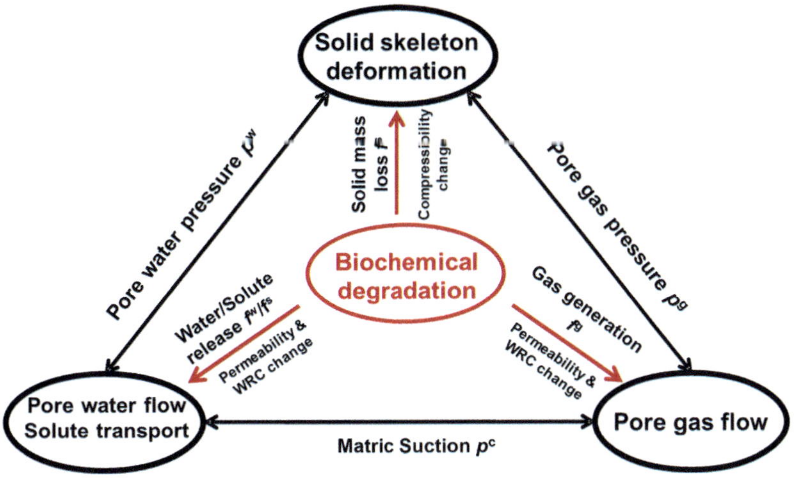

Fig. 5.1 Processes in the degradation–consolidation model for landfilled MSW (Chen et al. 2020)

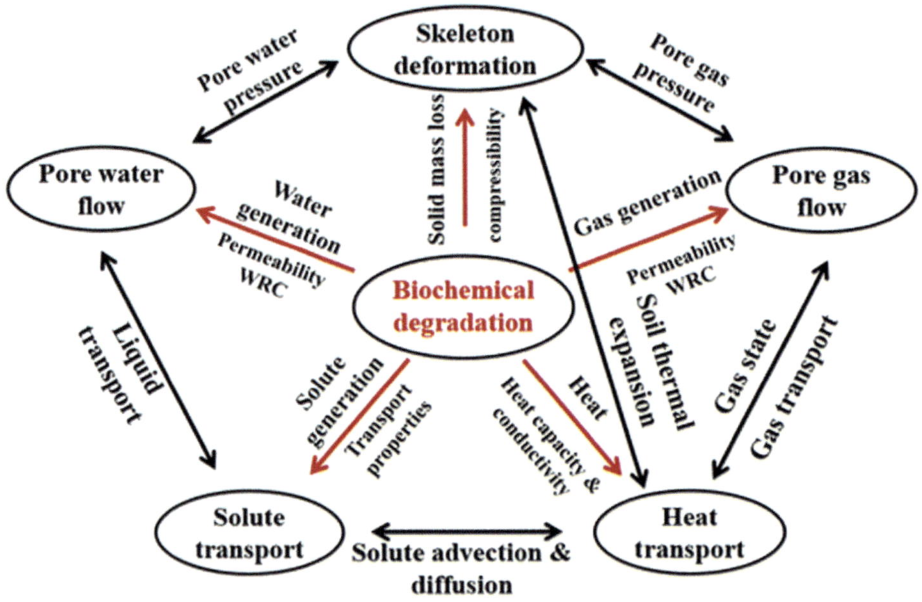

Fig. 5.2 Processes in the degradation–consolidation model for landfilled MSW considering heat transport effects (reprinted from Li et al. 2021, with permission from Elsevier)

(3) Degradation–consolidation model for NGHs

The coupled processes of HBSs primarily comprise NGH dissociation (decomposition), skeleton deformation, gas or water migration, and heat transfer. The coupled relationships associated with the exploitation of NGHs can be highlighted by simplifying Fig. 5.2 to give Fig. 5.3. NGH dissociation is an endothermic process, absorbing heat from the surroundings for solid hydrate dissociation and liquid water evaporation, which, in turn, affect soil stiffness and strength. Simultaneously, pore pressure variations resulting from gas and liquid release from NGHs alter the effective stresses and heat convective parameters of the soil skeleton, further influencing NGH dissociation. Additionally, temperature changes cause all three sediment phases to undergo thermal expansion and induce variations in the NGH dissociation process. Meanwhile, skeleton deformation reciprocally affects pore pressures. Thus, the degradation–consolidation model for NGHs makes the following assumptions: (1) Solid, hydrate, and liquid sediment phases can thermally expand. (2) Heat transfer processes include conduction and convection. (3) Darcy's law and Fick's law describe the seepage and diffusion processes of pore water and dry air. (4) Secondary NGH or ice formation does not occur during NGH dissociation (Ye et al. 2023).

(4) Governing equations

As depicted in Figs. 5.1, 5.2 and 5.3, the proposed degradation–consolidation models for skeleton-degradable soils encompass multiple processes governed by a momentum balance equation, a mass balance equation, and an energy balance equation, with biochemical or chemical degradation serving as the driving force.

(a) Momentum balance equation

According to the principle of effective stress, the momentum balance equation of a mixture considering water saturation, capillary pressure, and gas pressure is expressed as

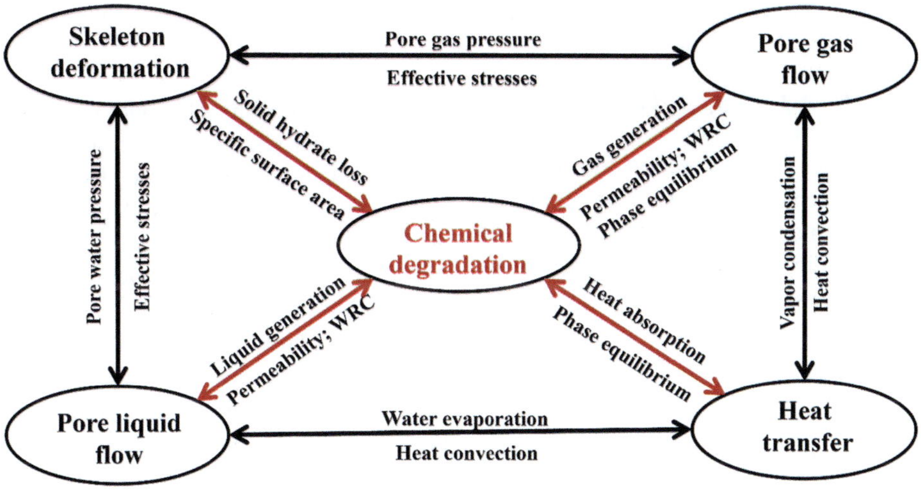

Fig. 5.3 Processes in the degradation–consolidation model for NGHs

$$\nabla(\sigma' - \alpha(p^g - \chi p^c)\boldsymbol{I}) + \rho \boldsymbol{g} = 0 \quad (5.1)$$

where $\boldsymbol{\sigma}'$, denoting the effective stress tensor, is defined as $\boldsymbol{\sigma}' = \boldsymbol{D}_s : \nabla \boldsymbol{u}$, with $\boldsymbol{D}_s$ representing the stiffness matrix. $\boldsymbol{u}$ denotes the displacement vector; α denotes the Biot coefficient; $\boldsymbol{I}$ denotes the unit vector; $\boldsymbol{g}$ represents the gravity vector; χ indicates the Bishop coefficient; and ρ signifies the density of the porous medium, which is defined as

$$\rho = (1 - n)\rho^s + nS_w\rho^w + n(1 - S_w)\rho^g \quad (5.2)$$

Here, ρ^s and ρ^w denote the densities of the solid and liquid phases, respectively; and ρ^g represents the density of gas phases and is defined as $\rho^g = \rho^v + \rho^a$, with ρ^v and ρ^a denoting the densities of vapor and dry air, respectively. Additionally, n represents the porosity of waste, and S_w denotes the liquid saturation of waste. The stress–strain relationship can be described by different constitutive models, as introduced in Chap. 3.

Capillary pressure (p^c), defined as the pressure difference between the gas and liquid phases, is expressed as follows:

$$p^c = \rho^g - \rho^w \quad (5.3)$$

(b) Mass balance equation

The mass balance equations for liquid water and vapor in a deformable porous medium are as follows:

$$\frac{\partial(nS_w\rho^w)}{\partial t} + \frac{\partial[n(1 - S_w)\rho^v]}{\partial t} + \nabla \cdot \boldsymbol{J}_w = Q_w - Q_{sw} \quad (5.4)$$

Similarly, the mass balance equation for dry air in a deformable porous medium is as follows:

$$\frac{\partial[n(1 - S_w)\rho^a]}{\partial t} + \nabla \cdot \boldsymbol{J}_a = Q_a \quad (5.5)$$

In the above equations, Q_w and Q_a denote the amounts of generated water and gas, respectively; Q_{sw} represents the quantity of water released from particles into the surrounding pore space; and $\boldsymbol{J}_w$ and $\boldsymbol{J}_a$ signify the fluxes of water and vapor and of dry air, respectively. In Eqs. (5.4) and (5.5), the terms of $\partial(nS_w\rho^w)/\partial t$ and $\partial[n(1 - S_w)\rho^a]/\partial t$ can be calculated according to the water retention curve of the soil. Details of the water retention curve of skeleton-degradable soils are introduced in Chap. 4.

Among these, the water and vapor fluxes ($\boldsymbol{J}_w$) are computed using the following equation:

$$\boldsymbol{J}_w = \boldsymbol{J}_{cw} + \boldsymbol{J}_{cv} + \boldsymbol{J}_{dv} \quad (5.6)$$

where $\boldsymbol{J}_{cw}$ denotes the convection flux of pore liquid water, $\boldsymbol{J}_{cv}$ represents the convection flux of vapor, and $\boldsymbol{J}_{dv}$ indicates the diffusion flux of vapor. The specific expressions for these quantities are as follows:

$$\boldsymbol{J}_{cw} = \rho^w \frac{\boldsymbol{k}k_{rl}}{\mu_l}(-\nabla p^g + \nabla p^c + \rho^w \boldsymbol{g}) \quad (5.7)$$

$$\boldsymbol{J}_{cv} = \rho^v \frac{\boldsymbol{k}k_{rg}}{\mu_g}(-\nabla p^g + \rho^g \boldsymbol{g}) \quad (5.8)$$

$$\boldsymbol{J}_{dv} = \boldsymbol{D}_g^v \nabla \rho^v \quad (5.9)$$

where $\boldsymbol{k}$ denotes the intrinsic permeability of waste; k_{rl} and k_{rg} represent the relative permeabilities of the liquid water and gas phases, respectively; μ_l and μ_g denote the viscosities of the liquid water and gas phases, respectively; and $\boldsymbol{D}_g^v$ signifies the diffusion coefficient of vapor.

Moving further, the dry air flux ($\boldsymbol{J}_a$) is computed using the following equation:

$$\boldsymbol{J}_a = \boldsymbol{J}_{ca} + \boldsymbol{J}_{da} \quad (5.10)$$

where $\boldsymbol{J}_{ca}$ denotes the convection flux of dry air, and $\boldsymbol{J}_{da}$ represents the diffusion flux of dry air. The specific expressions for these quantities are

$$\boldsymbol{J}_{ca} = \rho^a \frac{\boldsymbol{k}k_{rg}}{\mu_g}(-\nabla p^g + \rho^g \boldsymbol{g}) \quad (5.11)$$

$$\boldsymbol{J}_{da} = \boldsymbol{D}_g^a \nabla \rho^a \quad (5.12)$$

where $\boldsymbol{D}_g^a$ denotes the diffusion coefficient of dry air. The intrinsic and relative permeability can be varied with the compression or deformation of the soil. The constitutive model can be found in Chap. 4.

The transport process of dissolved solutes can be described by the mass conservation equation as follows:

$$\frac{\partial S_i}{\partial t} + \boldsymbol{J}_{cw} \nabla S_i - \boldsymbol{D}_l \nabla^2 S_i = Q_{S_i} \quad (5.13)$$

where S_i denotes the ith solute concentration in the pore water phase. In this case, two chemical components are considered as dissolved solutes, namely volatile fatty acids (VFAs) and methanogenic biomass (MB). $\boldsymbol{D}_l$ denotes the hydraulic diffusion coefficient, which varies with degradation-induced porosity, and Q_{S_i} represents the source term of the ith solute.

Similar to the dissolved solute, the transport of LFG in a dry air mixture comprising CO_2, H_2S, NH_3, CH_4, and H_2, can be modeled as follows:

$$\frac{\partial S_i}{\partial t} + \boldsymbol{J}_{ca} \nabla S_i - \boldsymbol{D}_g \nabla^2 S_i = Q_{S_i} \quad (5.14)$$

where $\boldsymbol{D}_g$ denotes the effective dispersion tensor for a multicomponent gas and is defined as $\boldsymbol{D}_g = \boldsymbol{D}_g^a = \boldsymbol{D}_g^v$. Furthermore, Q_{S_i} represents the source term of a dry gas mixture containing CO_2, H_2S, NH_3, CH_4, and H_2.

(c) Energy balance equation

The energy balance equation considering heat convection, heat conduction, and heat sourced from biochemical reactions is expressed below:

$$(\rho C_p)_{eff} \frac{\partial T}{\partial t} + \left(C_p^w \boldsymbol{J}_{cw} + C_p^g \boldsymbol{J}_{ca}\right) \cdot \nabla T - \nabla \cdot (\chi_{eff} \nabla T) = R_j \Delta H_j - \dot{m} \Delta H_v \quad (5.15)$$

where R_j denotes the reaction rate, ΔH_j represents the specific enthalpy of the chemical reaction j, $\dot{m}$ indicates the water evaporation mass rate, ΔH_v denotes the specific evaporation enthalpy, and $(\rho C_p)_{eff}$ and χ_{eff} respectively denote the effective specific heat and effective thermal conductivity tensor for a multiphase mixture. The calculation formulas for $(\rho C_p)_{eff}$ and χ_{eff} are as follows:

$$\left(\rho C_p\right)_{eff} = (1-n)\rho^s C_p^s + nS_w\rho^w C_p^w + n(1-S_w)\rho^g C_p^g \tag{5.16}$$

$$\rho\chi_{eff} = (1-\phi)\rho^s\chi_s + nS_w\rho^w\chi_w + n(1-S_w)\rho^g\chi_g \tag{5.17}$$

$$\dot{m} = \rho^w\frac{\partial(nS_w)}{\partial t}\boldsymbol{I}_0 + \boldsymbol{J}_{cw} \tag{5.18}$$

where χ_s, χ_w, and χ_g denote the thermal conductivities of the solid, liquid, and gas phases, respectively, and $\boldsymbol{I}_0$ represents the identity tensor.

(d) Biochemical reactions

Biochemical degradation drives the multifield processes within landfills. Previous studies have introduced an anaerobic two-stage degradation model to characterize MSW degradation in anaerobic conditions (Chen et al. 2020; Li et al. 2021). This anaerobic model encompasses 13 substances, including rapidly degradable cellulose (RDC) with intraparticle water; slowly degradable cellulose (SDC); sugars; proteins; lipids; pore water; VFAs; MB; and gases such as CO_2, H_2S, NH_3, CH_4, and H_2. Additionally, the model encompasses seven biochemical reactions, including the hydrolysis of RDC, SDC, sugars, proteins, and lipids; methanogenesis of VFA, and exhaustion of MB.

An aerobic degradation model has also been developed to characterize MSW degradation in an aerobic environment (Xiao et al. 2022). This model considers seven biochemical reactions, encompassing the aerobic hydrolysis of RDC, SDC, sugars, fat, and proteins; synchronous nitrification and denitrification; and MB decay. Furthermore, it includes 12 substances, including RDC; SDC; sugars; lipids; proteins; MB; sulfate radicals (SO_4^{2-}); gases such as NH_3, O_2, CO_2, and N_2; and water.

The biochemical reaction rates are dependent on the concentration of substances, water content, temperature. The models to describe the biochemical reaction rates are introduced in Chap. 2.

(e) Chemical reactions

As NGHs within HBSs disintegrate, they release CH_4 and water. The Kim–Bishnoi kinetic reaction model characterizes the chemical dissociation of NGHs induced by reduced pressure, increasing temperature, or other factors (Kim et al. 1987; Ye et al. 2023). This model encompasses solid NGHs, as well as the produced water and CH_4, which can exist in both liquid and gas phases.

Constitutive equations for the chemical reactions occurring under varying pressure and temperature conditions, deformation, and liquid–gas flows are outlined in Chaps. 2, 3, and 4, respectively, and will not be repeated here.

5.2.2 Finite Element Numerical Solution Framework

(1) Numerical solution to the degradation–consolidation model for landfilled MSW

The impacts of biochemical reactions on multi-field coupling processes within landfilled MSW can be reflected in several ways: (1) introducing source terms for water, gas, and heat; (2) changing the porosity and compressibility resulting from mass loss induced by biochemical degradation; and (3) modifying the engineering properties of landfilled MSW owing to changes in porosity. Additionally, the effects of temperature on these processes are evident in the (1) variations in water and gas source terms, accounting for evaporation and condensation, and in (2) changes in the porosity and saturation resulting from temperature variations.

To solve the finite element model, an integrated framework comprising the Galerkin and generalized trapezoidal methods is adopted. Here, the former addresses spatial integration, while the latter addresses time integration for the weak forms of the balance equations. Following spatial discretization using the isoperimetric formulation, a system of nonsymmetric, nonlinear, and coupled equations is obtained:

$$\begin{bmatrix} C_{gg} & C_{gc} & C_{gt} & C_{gu} \\ 0 & C_{cc} & C_{ct} & C_{cu} \\ 0 & C_{tc} & C_{tt} & C_{tu} \\ 0 & 0 & 0 & 0 \end{bmatrix} \frac{\partial}{\partial t} \begin{bmatrix} p^g \\ p^c \\ T \\ \boldsymbol{u} \end{bmatrix} + \begin{bmatrix} K_{gg} & K_{gc} & K_{gt} & 0 \\ K_{cg} & K_{cc} & K_{ct} & 0 \\ K_{tg} & K_{tc} & K_{tt} & 0 \\ K_{ug} & K_{uc} & K_{ut} & K_{uu} \end{bmatrix} \begin{bmatrix} p^g \\ p^c \\ T \\ \boldsymbol{u} \end{bmatrix} = \begin{bmatrix} f^g \\ f^c \\ f^T \\ f^u \end{bmatrix} \tag{5.19}$$

where the solid displacements ($\boldsymbol{u}$), gas and capillary pressures (p^g, p^c), and temperature (T) constitute the primary variables of the multiphase flow and deformation-coupled equations. Furthermore, $\boldsymbol{C}$ and $\boldsymbol{K}$ denote the storage constants and flux constants, respectively, and their specific expressions are detailed in Li et al. (2021).

Figure 5.4 illustrates the solution diagram of the degradation–consolidation model for landfilled MSW. Within this framework, solute transport, multiphase flow, deformation, and heat transport are solved using OGS. Meanwhile, biochemical degradation processes are handled using a special calculator developed by Zhejiang University (ZJU calculator).

During computation, material parameters and initial and boundary conditions are input into OGS. Meanwhile, governing equations for skeleton deformation, liquid–gas flows, solute migration, and heat transport processes are implemented in C+ + within the program. Furthermore, chemical reactions involved in degradation are scripted in Python using the ZJU Calculator (Chen et al. 2016). Notably, OGS interfaces with the ZJU biochemical degradation code to exchange water content and solute concentration information. Following calculations, results pertaining to water, gas, solute, and heat generation are relayed back to OGS. The influence of deformation and solid mass loss on porosity variations is meticulously considered. Notably, porosity directly influences the water retention curve, intrinsic permeability, and diffusion coefficient. Water, gas, and solutes, such as VFA and MB, generated during degradation are incorporated into OGS as source terms to model their impacts on the seepage process. Input and output files are updated at each calculation step.

(2) Numerical solution to degradation–consolidation model for NGHs.

The solution strategy for the degradation–consolidation model of NGHs is outlined in Fig. 5.5. This coupled model, comprising the momentum

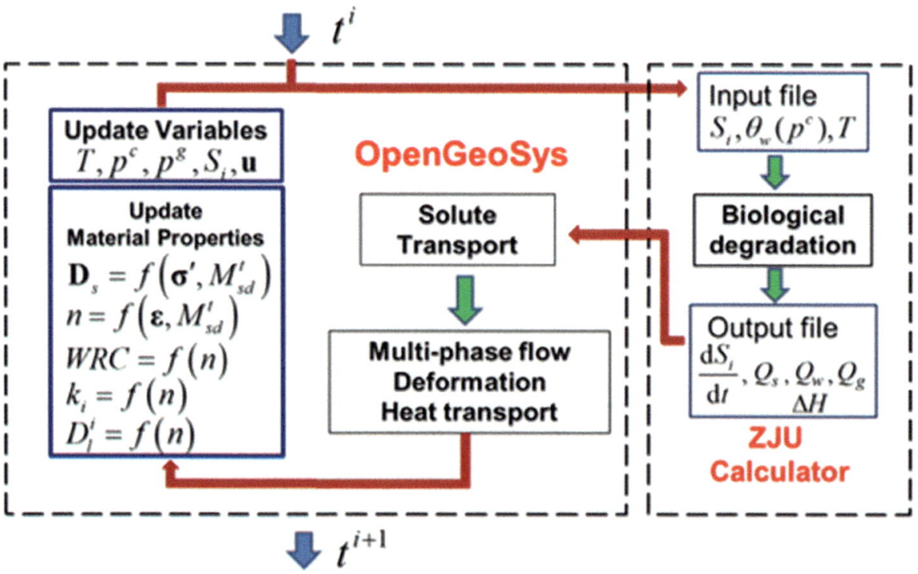

Fig. 5.4 Solution diagram of the degradation–consolidation model for landfilled MSW (Chen et al. 2020)

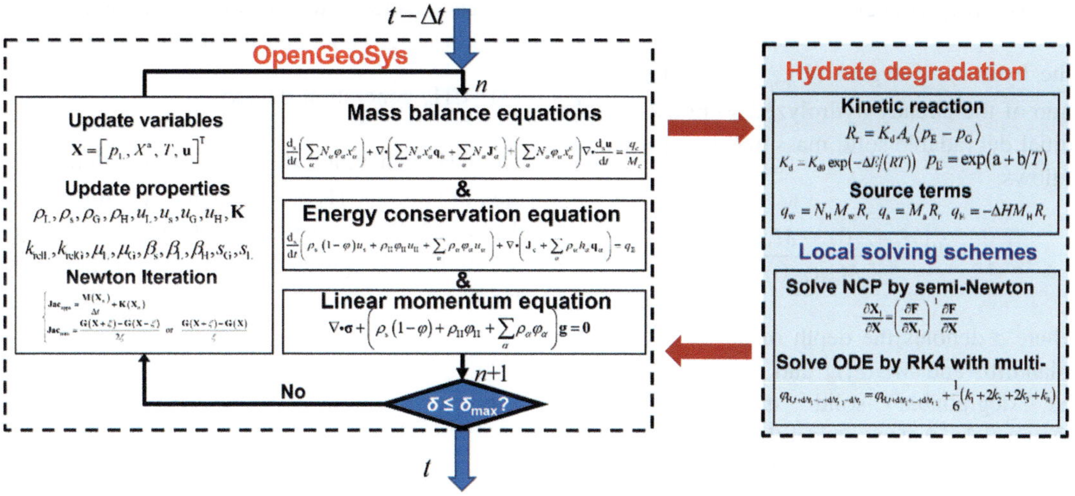

Fig. 5.5 Solution diagram of the degradation–consolidation model for NGHs

balance equation (Eq. (5.1)), mass balance equations (Eqs. (5.4) and (5.5)), and energy balance equation (Eqs. (5.15)), is solved using a monolithic scheme. The required nonlinear iteration is performed based on the Newton–Raphson method. The overall strategy is as follows: (1) Based on the hydrate kinetic reaction and phase equilibrium equations, the reaction rates of hydrates, liquids, and gases and energies are computed and adopted as the source terms of the mass and energy equations. (2) The pore water pressure (p_L) and the total molar fraction of the gas component (X^a) are considered as the primary variables. (3) The nonlinear complementary problem approach is adopted by establishing a semi-smooth algebraic equation system, which is solved using the semi-Newton–Raphson method. This system includes local state variables such as liquid saturation (s_L), molar fraction of water in the liquid-phase (x_L^w), and molar fraction of CH_4 or CO_2 in the gas phase (x_G^a). (4) Based on the mass balance equation of hydrates, their volumetric fraction (φ_H) is determined using the fourth Runge–Kutta method combined with multiple subtime steps, following which its impact on soil stiffness is considered. (5) A high computational parallel method with domain decomposition is adopted. This includes mesh partitioning using the Partmesh tool in OGS, setting the linear iteration method in the input file with a.prj extension, reading the subdomain mesh, and solving the equation system using the PETSc package (Wang et al. 2015; Ye et al. 2023).

5.3 Degradation–Consolidation Behavior of Landfilled Municipal Solid Waste

5.3.1 Stabilization Indicators

As landfilled MSW degrades, the loss in its solid mass triggers the production of water, gas, and solutes over time. These dynamic processes change the displacement, pore water pressure, pore gas pressure, and solute concentration field within the landfilled MSW.

Utilizing the degradation–consolidation model for landfilled MSW, five indicators are introduced to evaluate the anaerobic degradation process of landfilled MSW. These indicators encompass the normalized ratio between cellulose and lignin, hydrolysis degree, intraparticle water release degree, CH_4 generation degree, and consolidation degree.

(1) Hydrolysis degree

The hydrolysis degree (U_{hyd}^{t}) is defined as the ratio of the already hydrolyzed solid mass to the initial degradable solid mass and is expressed as follows:

$$U_{hyd}^{t} = \frac{\int_0^z [M_{sd}(z,0) - M_{sd}(z,t)]dz}{\int_0^z M_{sd}(z,0)dz} \quad (5.20)$$

where z denotes the depth of an MSW layer in a landfill, and $M_{sd}(z,t)$ and $M_{sd}(z,0)$ represent the degradable solid masses (excluding intraparticle water) at depth z at time t and the initial time, respectively. The degradable solid mass (M_{sd}) represents the total dry mass of five degradable compositions. According to this equation, the hydrolysis degrees of fresh MSW is 0%, whereas that of fully decomposed MSW is 100%. This indicator considers the influence of sugars, proteins, and lipids.

(2) Water release degree.

The water release degree (U_{water}^{t}) is defined as the ratio of the current volumetric water content to the maximum volumetric water content of fresh MSW. Here, the maximum volumetric water content is determined by the sum of the initial water content and intraparticle water content. Mathematically, this is expressed as follows:

$$U_{water}^{t} = \frac{\int_0^z [S_6(z,t)]dz}{\int_0^z \left[S_6(z,0) + \lambda_w(z,0)S_1(z,0)\frac{M_{water}}{M_{RDC}}\right]dz} \quad (5.21)$$

where $S_6(z,0)$ and $S_6(z,t)$ denote the volumetric water contents at depth z at time t and the initial time, respectively. $\lambda_w(z,0)$ represents the intraparticle water factor of fresh MSW, and M_{water} and M_{RDC} denote the molar masses of water and RDC, respectively. Given that the intraparticle water content of HFWC MSW significantly exceeds that of LFWC MSW, the water release degree quantifies the release potential of intraparticle water generated in landfilled MSW.

(3) CH_4 generation degree.

The CH_4 generation degree ($U_{CH_4}^{t}$) is defined as the ratio of the CH_4 produced and accumulated to the initial CH_4 generation potential, thus quantifying the CH_4 production potential of landfilled MSW.

$$U_{CH_4}^{t} = \frac{\int_0^Z Q_{CH_4}(z,t)dz}{\int_0^Z P_{CH_4}(z,0)dz} \quad (5.22)$$

$$P_{CH_4}(z,0) = 3M_{CH_4}\left[\frac{S_1(z,0)}{M_{RDC}} + \frac{S_2(z,0)}{M_{SDC}} + \frac{S_3(z,0)}{M_{sugar}} + \frac{S_4(z,0)}{M_{protein}} + \frac{S_5(z,0)}{M_{lipids}}\right] \quad (5.23)$$

where $Q_{CH_4}(z,t)$ denotes the CH_4 generated and accumulated in MSW at depth z until time t, and $P_{CH_4}(z,0)$ represents the CH_4 generation potential of fresh MSW at depth z. This potential can be calculated based on the chemical compositions of fresh MSW. Furthermore, M_{CH_4}, M_{RDC}, M_{SDC}, M_{sugar}, $M_{protein}$, and M_{lipids} signify the molar masses of various chemical components.

(4) Consolidation degree

The consolidation degree (U_{set}^{t}) is defined as the ratio of the settlement of MSW at a certain time to the final settlement of completely degraded MSW:

$$U_{set}^{t} = \frac{u^t}{u^{end}} \times 100\% \quad (5.24)$$

where u^t denotes the surface settlement at time t, and u^{end} represents the surface settlement under complete MSW degradation and dissipation of excess pore pressure (EPP). Final settlement occurs when the gas pressure within the MSW body equals atmospheric pressure and the water content is at or below the field water capacity. The consolidation degree of landfilled MSW resembles that of conventional soils, with the primary difference being degradation-induced deformation, which significantly influences the consolidation process.

(5) Normalized ratio between cellulose and lignin

The normalized ratio between cellulose and lignin (β) is defined as follows:

$$\beta = 1 - \frac{\int_0^z R_{C/L}(z,t)dz}{\int_0^z R_{C/L}(z,0)dz} \quad (5.25)$$

where $R_{C/L}$ denotes the ratio between cellulose and lignin, and $R_{C/L}(z,0)$ and $R_{C/L}(z,t)$ represent $R_{C/L}$ values at depth z for fresh and aged MSW, respectively. Given the difficulty of lignin decomposition, $R_{C/L}$ is often used to quantify the degradation and stabilization state of landfilled MSW. This ratio typically ranging from three to four for fresh MSW. However, owing to variations in the initial compositions of fresh MSW across different regions, the value of $R_{C/L}$ also varies, rendering standardized comparisons impractical. Following normalization, the value of β ranges from zero for fresh MSW to one for completely degraded MSW. The advantage of using β over other stabilization indicators lies in its easy measurement using the initial and current $R_{C/L}$ values, which can be determined in a laboratory using in-situ-sampled MSW and fresh MSW.

5.3.2 Anaerobic Stabilization Behavior

Chen et al. (2020) investigated the anaerobic degradation stabilization behavior of representative HFWC and LFWC MSWs using the above defined indicators. The chemical compositions of fresh MSWs are listed in Table 5.1. Given the exclusive dependence of the degradation process on individual degradable chemical compositions and the degradation environment, biochemical degradation-related parameters for both HFWC and LFWC MSWs are identical, as indicated in Table 5.2. To evaluate the degradation and stabilization behaviors of HFWC and LFWC MSWs and to focus on the influence of different initial contents on anaerobic degradation and stabilization, the values of all other parameters in the degradation–consolidation model remain consistent for both MSW types, except the intraparticle water content.

In landfills, waste is typically disposed of in a layer-by-layer manner, resulting in an initial thickness of approximately 5–10 m per layer. The waste contained in each layer has a similar age and level of degradation. Hence, a one-dimensional model with a height of 10 m was established. The initial and boundary conditions of this model are depicted in Fig. 5.6. Initially, the gas pressure was set to 0 Pa, while the water content was set to 0.283. The top boundary was permeable to gas but impermeable to liquid, and the gas pressure was fixed at 0 Pa. Meanwhile, the bottom boundary was impermeable to gas but permeable to liquid. No additional loading was applied to the top boundary, and the domain deformed under its self-weight stress. The simulation duration was set to 50 years.

Figure 5.7 compares the anaerobic degradation and stabilization indicators of HFWC and LFWC MSWs. As depicted in Fig. 5.7a, the hydrolysis degree of HFWC MSW surpasses 80% in the initial three years. In contrast, under

Table 5.1 Chemical compositions (mass per unit volume) of landfilled HFWC and LFWC MSW (kg/m^3) (Chen et al. 2020)

Category	RDC	SDC	Sugars	Proteins	Lipids	Inert	Intraparticle water	Interparticle water
HFWC MSW	31	20	66	5.5	5.5	130	442	283
LFWC MSW	137	176	28	2.3	2.3	167	111	37

Table 5.2 Parameters of the degradation–consolidation model for landfilled MSW under anaerobic conditions (Chen et al. 2020)

Intraparticle water release factor		Hydrolysis constant factor					Interpore water content impact parameters for hydrolysis	
λ_w		k_{H1} (day^{-1})	k_{H2} (day^{-1})	k_{H3} (day^{-1})	k_{H4} (day^{-1})	k_{H5} (day^{-1})	θ_{min}	θ_{max}
HFWC MSW	LFWC MSW							
130	10	3×10^{-3}	2×10^{-4}	3×10^{-3}	3×10^{-3}	3×10^{-3}	12%	45%
Acid inhibition parameters for hydrolysis			Methanogenesis parameters			Acid inhibition parameters for methanogenesis		Biomass death factor
K_h (kg m^{-3})		n_h	Y	k_{mmax} (day^{-1})	K_S	K_m (kg m^{-3})	n_m	K_d (day^{-1})
10		4	0.08	0.05	4	5	3	0.005
Compression parameters					Porosity	Intrinsic permeability	WRC parameters	
C'_{c0}		$C'_{c\infty}$	$\varepsilon_{s\infty}$ (σ'_0)	σ'_0 (kPa)	n	k_i (m^2)	a_v	n_v
0.2075		0.1	0.38	0.5	0.55	10^{-12}	0.88	1.58
Free diffusion coefficient					Tortuosity			
D_0 (m^2 s^{-1})					τ			
10^{-5}					1			

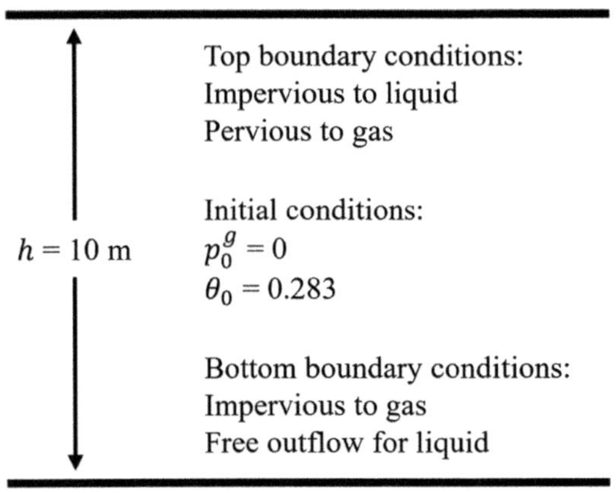

Fig. 5.6 Model depicting the stabilization behavior of a landfilled MSW layer (Chen et al. 2020)

identical conditions, the hydrolysis degree of LFWC MSW reaches only approximately 50–60%, remaining below 80% even after 15 years.

In terms of the water release degree (Fig. 5.7b), HFWC MSW releases its entire intraparticle water content in the initial year. However, LFWC MSW releases a considerably lower amount of intraparticle water compared to HFWC MSW, although specific values are not provided.

As depicted in Fig. 5.7c, the CH_4 generation degrees of HFWC and LFWC MSWs demonstrate varying trends. In both cases, CH_4 generation commences once acid inhibition subsides. For LFWC MSW, CH_4 production increases almost linearly over a logarithmic time scale. In contrast, for HFWC MSW, the majority of CH_4 generation occurs within 2 years owing to the higher proportion of rapidly degradable components within HFWC MSW. The CH_4 generation degree of HFWC MSW reaches 0.8 after 3 years before slowing down significantly. Additionally,

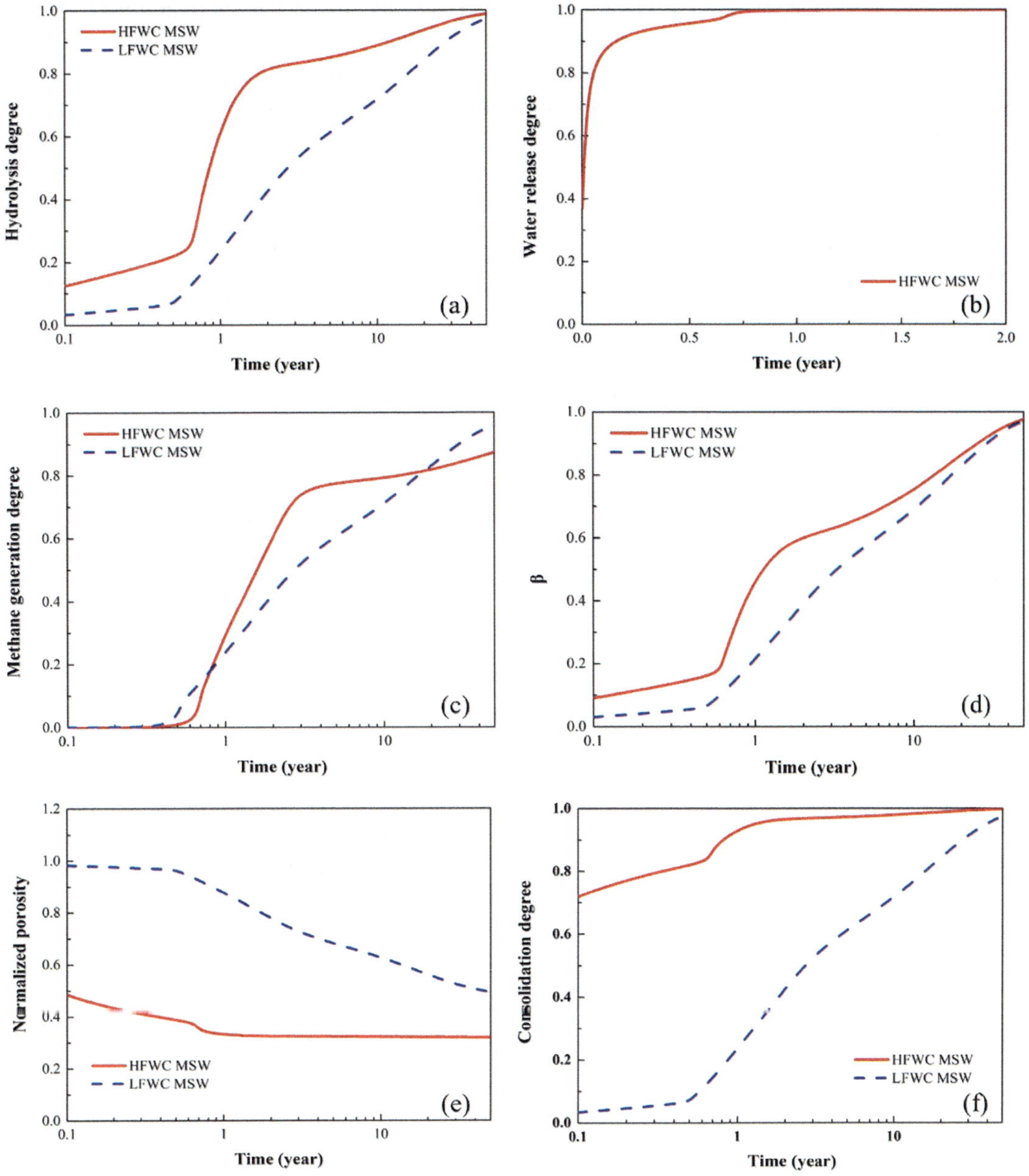

Fig. 5.7 Computed stabilization indicators of HFWC and LFWC MSWs under anaerobic degradation conditions: **a** hydrolysis degree, **b** water release degree, **c** CH_4 generation degree, **d** β value, **e** uniformized porosity, and **f** consolidation degree (Chen et al. 2020)

the CH_4 generation degree of HFWC MSW remains below 0.9 after 50 years, whereas for LFWC MSW, it approaches 1.0. This disparity suggests that over 10% of the CH_4 generation potential is lost during the degradation of HFWC MSW, possibly owing to the influence of the bottom drainage boundary. During the initial degradation phase, substantial amounts of leachates are produced and drained through the bottom boundary, carrying hydrolysis products such as VFAs. If retained within the MSW body, these VFAs can further transform into CH_4. Hence, owing to VFA drainage through leachate systems, the CH_4 generation potential of HFWC MSW can be lower than the initial potential.

As shown in Fig. 5.7d, the β values and hydrolysis degrees of both HFWC and LFWC MSWs demonstrate a correlation. This alignment arises because the values of both β and the hydrolysis degree are obtained based on the mass loss of degradable solid components. However, β does not consider sugars, which constitute a significant portion of HFWC MSW. Hence, β reaches a value of 0.6 after 2 years, a figure lower than the hydrolysis degree of HFWC MSW.

Porosity significantly influences the multifield coupling processes within landfilled MSW. In particular, hydraulic characteristics such as permeability and water retention are directly influenced by porosity. Figure 5.7e depicts the normalized porosities of both HFWC and LFWC MSWs. The normalized porosity (n^t_{norm}) is defined as follows:

$$n^t_{norm} = \frac{n^t}{n^0} \quad (5.26)$$

where n^t denotes the porosity at time t, and n^0 represents the initial porosity. Throughout degradation, the loss of solid mass and consolidation lead to an increase and decrease in porosity, respectively. However, the documented results reveal that porosity decreases initially, indicating that degradation-induced consolidation more significantly influences porosity change compared to solid mass loss. Consolidation dominates the porosity change. Owing to the rapid consolidation of HFWC MSW, its porosity decreases to a relatively stable level after 2 years. Conversely, LFWC MSW requires almost 50 years to undergo similar consolidation phenomena.

The findings displayed in Fig. 5.7f reveal that HFWC MSW reaches a relatively stable state after 3 years, much earlier than LFWC MSW. Degradation-induced consolidation predominantly results from solid mass decomposition. As suggested by the hydrolysis degree results (Fig. 5.7a), over 80% of the degradable solid mass in HFWC MSW decomposes within the first 3 years, considerably faster than in LFWC MSW. Notably, the model includes a drainage boundary for leachates at the bottom and for LFG at the top. The water table or pore pressure significantly influences settlement development. The outcomes indicate that with leachate draining out of the landfilled MSW layer, the settlement or consolidation degree of HFWC MSW increases rapidly within the initial 2–3 years. However, for LFWC MSW, owing to the lower water content of fresh MSW, the variation trend of the consolidation degree mirrors that of the hydrolysis degree.

5.3.3 Aerobic Stabilization Behavior

Landfilled MSW usually exists in an anaerobic reaction environment. To expedite the stabilization of aged MSW, containing high proportions of slowly degradable components, O_2 ventilation can be employed to convert the anaerobic degradation environment into an aerobic environment. Here, O_2 acts as an oxidant to hydrolyze the slowly degradable components of aged MSW into small dissolved solutes (Zhan et al. 2017a). Xiao (2022) adopted the degradation–consolidation model considering heat transport effects and the aerobic degradation model to simulate the aerobic degradation of aged MSW in a column.

As depicted in Fig. 5.8, the MSW column had a height of 0.9 m, width of 0.4 m, and thickness of 0.31 m. Its side boundaries were impermeable to gas and water. The gas pressure at the column top was equal to atmospheric pressure, while that at the column bottom was set to 100.5 kPa. Initially, the aged MSW sample had a temperature of 19 °C, porosity of 0.47, and moisture content of 43.14%. Air was injected from the column

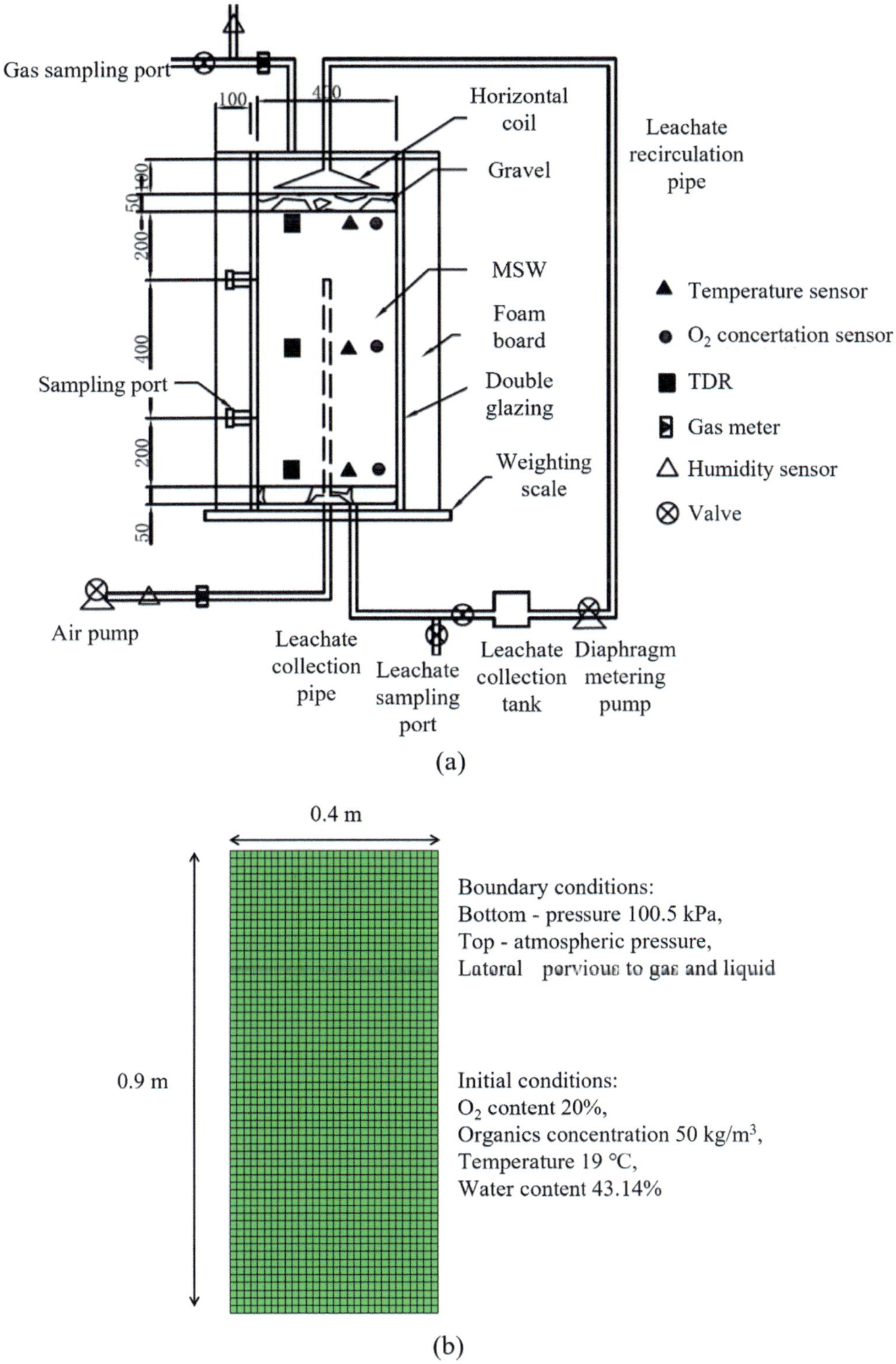

Fig. 5.8 Diagram of **a** aerobic column and **b** the aged MSW simulation column (Xiao 2022)

Table 5.3 Chemical compositions (mass per unit volume) of the aged MSW (kg/m^3) (Xiao 2022)

Parameter	RDC	SDC	Sugars	Proteins	Lipids	MB	NH_3
Value	0	50	0	0	0	0.01	10

Table 5.4 Parameters of the degradation–consolidation model for landfilled MSW under aerobic conditions (Xiao 2022)

k_{H1} (day^{-1})	k_{H2} (day^{-1})	k_{H4} (day^{-1})	k_{H5} (day^{-1})	k_N (day^{-1})	k_d (day^{-1})	ϕ_0 (-)	$\boldsymbol{k}$ (m^2)
0.1	0.05	0.1	0.1	0.1	0.1	0.61	1×10^{-10}
α_v (–)	n_v (–)	χ_s (W/m/K)	χ_w (W/m/K)	χ_g (W/m/K)	C_p^s (J/kg/K)	C_p^g (J/kg/K)	Density (kg/m^3)
0.88	1.58	0.184	0.58	0.03	1,333	2,215.2	1,600

bottom at a constant rate of 0.75 m^3/h and discharged from the column top. Table 5.3 lists the chemical compositions of the aged MSW sample, indicating a high concentration of SDC at 50 kg/m^3. Table 5.4 details the model parameters adopted for the simulation of landfilled MSW under aerobic conditions.

Figure 5.9 compares the simulated results with the experimental data. As depicted in Fig. 5.9a, the simulated content of SDC closely aligns with the experimental measurements. Initially, the SDC concentration decreases from 50 kg/m^3 to 11.42 kg/m^3 over 4 months, exhibiting rapid degradation followed by a gradual slowdown. This pattern can be attributed to the greater concentration of degradable substances in the initial stages, resulting in a faster reaction rate. As organic matter depletes, the degradation rate progressively decreases, eventually approaching zero.

Figure 5.9b illustrates the temporal variation of NH_4-N concentration. Notably, the numerical simulation accurately captures the degradation pattern of NH_4-N under aerobic conditions. NH_4-N concentration drops below 20 mg/L in slightly over half a month, subsequently stabilizing at a relatively low level. This decline occurs primarily owing to the simultaneous nitrification and denitrification of NH_4-N in the liquid-phase upon O_2 ventilation, producing N_2. Thereafter, subsequent NH_4-N generation via solid-phase hydrolysis and N_2 generation through liquid-phase consumption maintain a dynamic equilibrium process.

Figure 5.9c depicts the temporal variation in residual water content. The numerically simulated residual water content aligns with experimental observations. Initially, the total water volume in the column is 36 kg. With ventilation, a significant amount of water evaporates and is discharged with the gas, resulting in progressive water loss. Consequently, by the end of the experiment, remaining water in the reactor progressively decreases to 21 kg.

Real-time temperature readings from sensors positioned at the top, middle, and bottom of the column are presented in Fig. 5.9d. The numerical simulation effectively captures temperature fluctuations resulting from biochemical heat release and variations in water evaporation and transport. From an initial value of 19 °C, the reactor temperature rapidly surged to 29 °C within 2 months with the onset of aerobic reactions. This peak temperature was maintained for the subsequent two months, driven by the abundant degradable organic matter fostering microbial growth and water evaporation. In later stages, as organic matter degradation becomes

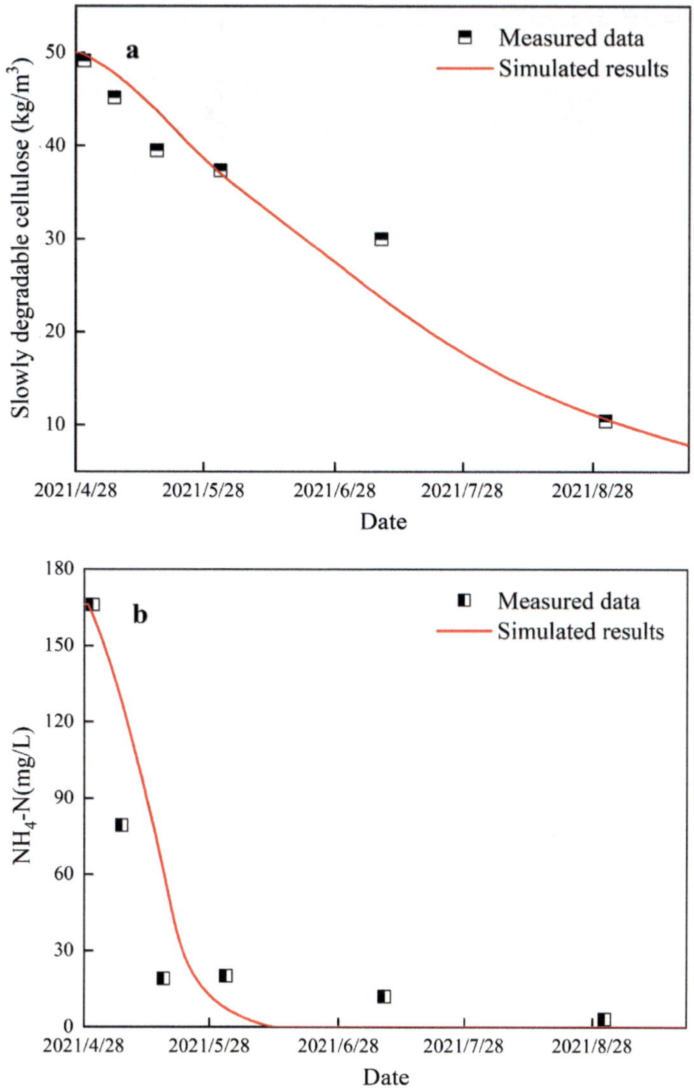

Fig. 5.9 Simulated aerobic degradation of aged MSW (Xiao 2022)

gradual, heat release primarily sustains microbial growth and water evaporation, resulting in minimal temperature fluctuations. Notably, distinct temperature variations at different locations reflect airflow dynamics. For instance, the uppermost position exhibits the highest peak temperature owing to increased O_2 concentrations from the upward airflow, which promote more complete aerobic reactions, and reduced moisture contents from water evaporation, resulting in reduced MSW heat capacity.

Fig. 5.9 (continued)

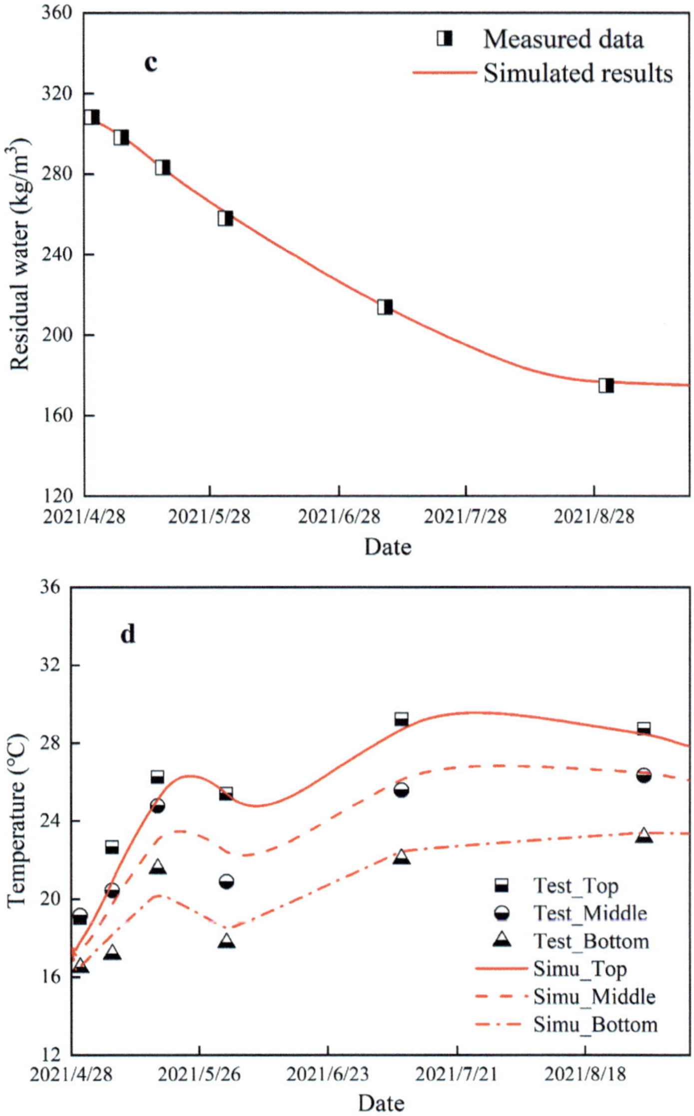

5.4 Large-Scale Bioreactor Experiment and Field Survey Verification

To date, numerous indoor and onsite experiments have been conducted worldwide to examine the biochemical degradation, hydraulic dynamics, and mechanical characteristics of landfilled MSW. However, large-scale experiments specifically focusing on HFWC MSW are lacking.

To address this gap, researchers at Zhejiang University, China, performed two large-scale bioreactor experiments on HFWC MSW (Zhan et al. 2017a, b, 2019; Xu et al. 2019; Xu 2016). These experiments included a cell with vertical one-dimensional liquid and gas flow (labeled as CELL1) and another cell with axisymmetric two-dimensional liquid and gas flow (labeled as CELL2), with dimensions of 5.0 m × 5.0 m 7.5 m (width × length × height) and 5.0 m 4.0 m × 7.5 m (width × length × height), respectively. Fresh HFWC MSW sourced from

Hangzhou was subjected to various treatments, including vertical loading, multiple leachate drawdowns and refills, matured leachate refilling, warm leachate recirculation, and ventilation. The measurement parameters included temperature, earth pressure, settlement, moisture content, and leachate levels. Leachate measures included leachate quantities and biochemical properties (pH, COD, biochemical oxygen demand (BOD), VFAs, ammonia nitrogen, alkalinity, etc.). Meanwhile, LFG measurements examined gas pressures, fluxes, and component concentrations (O_2, CH_4, CO_2, etc.). Furthermore, the permeability coefficients of the liquid and gas phases were tested. The monitoring duration for CELL1 and CELL2 were 1,050 days and 710 days, respectively.

Adopting the CELL1 experiment as an example, subsequent sections detail the setup, procedure, and monitoring results of the experiment conducted at Zhejiang University (Zhan et al. 2017a, b; Xu 2016). Furthermore, the section describes the results of a numerical verification analysis conducted over the first 400 days of the experiment using the degradation–consolidation model outlined in Sect. 5.2.

5.4.1 Experimental Setup

Figure 5.10a depicts a schematic profile of the experimental cell, while Fig. 5.10b illustrates the layout of transducers. The experimental cell was constructed as a steel box with an impermeable bottom and side walls. To minimize heat loss, a 0.2-m-thick insulation board was installed along the inner wall, while a 2-mm-thick high-density polyethylene (HDPE) membrane functioned as an impermeable layer between the insulation board and waste. The cell structure comprised a leachate drainage layer, MSW layer, leachate recirculation and LFG collection layer, and loading layer, arranged from bottom to top.

Leachate drainage layer: The base of the cell was filled with approximately 0.3 m of gravel for leachate drainage. The leachate drainage layer from bottom to the top comprised an HDPE membrane (with a thickness of 1.5 mm), a nonwoven geotextile (with a specification of 400 g/m^2), a drainage pipe network within a washed crushed stone layer (with a particle size of 2–4 cm), a woven geotextile (with an effective pore size of 0.3–1.0 mm), a leachate drainage and refilling pipeline within a crushed stone layer (with a thickness of 60 cm), a nonwoven geotextile (with a specification of 400 g/m^2), and an HDPE geomembrane (with a thickness of 1.5 mm).

MSW layer: Waste from a transfer station in Hangzhou, China, was transported by sanitation trucks. This unprocessed waste was filled in 1.0-m-thick lifts. The waste was weighed and disposed into the cell using an elevator bucket equipped with a crane scale. Subsequently, it was manually distributed to achieve a leveled surface and compacted using a 4.0 t steel block with a basal area of 1.2 m^2. Following waste filling, the cell initially contained 91.3 t of waste, with an initial waste thickness of 5.3 m and a unit weight of 8.0 kN/m^3. This waste comprised 59.4% of food waste, and its initial moisture content was 70.2%, categorizing it as HFWC MSW. Following 1 year of degradation, the food waste content in CELL1 decreased to 22.6%, whereas its paper content increased to 22.9%, approaching that of LFWC MSW.

Leachate recirculation and LFG collection layer: Above the waste layer, a 0.6-m-thick gravel layer was deposited for leachate recirculation and LFG collection, exerting a surcharge loading of 7.6 kPa. Furthermore, between Day 15 and Day 16, an additional surcharge load of 25.8 kPa was applied by the gravel layer. The total vertical loading reached 33.4 kPa, simulating the overburden pressure of the upper waste layer.

As depicted in Fig. 5.10b, transducers were positioned in seven layers designated as L_0–L_6. With the datum point set at the base of the waste layer, the initial elevations of layers L_0–L_6 were −0.3, 0.0, 0.9, 1.9, 3.0, 4.4, and 5.3 m, respectively. Two pairs of earth pressure cells (BGK4800-350 kPa, Geokon Instruments Co., Ltd., China) were installed within layers L_1 and L_4 to record the total vertical and horizontal stresses within the waste body. Meanwhile, two

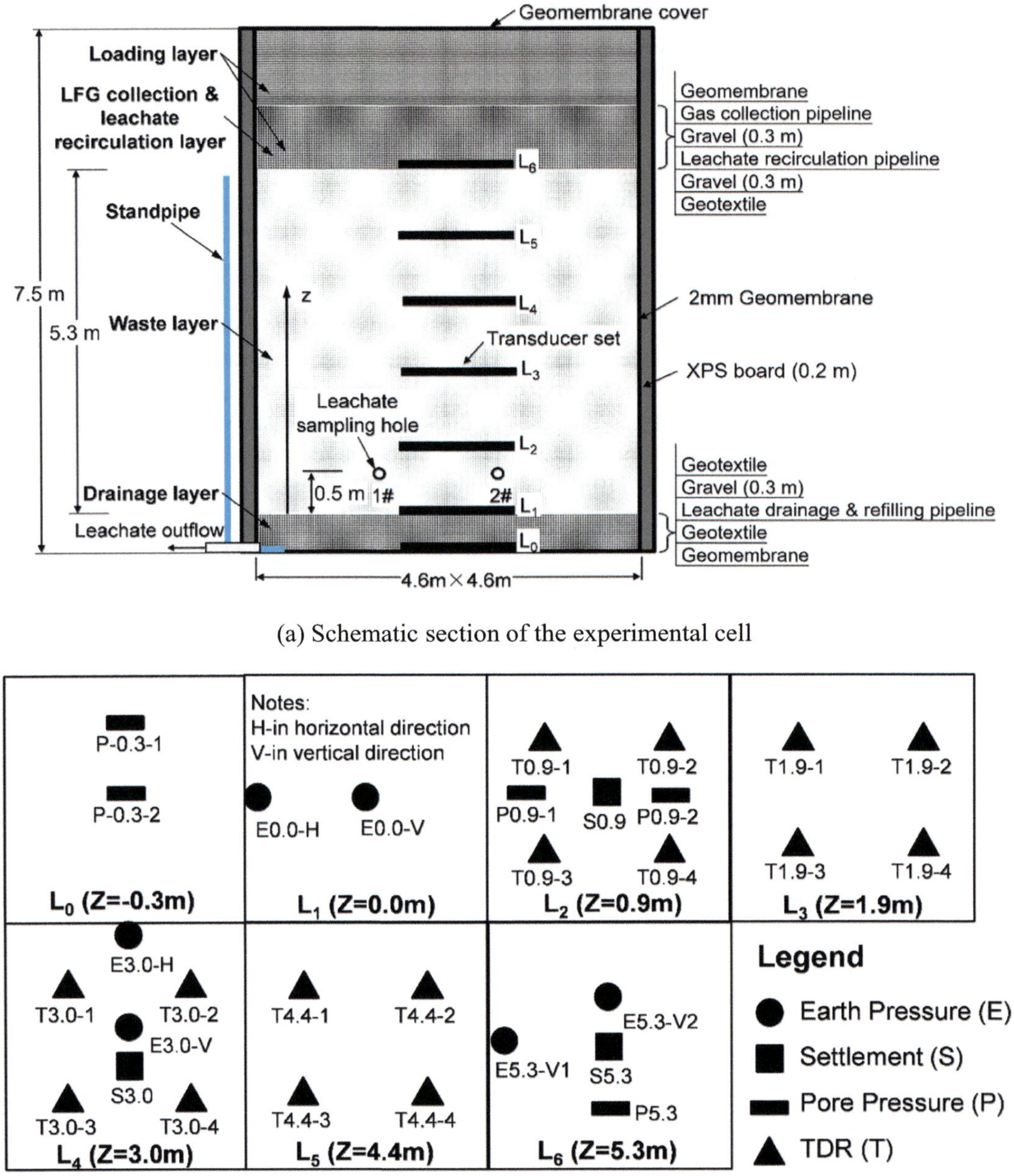

(a) Schematic section of the experimental cell

(b) Layout of transducers

Fig. 5.10 Experimental setup (reprinted from Zhan et al. 2017a, with permission from Elsevier)

cells were placed within layer L_6 to measure the total vertical stress acting on the waste surface. Subsequently, approximately 5 cm of coarse sand was backfilled above the active face of the cells. Three settlement plates (GK4650-1–70 kPa, Geokon Inc., America) were installed within layers L_2, L_4, and L_6 to examine waste compression behaviors. Furthermore, two pore water pressure transducers (BGK4500AL-70 kPa, Geokon Instruments Co., Ltd., China) were installed within layer L_2 to quantify the head of the leachate mound within the waste body. Meanwhile, layer L_0 and layer L_6 were equipped with two and one pore water pressure transducers to measure the hydraulic head in the drainage layer and recirculation layer, respectively. For waste temperature measurements, thermistors were incorporated into each earth pressure cell, settlement plate, and pore water pressure transducer. These transducers were ultimately connected to an automatic logging system (DT80G, Thermo Fisher Scientific Inc., Australia), recording data four times per day. Additionally, 16 time-domain reflectometry (TDR) probes were installed within layers L_2, L_3, L_4, and L_5, distributed in a rectangular pattern with four probes per layer. All TDR probes were coated with plastic tubes to minimize signal attenuation caused by the high electrical conductivity of the waste. Leachate levels within the waste were also recorded using a standpipe inserted into the side of the cell.

5.4.2 Experimental Procedure and Results

(1) Experimental procedure

Waste filling commenced on June 8, 2013, and lasted for seven days. The day following the completion of waste filling was designated as the datum time. The experimental operations, their durations up to June 2016, and objectives are listed in Table 5.5.

Leachate drawdown and refilling (hydraulic test): By June 2016, CELL1 conducted seven cycles of leachate drawdown and refilling. A schematic view of the hydraulic test is illustrated in Fig. 5.11. Notably, the principle of this hydraulic test resembles that of a variable head test. The bottom waste layer with a thickness of L was utilized as the sample section, while the top waste layer acted as the variable head pipe. The leachate drawdown test involved free drainage under gravity, reducing the leachate level within the waste body to a specific height. Notably, the leachate level was lowered in stages from the top to the bottom of the waste body. The target height of leachate outflow was regulated by adjusting the elevation of a three-way pipe. During the leachate refilling test, leachate was introduced from the bottom of the waste body to increase the leachate level to a certain height. The leachate refilling test also proceeded in stages, and the target leachate level was regulated using a constant-head recharge leachate tank. During the experiment, the leachate flow rate (Q), leachate level (h_w) and corresponding time (t) were recorded, and the permeability coefficient was calculated according to Darcy's law.

Intermittent aeration (ventilation test): The objective of this test was to quantify the vertical gas-phase permeability coefficient of landfilled MSW and examine the effect of O_2 on anaerobic degradation. The experimental setup schematic is illustrated in Fig. 5.12. The ventilation commenced when the leachate level was lowered to the bottom of the drainage layer. High-pressure gas from the air compressor was regulated to a set pressure using a pressure-regulating valve, following which it was evenly diffused into the drainage layer through a pipe network. This gas then ascended into the waste layer and entered the LFG collection layer, ultimately being discharged through the LFG collection pipeline network. Pressure sensors and gas flow meters monitored gas pressure and flow rates at the bottom inlet and top outlet, while U-tubes recorded the gas pressure at a distance of $h_2 = 0.45$ m from the bottom of the waste layer. Concurrently, to investigate the impact of the aeration process on anaerobic degradation, gases were collected from the top outlet for concentration measurements, while leachate was collected from the bottom drainage layer for quality assessments.

Mature leachate recirculation: To expedite the methanogenesis of landfilled MSW, mature leachate sourced from the Tianziling Landfill in Hangzhou was reintroduced during the second and first cycles of leachate drawdown and refilling. Table 5.6 summarizes the typical quality characteristics of the mature leachate, indicative of the methanogenesis stage.

Table 5.5 Experimental operations (reprinted from Zhan et al. 2017a, with permission from Elsevier)

Label	Elapsed time (days)	Operations	Objectives
a	1–6	1st loading: 7.6 kPa	To simulate the landfilling process under inadequate leachate drainage
b	15–16	2nd loading: 33.4 kPa	
c	33–69	1st cycle of leachate drawdown and refilling; manual saturation of waste first began on Day 65	To test the hydraulic conductivity
d	125–141	2nd cycle of leachate drawdown and refilling: refilled with 3.88 m^3 of mature leachate (1.88 m^3 was added to the drainage layer)	To test the hydraulic conductivity and expedite waste degradation
e	216–256	Heated leachate recirculation: from the bottom to the top of the waste layer, alternating with a flow rate of 1.0 m^3/day	To increase waste temperature and accelerate waste degradation
f	346–350	3rd cycle of leachate drawdown and refilling	To test the hydraulic conductivity
g	353–371	Waste sampling: at a depth of 1–2 m	To characterize one-year-old waste
h	417–468	1st intermittent aeration: flow rate of 13–16 m^3/h and duration of 3–5 h/day	To examine the effect of O_2 on anaerobic degradation
i	853–861	4th cycle of leachate drawdown and refilling	
j	894–897	5th cycle of leachate drawdown and refilling	
k	930–996	3rd loading: 97.2 kPa 6th cycle of leachate drawdown and refilling 4th loading: 127.2 kPa 2nd intermittent aeration	
l	997-	5th loading: 247.2 kPa 7th cycle of leachate drawdown and refilling 3rd intermittent aeration	

Warm leachate recirculation: Due to low winter temperatures, the average temperature within the waste body remained below 20 °C, hindering methanogenesis. To mitigate the inhibitory effect of this waste temperature reduction on biochemical degradation, the leachate was heated to 40–45 °C using a heating system before being recirculated through the bottom drainage layer or the top LFG collection layer.

(2) Experimental Results

Amount of collected leachate: Figure 5.13 illustrates the variations in the amount of collected leachate. Within the first two months, 26.8 m^3 of leachates were collected from the 91.3 t HFWC MSW. Over the final 10 months, the effective stress remained almost constant, and leachate production increased by 1.1 t. After 1 year of landfilling, with increments in the effective stress, leachate production further escalated. Under constant stress conditions, the majority of leachate production occurred within the first 2 months after landfilling. Remarkably, these leachate production data have been incorporated in the 2019 update of the 2006 IPCC guidelines for national greenhouse gas inventories (IPCC 2019).

Liquid-phase permeability coefficient: Figure 5.14 presents the measurement results of the liquid-phase permeability coefficient during the leachate drawdown and refilling phase. Notably,

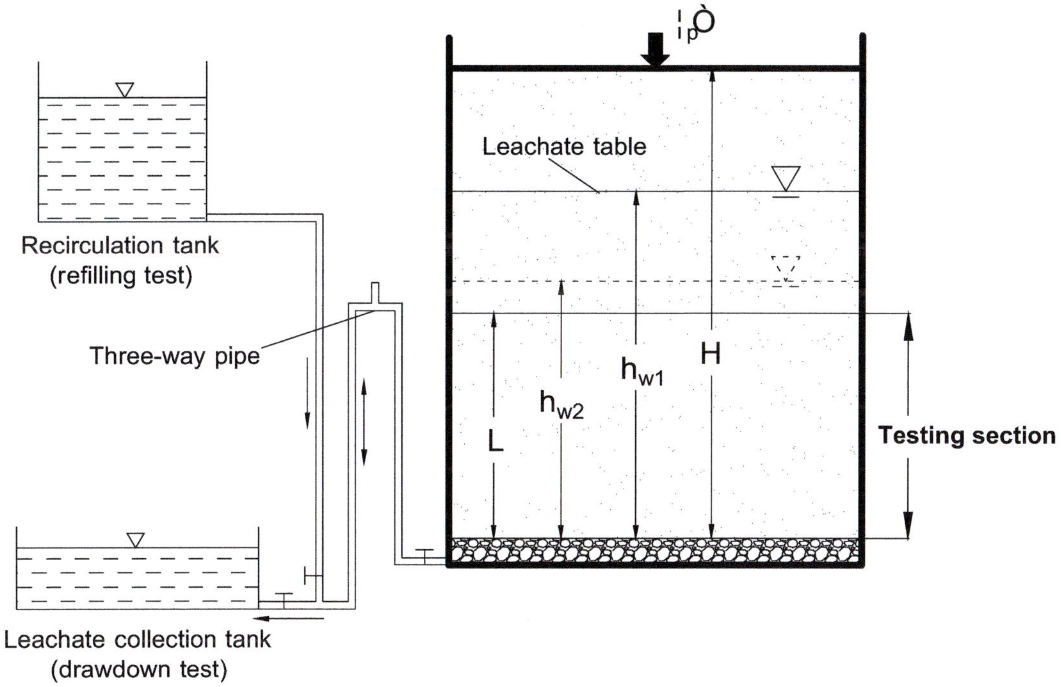

Fig. 5.11 Schematic view of the leachate drawdown and refilling test (reprinted from Zhan et al. 2017b, with permission from Elsevier)

Fig. 5.12 Schematic view of the intermittent aeration process (Xu 2016)

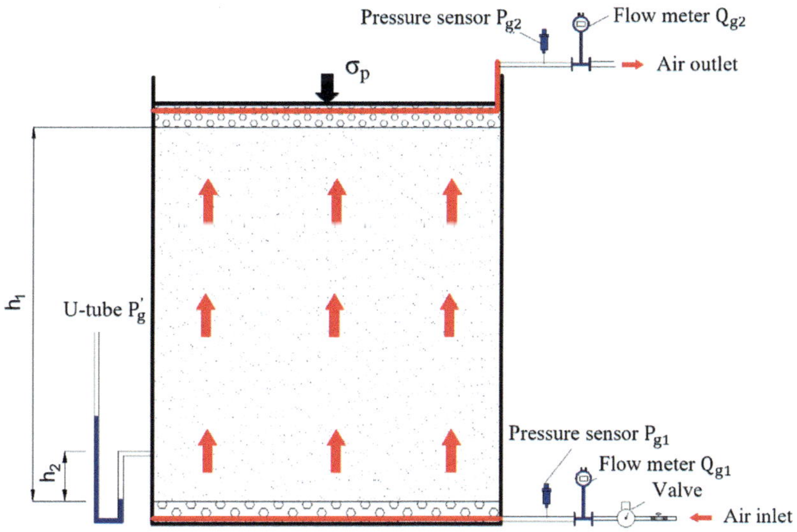

Table 5.6 Quality of mature leachate

Indicators	pH	COD (mg/L)	VFA (mg/L)	Alkalinity (mg/L)
Value	7.91	4,043	1,736	4,169

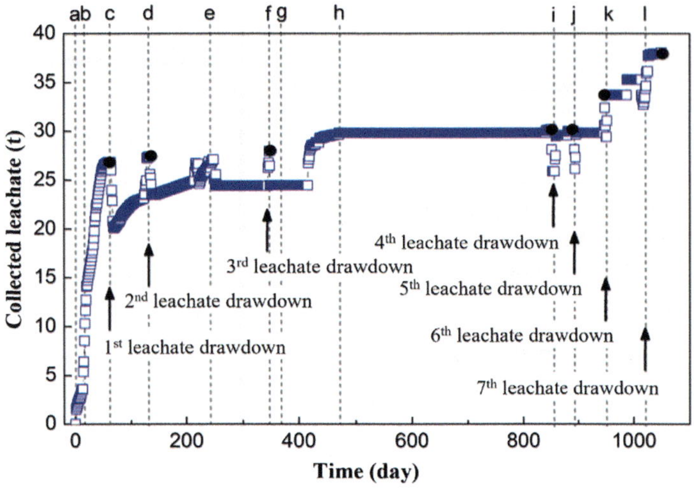

Fig. 5.13 Analysis of the collected leachate (Xu 2016)

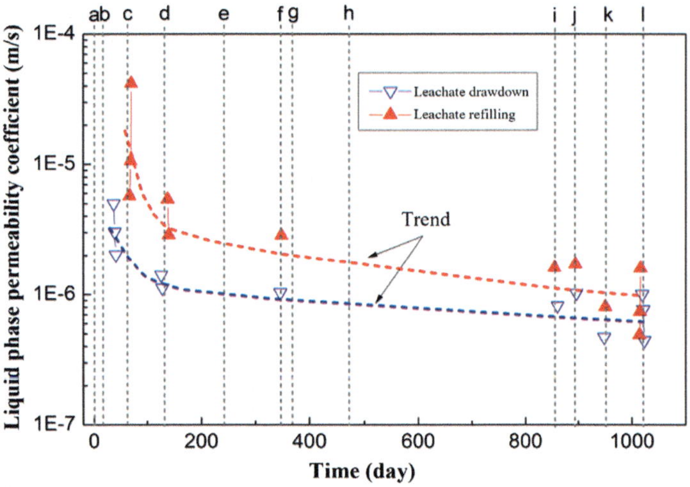

Fig. 5.14 Measurement results of the liquid-phase permeability coefficient (Xu 2016)

hollow triangular points in the figure denote results obtained from the leachate drawdown test, while solid triangular points denote results obtained from the leachate refilling test. The liquid-phase permeability coefficient decreased as the leachate level declined and increased as the leachate level escalated. Within the same leachate drawdown and refilling cycle, the liquid-phase permeability coefficient derived from the leachate refilling test exceeded that obtained from the leachate drawdown test. Ultimately, both coefficient values converged as the waste body underwent compression. Furthermore, despite similar waste layer thickness and porosity values, the values of the liquid-phase permeability coefficient obtained from distinct tests within the same waste section demonstrated significant variations, potentially owing to gas generation and entrapment beneath the leachate level.

Gas-phase permeability coefficient: The values of the gas phase permeability coefficient ranged from 2.84×10^{-6} to 1.22×10^{-5} m/s. Furthermore, the gas-phase permeability coefficient decreased significantly with increasing overlying stress. Figure 5.15 depicts the variation in the gas-phase permeability coefficient with volumetric water content, presenting a notable decrease in the gas-phase permeability coefficient with increasing volumetric water content.

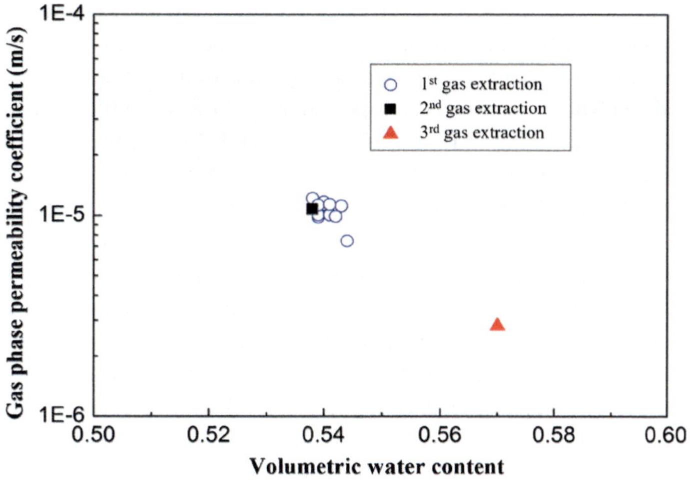

Fig. 5.15 Measurement results of the gas-phase permeability coefficient (Xu 2016)

Waste temperature: Built-in thermocouples monitored temperatures at various points within the waste body.

Figure 5.16 illustrates the temporal changes in waste temperature, alongside recorded air temperatures. Here, $T_{0.0}$, $T_{0.9}$, $T_{3.0}$, and $T_{5.3}$ represent the average temperatures recorded by the thermistors installed within layers L_1, L_2, L_4, and L_6, respectively. Notably, waste temperatures at different depths exhibited similar patterns, with a slight lag behind fluctuations in air temperature. This delay originated from the insulating effect of the lined insulation board, which retained heat within the waste. Overall, waste temperatures ranged from 14.5 °C to 39.2 °C, closely mirroring summer air temperatures and surpassing winter temperatures. Notably, temperature variations near the top and bottom (i.e., $T_{0.0}$ and $T_{5.3}$) of the waste body were more responsive to seasonal air temperature changes compared to those near the body center (i.e., $T_{0.9}$ and $T_{3.0}$). This disparity was attributed to the lack of insulation boards along the bottom and top boundaries of the waste body.

Initially, waste temperature increased rapidly following waste placement. Specifically, temperatures $T_{0.0}$, $T_{0.9}$, $T_{3.0}$, and $T_{5.3}$ increased from initial values of 23.5 °C, 22.8 °C, 23.6 °C, and 22.2 °C to peak values of 33.9 °C, 37.5 °C,

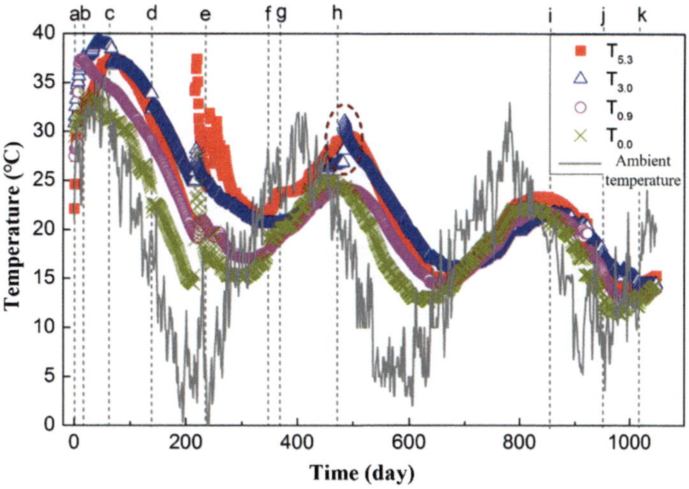

Fig. 5.16 Measurement results of waste temperatures at various buried depths (Xu 2016)

39.2 °C, and 37.6 °C after 19, 16, 45, and 69 days, respectively. This increase was attributed to the heat released from aerobic reactions and elevated summer temperatures. Subsequently, the values of $T_{0.0}$, $T_{0.9}$, $T_{3.0}$, and $T_{5.3}$ gradually decreased to 14.5 °C, 19.8 °C, 24.9 °C, and 20.7 °C, respectively, owing to a drop in air temperature. To prevent a further reduction in the waste temperature, the leachate was heated to 40–45 °C before being recirculated. During the recirculation of this warm leachate, waste temperatures at different depths increased such that the average value increased by 3–7 °C. During the period between Day 418 and Day 468, intermittent aerobic operations were conducted after leachate draining. Despite low air temperatures, the waste temperature increased rapidly. Specifically, over 50 days, the values of $T_{0.0}$, $T_{0.9}$, $T_{3.0}$, and $T_{5.3}$ increased by 3.9 °C, 4.8 °C, 2.4 °C, and 2.5 °C, respectively. Following this period, the waste temperature further continued to increase for weeks. Notably, a rapid increase in temperature $T_{3.0}$ was observed in the period between Day 480 and Day 495, potentially owing to the localized release of heat from methanogenesis reactions.

Overall, waste temperatures exhibited periodic fluctuations corresponding to seasonal changes in ambient temperature. Additionally, within each periodic fluctuation cycle, both the lowest and highest temperatures decreased over time. This may be related to the consistent degradation of waste and the ongoing reduction in heat generation capacity.

Leachate quality: Figure 5.17 illustrates the temporal variations in the pH, COD, BOD, and VFA concentrations in the leachate. Notably, the variation patterns of COD, BOD, and VFA concentrations were consistent among themselves but opposite to the variation trends of pH. Overall, the ammonia nitrogen concentration and alkalinity of the leachate demonstrated a consistent upward trend. During the intermittent aeration period between Day 418 and Day 468, the ammonia nitrogen concentration and alkalinity of the leachate decreased significantly, primarily owing to the occurrence of nitrification and denitrification within the waste layer. Subsequently, these values increased gradually.

LFG composition: As depicted in Fig. 5.18, based on LFG compositions, the degradation process can be divided into three stages—the

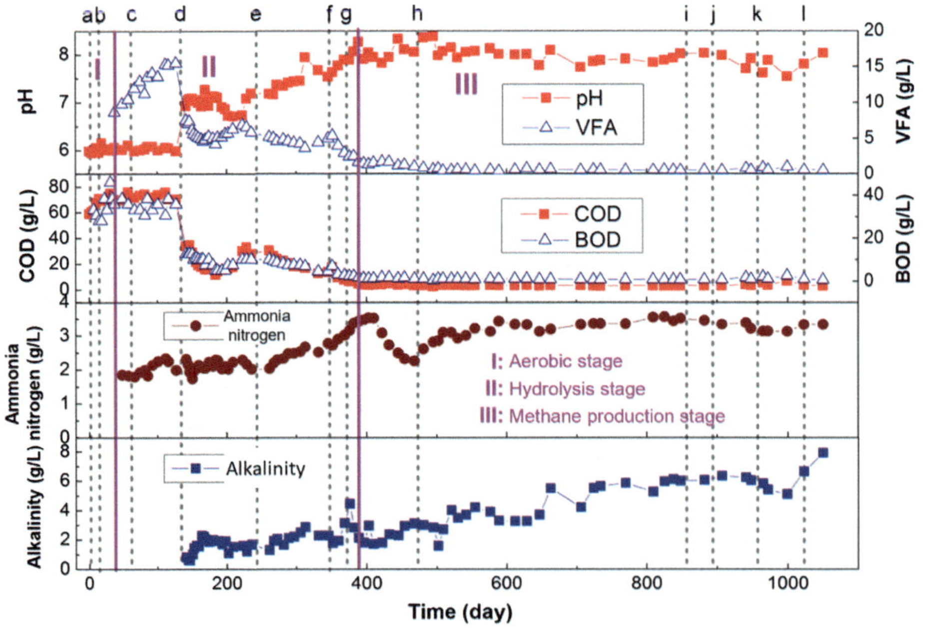

Fig. 5.17 Measurement results of leachate quality (Xu 2016)

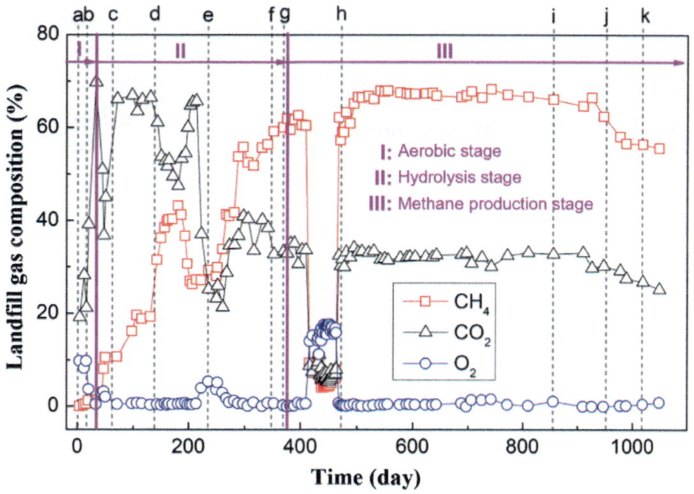

Fig. 5.18 Measurement results of LFG composition (Xu 2016)

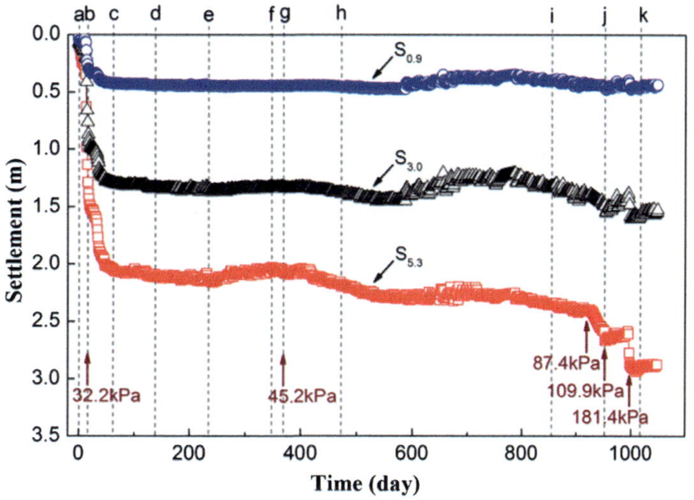

Fig. 5.19 Measurement results of settlements at various burial depths (Xu 2016)

initial aerobic stage, hydrolysis stage (including acidogenesis and acetogenesis), and stable methanogenesis stage. The end of the aerobic stage is marked by the depletion of O_2, reaching a concentration of approximately 0%. Meanwhile, the beginning of the stable methanogenesis stage is marked by the relatively stabilization of CH_4 and CO_2 concentrations,and the hydrolysis stage occurs between these two stages. The figure indicates that the initial aerobic,hydrolysis, and stable methanogenesis stages lasted approximately 1 month, 11.5 months, and 22.5 months, respectively.

Settlement: Figure 5.19 illustrates the temporal variations in settlement at different burial depths of CELL1. Results reveal that under an overlying load of 32.2 kPa, the settlement strain within 2 months reaches 0.39–0.47, approximately 95% of the total strain within 1 year. Furthermore, the primary compression ratio of fresh MSW is 0.381. This value reduces to 0.215 after 2.5 years.

5.4.3 Numerical Verification and Analysis

The degradation and consolidation process of waste over the first 400 days of the CELL1 experiment was numerically simulated using the degradation–consolidation model detailed in

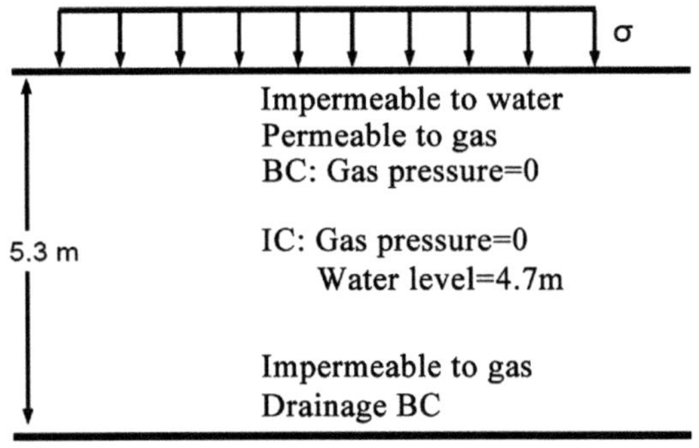

Fig. 5.20 Schematic of the numerical model of CELL1 (reprinted from Li et al. 2023, with permission from Elsevier)

Sect. 5.2. To this end, a 5.3-m-high one-dimensional model was established, with initial and boundary conditions as depicted in Fig. 5.20.

The bottom boundary of the model featured a drainage mechanism but was impervious to gas. The flow boundary condition was determined based on the volume of collected leachate. Meanwhile, the top boundary of the model was impermeable to water but permeable to gas. Due to potential leakage from the HDPE membrane, no significant positive gas pressure was detected during testing; hence, the gas pressure boundary was fixed at 0 Pa. The top load application scheme was as follows: From Day 0 to Day 14, Day 15 to Day 369, and Day 370 to Day 400, the pressures were 7.3 kPa, 32.2 kPa, and 45.2 kPa, respectively. The temperatures measured at the top and bottom of the MSW body were adopted as boundary conditions. Details pertaining to the chemical composition of fresh MSW and model parameters are provided in Tables 5.7 and 5.8, respectively.

(1) VFA concentration

Figure 5.21a compares the VFA concentrations in the leachate obtained from the model simulation and experiment. The red and blue lines represent simulated VFA concentrations in the presence and absence of mature leachate recirculation and heating, respectively. The red line effectively demonstrates VFA accumulation in the initial stages of degradation, followed by a rapid decline after leachate recirculation and heating. Conversely, the blue line indicates that in the absence of these processes, the MSW does not enter the stable CH_4 production phase within 400 days.

(2) Ratio between cellulose and lignin

Figure 5.21b compares the cellulose and lignin (C/L) ratios obtained from the model simulation and experiment. The red and blue lines represent simulated C/L ratios in the presence and absence of mature leachate recirculation and heating, respectively. Initially, both lines demonstrate good alignment with experimental data. However, after one year, discrepancies emerge as the calculated C/L values surpass the experimental values (by 21% for the red line and 29% for the blue line). Notably, the C/L ratio, in conjunction with indicators such as leachate and gas production potential, quantifies landfill stabilization (Chen 2014; Liu et al. 2016). Both Fig. 5.21a and b indicate that the recirculation of mature leachate effectively accelerates degradation and stabilization.

(3) CH_4 generation

Figure 5.21c compares the CH_4 generation amounts obtained from the model simulation and experiment. The red and blue lines represent the simulated CH_4 amounts in the presence and absence of mature leachate recirculation and heating, respectively. Both lines follow similar trends before operation III. However, subsequently, the red line exhibits a much pronounced

Table 5.7 Chemical composition of fresh MSW (kg/m^3) (reprinted from Li et al. 2023, with permission from Elsevier)

Parameter	Rapidly degradable cellulose	Slowly degradable cellulose	Sugars	Proteins	Lipids
Value	26.95	20.32	49.95	4.16	4.16

Table 5.8 Parameters of the degradation–consolidation model for landfilled MSW considering heat transport effects (reprinted from Li et al. 2023, with permission from Elsevier)

λ_w (–)	k_{H1} (day^{-1})	k_{H2} (day^{-1})	k_{H3} (day^{-1})	k_{H4} (day^{-1})	k_{H5} (day^{-1})	θ_{min} (–)	θ_{max} (–)
130	0.004	0.0005	0.004	0.004	0.004	0.12	0.45
K_h (–)	n_h (–)	Y (–)	k_{mmax} (day^{-1})	K_s (–)	K_m (kg/m^3)	n_m (–)	K_d (day^{-1})
12	4	0.05	0.06	30	25	3	0.005
Deformation						Heat transport	
C_{c0}' (–)	$C_{c\infty}'$ (–)	$\varepsilon_{s\infty}$ (–)	ϕ_0 (–)	ρ_w (kg/m^3)	ρ_s (kg/m^3)	χ_{eff} (W/m/K)	C_{peff} (J/kg/K)
0.38	0.1	0.38	0.47	1,000	820	0.184	1,333
Thermal conductivity of landfill gas (W/m/K)		C_p^g (J/kg/K)		Thermal conductivity of leachate (W/m/K)		C_p^w (J/kg/K)	
0.03		2,215.2		0.58		4,200	
Liquid and gas flow							
D_{l7} (m^2/s)	D_{l8} (m^2/s)	D_{g10} (m^2/s)	D_{g11} (m^2/s)	D_{g12} (m^2/s)	D_{g13} (m^2/s)		
1.5×10^{-9}	1.5×10^{-9}	1.5×10^{-5}	1.5×10^{-5}	1.5×10^{-5}	1.5×10^{-5}		
n (–)	θ_r (–)	m_v (–)	γ_v (–)	μ_l (kg/m/s)	μ_g (kg/m/s)		
0.78	0.33	0.36	6.2	0.001	0.00001		

rapid increase. By Day 400 of the large-scale experiment, the calculated CH_4 generation amount reaches 6.85 kg/m^3, while the corresponding initial mass of degradable carbon-containing substances is 105.54 kg/m^3. This indicates that a substantial portion of the CH_4 production potential is realized within the first 400 days of degradation, resulting in a subsequent swift decline in the gas production potential.

(4) Ratio between CH_4 and CO_2

Figure 5.21d compares the CH_4 to CO_2 (CH_4/CO_2) ratios obtained from the model simulation and experiment. Notably, the model calculation effectively captures the evolving trend of the CH_4/CO_2 ratio observed in the experiment, initially increasing before stabilizing. By Day 400, the volume increment ratio of CH_4 to CO_2 approaches two, indicating the transition of the degradation process to a stabilization stage.

(5) Temperatures at various burial depths

Figure 5.21e compares the temperatures derived from the model simulation and experiment. Notably, aerobic degradation caused an initial rapid increase in temperature owing to the release of heat from the MSW. As the model disregarded aerobic degradation, initial temperature changes were determined by adopting the temperature values measured over seven days as boundary conditions. Subsequently, with O_2 depletion, anaerobic degradation commenced, and waste temperatures gradually decreased due to low winter temperatures. Owing to the lack of insulation at the bottom of the cell, temperature changes here were primarily influenced by the boundary. During leachate recirculation periods, waste temperatures increased variably, increasing by an average of 3–7 °C. Notably, temperatures at the bottom decreased slightly during water level rise tests, likely owing to the lower temperature of the recharged leachate compared to

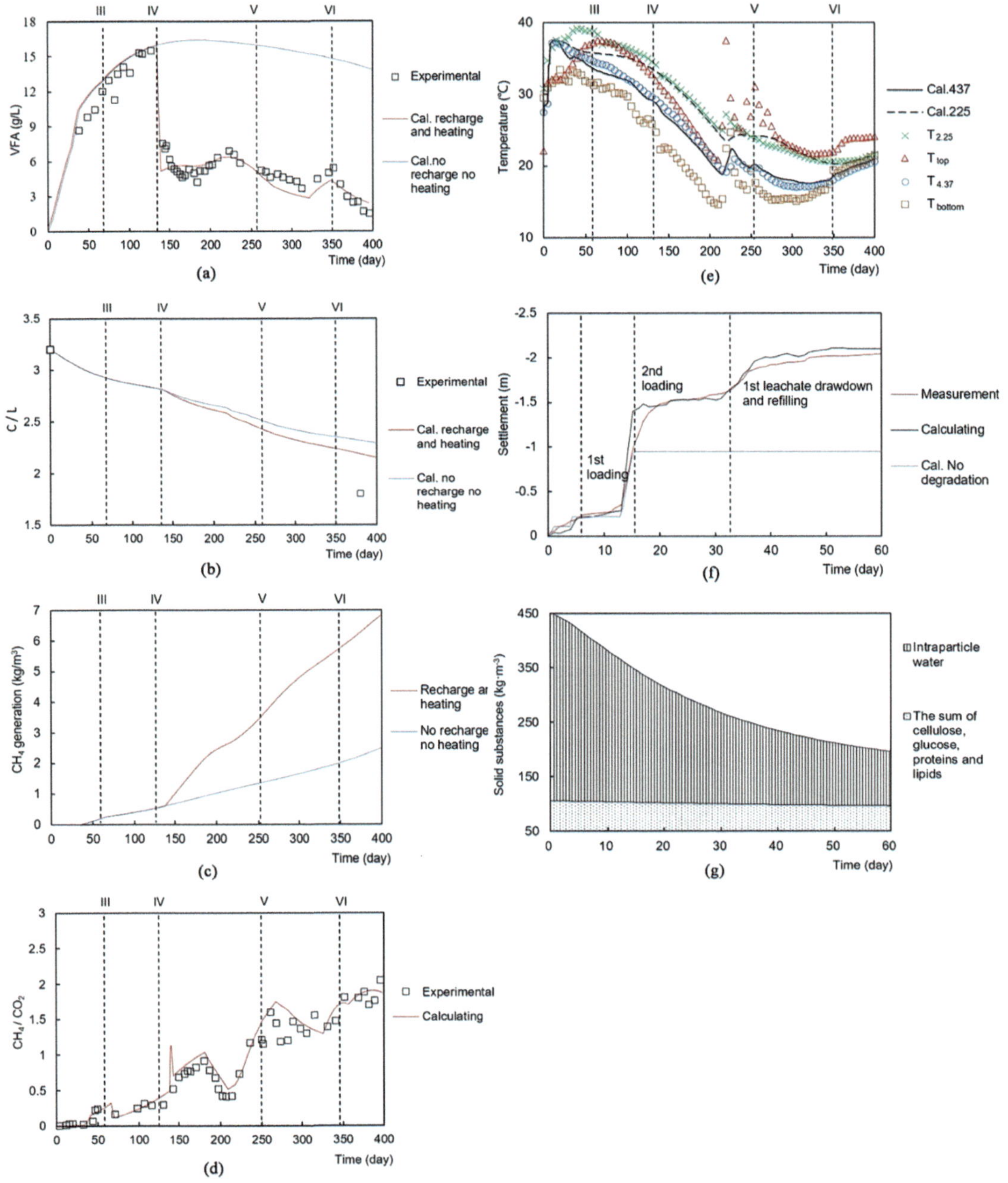

Fig. 5.21 Comparison between the calculated and experimental **a** VFA concentrations, **b** C/L ratios, **c** CH_4 generation amounts, **d** CH_4/CO_2 ratios, **e** temperatures, **f** settlement, **g** Comparison between the concentrations of intraparticle water and other degradable components (reprinted from Li et al. 2023, with permission from Elsevier)

that of the waste. These findings suggest that inadequate insulation at the cell bottom prevents significant temperature increments during methanation.

(6) Settlement within 60 days

Figure 5.21f compares the settlement at a depth of 0.9 m derived from the model simulation and the experiment. Meanwhile, Fig. 5.21g compares the calculated intraparticle water concentration with the sum of calculated concentrations of cellulose, glucose, proteins, and lipids within the first 60 days. Notably, the settlement reached 2 m within 60 days, constituting 91% of the total settlement. In Fig. 5.21f, the blue line depicts the deformation values computed under identical boundary conditions considering MSW as a completely nondegradable soil. The results reveal that loading primarily influenced settlement before Day 15, while rapid MSW degradation dominated between Day 15 and Day 60. Meanwhile, the concentration results reveal a decrease in degradable matter content from 452.30 kg/m^3 to 195.26 kg/m^3, along with a decrease in the total concentration of cellulose, glucose, proteins, and lipids from 105.56 kg/m^3 to 96.14 kg/m^3 from the experiment onset to Day 60. The disparity between the two represented the intraparticle water content, which decreased from 346.74 kg/m^3 to 99.12 kg/m^3. Notably, the mass change of intraparticle water accounted for 96.34% of the mass change of degradable matter within the MSW, suggesting that settlement within the initial 60 days was primarily attributable to intraparticle water release.

Overall, the simulation results of various biochemical, hydraulic, and mechanical indicators derived from the degradation–stabilization model demonstrate good agreement with the experimental data. This underscores the potential applicability of the degradation–stabilization model in examining the biochemical–hydraulic–mechanical interactions within HFWC MSW landfills.

5.4.4 Field Survey and Evaluation of the Degree of Stabilization in Landfills

The distribution of the normalized ratio between cellulose and lignin (β) can be derived by sampling and testing the cellulose and lignin contents of MSWs at different burial depths. Figure 5.22 compares β values obtained from model simulations and an onsite experiment conducted on samples from HFWC MSW landfills in China (Chen et al. 2019; Liu 2016). The simulated results align closely with those of the onsite measurements. Within the first 2–3 years, degradation progresses rapidly. Hence, the value of β reaches 0.6 and subsequently increases gradually over time.

So far, Sect. 5.3.1 presented several indicators describing the degradation process of landfilled MSW from various perspectives. Based on these indicators, Sect. 5.3.2 subsequently delineated the anaerobic stabilization stages of HFWC MSW landfills (Chen et al. 2020). As depicted in Fig. 5.23, the anaerobic stabilization period of HFWC MSW landfills can be categorized into three stages: (I) rapid degradation, (II) slow degradation, and (III) post-stabilization.

The transition point between the rapid and slow degradation stages varies based on different stabilization indicators. In particular, the water release and consolidation degrees exhibit an inflection within 2 years, while the hydrolysis degree and β value demonstrate a transition from rapid to slow degradation after 2 years. Methanogenesis follows solid mass hydrolysis; hence, the CH_4 generation degree presents a lag.

During rapid degradation, 80% of the solid mass hydrolyzes, and all intraparticle water is released. Simultaneously, 60% of LFG is generated, and 80% compression is achieved over 2–3 years, resulting in a β value of 0.6. Following this, slow degradation commences, demonstrating over 90% solid mass hydrolyzation, 90% compression, and 80% LFG generation over 20–30 years, achieving an approximate β value of 0.9, indicating near-complete stabilization.

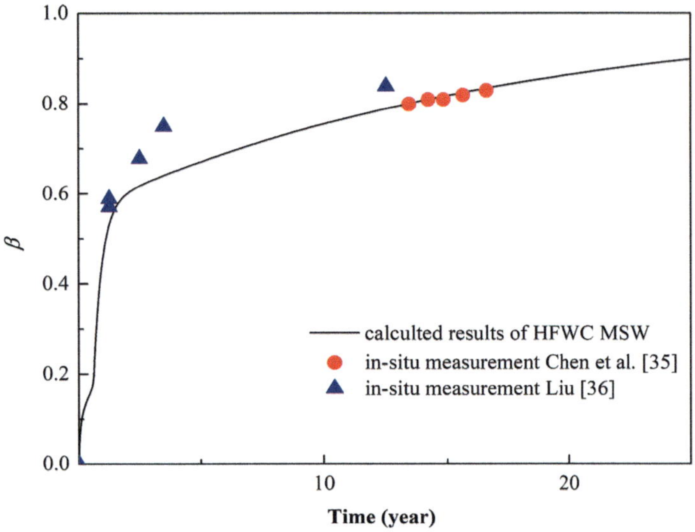

Fig. 5.22 Comparison between the C/L ratios derived from the model simulation and experiment (Chen et al. 2020)

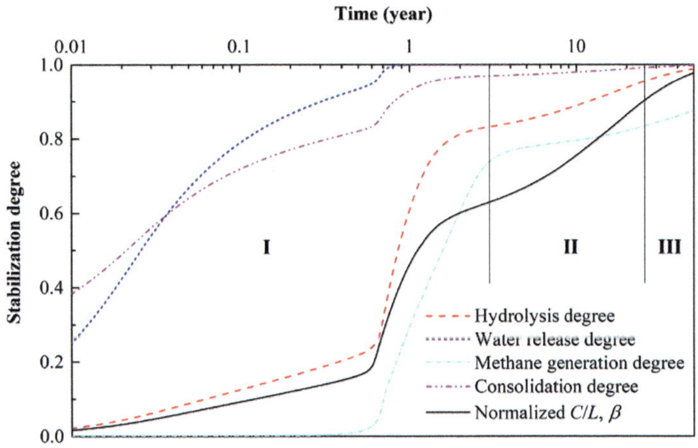

Fig. 5.23 Anaerobic stabilization stages of HFWC MSW landfills (Chen et al. 2020)

Different stabilization stages of landfilled HFWC MSW result in distinct degradation behaviors. For instance, during the rapid degradation stage, substantial amounts of leachates are released, accompanied by the rapid generation of LFG. These high leachate levels increase the risk of slope failure and reduce LFG collection efficiency, necessitating specific engineering measures. Meanwhile, in the slow degradation stage, residual organic substances undergo gradual degradation. Transforming the anaerobic degradation environment to an aerobic degradation environment with air ventilation can significantly accelerate degradation and shorten the duration of this stage. Finally, in the post-stabilization stage, HFWC MSW undergoes near-complete degradation, exhibiting minimal LFG generation and negligible landfill surface settlement, thus facilitating potential field or MSW reuse. However, comprehensive environmental assessments are imperative before considering such reuse.

5.5 Analysis of Two-Dimensional Landfill Slopes

According to the previous analysis, it is known that HFWC MSW landfills produce a large amount of LFG during the rapid degradation stage. In landfill engineering practices, this often

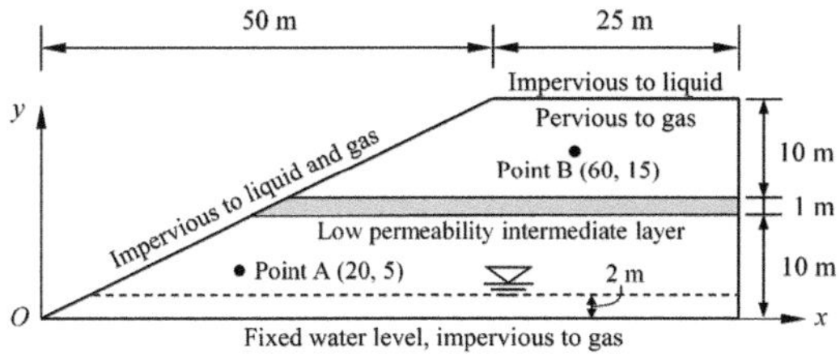

Fig. 5.24 Schematic view of a typical two-dimensional landfill slope (Ma 2022; Ma et al. 2024)

results in high LFG pressure accumulation in the waste body. Previous coupling simulations of MSW mass, such as McDougall (2007) and Chen et al. (2020), have primarily been conducted under one-dimensional conditions. However, these simulations fail to capture behaviors of MSW such as the lateral deformation of MSW.

A coupling simulation using an improved degradation–consolidation model was performed on a two-dimensional MSW slope (see Fig. 5.24), and then the coupling behaviors in it, especially the lateral deformation change rules, was emphatically discussed. Typically, a leachate drainage layer is positioned at the bottom of MSW mass. There is a certain height of leachate and high saturation in this layer, which significantly reduces its gas permeability.

To reflect this scenario, the bottom boundary was set as a fixed water level of 2 m, impermeable to gas. Additionally, the top and slope boundaries were both set to be impervious to water, simulating the effects of covering membranes, with gas pressure at the top boundary equated to atmosphere (pervious to gas). A low-permeability intermediate layer was introduced in the middle of the slope. Under its influence, the landfill gas generated in the bottom layer of MSW migrated upward slowly, leading to LFG accumulation and increased pore gas pressure in the bottom layer.

The parameter values of anaerobic degradation and multi-phase flow models used in this simulation are shown in Table 5.9. The skeleton deformation of MSW was calculated by the constitutive model proposed in Sect. 3.6. The parameter values of constitutive model of MSW are shown in Table 5.10.

5.5.1 Bio-Hydro-Mechanical Coupling Behaviors in the MSW Slope

Point A, which is close to the slope surface in the bottom layer, was selected as a typical point to analyze the variation of some physical quantities with degradation time. As illustrated in Fig. 5.25a and b, during the first 240 days of

Table 5.9 Parameters of the degradation–consolidation model for landfilled MSW (Ma 2022; Ma et al. 2024)

λ_w (–)	k_{H1} (day^{-1})	k_{H2} (day^{-1})	k_{H3} (day^{-1})	k_{H4} (day^{-1})	k_{H5} (day^{-1})	θ_{min} (–)	θ_{max} (–)
130	0.006	0.006	0.006	0.006	0.006	0.12	0.48
K_h (kg/m^3)	n_h (–)	Y (–)	k_{mmax} (day^{-1})	K_s (–)	K_m (kg/m^3)	n_m (–)	K_d (day^{-1})
12	3	0.04	0.1	30	25	4	0.005
D_0 (kg/m^3)	τ (–)	k_i (m^2)	α_v (–)	α_n (–)			
1e − 5	1	1e − 12	0.88	1.58			

Table 5.10 Parameter values of the constitutive model of MSW in the coupled simulation (Ma 2022; Ma et al. 2024)

λ	κ	υ	M	$p_{b-\max}$ (kPa)	β
0.5	0.05	0.3	1.5	1000	2.9

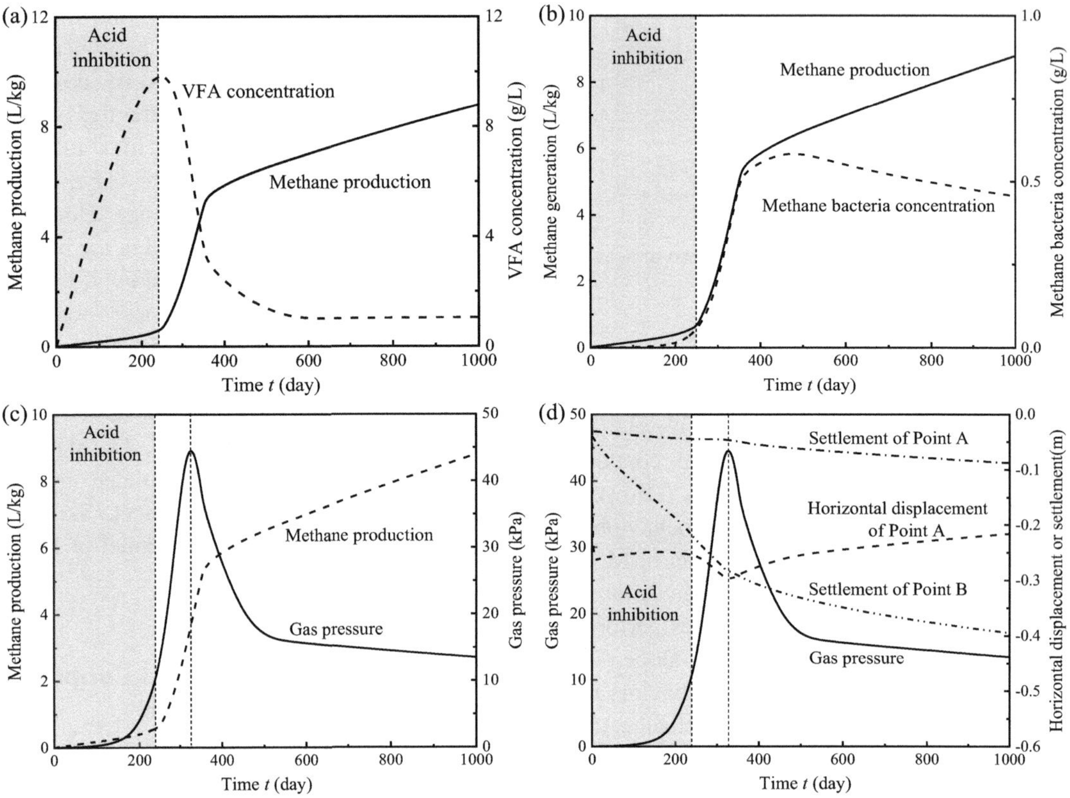

Fig. 5.25 Variation rules with degradation time of **a** VFA concentration and CH_4 production, **b** CH_4 production and methane bacteria concentration, **c** CH_4 production and gas pressure, and **d** LFG pressure and horizontal displacement of Point A, and vertical displacements of Points A and B (Ma 2022; Ma et al. 2024)

degradation, the concentration of VFA increased rapidly due to the hydrolysis of degradable substances. High VFA concentration could decrease the pH of leachate, then inhibit the growth of methane bacteria, a phenomenon known as acid inhibition (Chen et al. 2016). In this stage, the rates of methane production and methane bacteria growth rate were very low. With the increase in the concentration of methane bacteria, the acid inhibition was removed, and MSW entered the methanation stage, marked by the rapid decrease in VFA concentration and the increase in the methane production. In this example, the peak of gas generation rate approximately appeared on the 360th day. The limiting substrate of methanation is VFA. Therefore, the low VFA concentration resulted in the decrease in gas production rate and methane bacteria concentration after 400 days.

The variation of gas pressure, horizontal displacement, and settlement of the MSW at Point A are shown in Fig. 5.25c and d. The settlement of Point B located in the top layer is also plotted in Fig. 5.25d. Under the effect of gravity, the MSW close to the slope toe exhibited lateral bulge deformation (i.e., negative horizontal displacement) at the beginning of simulation. Subsequently, the deformation of MSW was influenced by the skeleton weakening induced by the degradation of waste and gas pressure. During the acid inhibition phase, the gas pressure was relatively low since the methane production was significantly inhibited. In this period, the displacement of MSW was mainly influenced by the degradation-induced skeleton weakening, leading to lateral contraction deformation. Consequently, the horizontal displacement increment of Point A was in the positive x-axis direction.

With the removal of acid inhibition, methane production and gas pressure increased rapidly. Dominated by the increasing gas pressure, the horizontal deformation increment of Point A shifted to the negative x-axis direction (i.e., lateral bulge deformation). After the 330th day, as gas pressure decreased and stabilized due to reduced methane production, degradation once again became the dominant factor influencing deformation. Hence, the deformation mode at Point A in this period mirrored that of the acid inhibition period.

The settlement of MSW is closely related to the effective stress; thus, the settlement of Point A hardly developed in the gas pressure increase stage. The gas pressure in the top layer of the slope remained low throughout degradation due to the top boundary being pervious to gas, resulting in continuous settlement in this region (e.g., Point B) under the influence of degradation.

5.5.2 Effective Stress Paths of MSW

The deformation mechanisms of MSW were further discussed combing the effective stress paths. As shown in Fig. 5.26a, due to the influences of the gas pressure, the effective stress path of the MSW at Point A is close to a reciprocating horizontal line on the p–q plane. When the gas pressure is increasing, the effective mean stress decreased, and the stress path extended horizontally to the left. When the gas pressure decreased gradually, the effective mean stress increased, and the stress path extended to the right. In this process, the deviatoric stress changed slightly. The stress state of Point B did not change obviously because the gas pressure was low in the top layer.

Whether the MSW was in elastic or plastic state could be judged by the following equation

$$f^{\mathrm{trail}} = q^2 + M^2 p^2 - M^2 p(p_{\mathrm{xb0}} - p_b) \quad (5.27)$$

If $f^{\mathrm{trail}} > 0$, the MSW is in the plastic state; otherwise, it is in the elastic state. As shown in Fig. 5.26b, during the acid inhibition, the f^{trail} of Point A was slightly larger than zero. With the increase in the gas pressure, the MSW at Point A was in elastic state ($f^{\mathrm{trail}} < 0$) because the horizontal stress path extended to the left and pointed the inside of the yield surface (close to the path of $\alpha = 180°$ in Fig. 3.29). Later, as the gas pressure decreased, the MSW was reloaded and entered the plastic state after reaching a yield point with $f^{\mathrm{trail}} = 0$ (similar to the path in Region III in Fig. 3.29). Therefore, in this simulation, the lateral bulge deformation of Point A produced by the increasing gas pressure was induced by the elastic rebound, while the contraction deformation was mainly the plastic deformation induced by the degradation. In addition, the MSW at Point B was always in

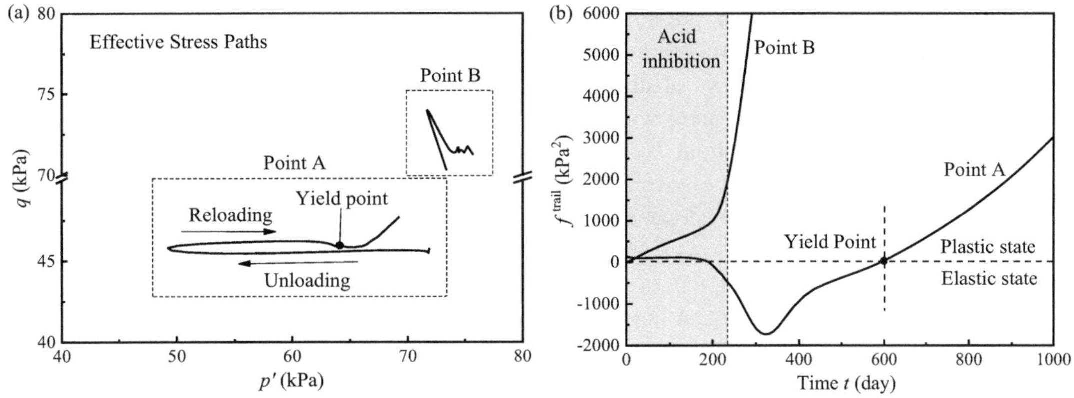

Fig. 5.26 **a** Effective stress paths and **b** Loading judgements of Points A and B (Ma 2022; Ma et al. 2024)

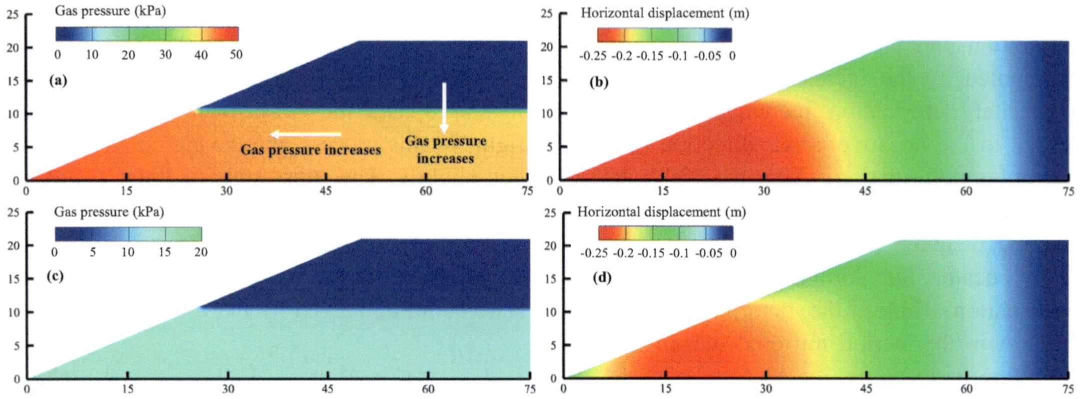

Fig. 5.27 Spatial distribution of **a** LFG pressure and **b** horizontal displacement of 330 days and **c** LFG pressure and **d** horizontal displacement of 800 days (Ma 2022; Ma et al. 2024)

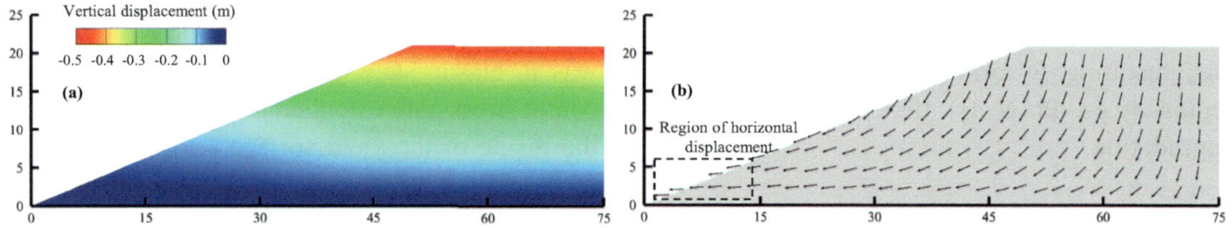

Fig. 5.28 **a** Spatial distribution of settlement and **b** deformation vector diagram of the MSW slope of 330 days (Ma 2022; Ma et al. 2024)

plastic state during the degradation process as the effective mean stress did not change obviously, being close to the path of $\alpha = 90°$ in Fig. 3.29.

5.5.3 Global Deformation Mode of the MSW Slope

The spatial distributions of gas pressure and horizontal displacement in the MSW slope are shown in Fig. 5.27. After 330 days of degradation, the gas pressure in the top layer was low, while the gas pressure in the bottom layer was larger than 40 kPa. In addition, the gas pressure increased as the distance to the slope surface decreased because of the impervious slope surface. Corresponding to this, the toe of MSW slope was the region with the largest horizontal displacement (larger than 0.25 m). After 800 days, the gas pressure in the bottom layer decreased to 10–15 kPa, and the horizontal displacement of the slope toe decreased to less than 0.2 m.

The spatial distribution of settlement of the MSW slope at the 330th day is shown in Fig. 5.28a. The settlement decreased with the depth. The settlement close to the slope surface was smaller than that of the same depth, which might result from the higher gas pressure in that region. The deformation vector diagram of the slope was plotted in Fig. 5.28b. The deformation directions of the MSW in slope top and toe were mainly horizontal and vertical, respectively. In the middle of the slope, there was a transition region of inclined direction of deformation. It could be imagined that if the MSW in the slope toe produces continuous horizontal displacement, the downward displacement of MSW in the middle of the slope could be exacerbated, and then a continuous sliding surface might be formed in the MSW slope.

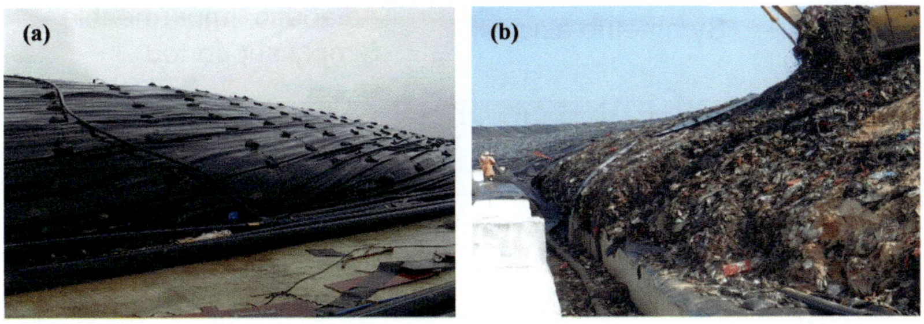

Fig. 5.29 **a** Lateral bulge deformation of the MSW slope in the Xingfeng Landfill, Guangzhou, China and **b** deformed slope after lifting the membrane (Ma 2022; Ma et al. 2024)

The bulge deformation even sliding of MSW mass due to high gas pressure was not rare in landfills worldwide. For example, a serious sliding occurred in the Payatas Landfill, Philippines in 2000, which buried more than 330 persons. Merry et al. (2005) found that this failure was closely related to the excessive pore pressure (especially gas pressure). Furthermore, the authors observed obvious lateral bulge deformation in the toe of the MSW slope in the Xingfeng Landfill, China (see Fig. 5.29). Ma et al. (2019) conducted a systematic field investigation in this landfill and found that the gas pressure at 15 m depth below the +135 m platform was approximately 50 kPa larger than static water pressure at the same depth, which is close to the calculated peak gas pressure of the MSW at Point A. Note that the measured high gas pressure located in the bottom layer and was close to the slope surface, that is, its relative position in the MSW slope of Xingfeng Landfill was close to that of Point A in the MSW slope of this simulation. The deep horizontal displacement of MSW slope in the Xingfeng Landfill was measured by inclinometer pipes. The measured horizontal displacement rate at the depth of 20 m in the high gas pressure region was stable at 0.47 mm/day after 19 days of the inclinometer pipe being placed. According to Fig. 5.25d, the average rate of horizontal displacement of Point A was calculated as 0.55 mm/day during the 310th-330th days (the gas pressure exceeded 40 kPa).Therefore, for the region in the Xingfeng Landfill where the (peak) gas pressure and relative position were similar to Point A, there was no large difference between the measured horizontal displacement rate of it and the calculation value of Point A in this simulation, which reflects the reliability of the simulation results to some extent.

5.6 Decomposition and Consolidation Behavior of Natural Gas Hydrates

5.6.1 Model Verification

(1) Comparison with unit cell tests

To validate the established degradation–consolidation model, it is compared with existing hydrate depressurization experiments (Masuda et al. 1999) and numerical simulations (Sun et al. 2019) utilizing the depressurization method. The mesh partition and boundary conditions of the model are illustrated in Fig. 5.30. Specifically, the initial temperature and pressure are 275.45 K and 3.75 MPa, respectively, with an outlet boundary pressure of 2.84 MPa. Furthermore, the initial porosity is 0.182, and the constitutive relationship of the sediment follows the modified elastic model as indicated in Eq. (3.43). The model parameters include E = 300 MPa, c_H = 1, E_H = 1,350 MPa, and μ = 0.35. Additionally, the initial liquid, gas, and hydrate saturations are 0.443, 0.206, and 0.351, respectively. The soil water retention curve (SWRC) follows the Brooks–Corey model with an inlet pressure value of 0.1 MPa, while the relative permeabilities of the liquid and gas phases follows the Corey

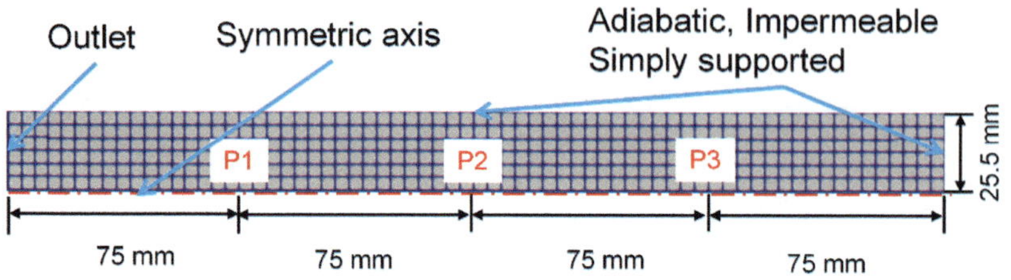

Fig. 5.30 Mesh and boundary conditions adopted in Masuda's test (Ye et al. 2023)

Table 5.11 Model parameters for the hydrate decomposition experiment (Ye et al. 2023)

Parameter	Symbol	Unit	Solid-phase	Hydrates	Liquid-phase	Gas-phase
Density	ρ^{π}	kg/m^3	2,600	910	1,000	Ideal gas equation
Thermal conductivity	k^{π}	W/m/K	8.8000	0.3930	0.5560	0.0335
Specific heat capacity	C^{π}	J/kg/K	800	2000	4,200	2,100
Viscosity coefficient	μ^{π}	Pa·s	–	–	1×10^{-3}	7×10^{-6}
Permeability characteristic constant	n^{π}	–	–	–	3	3
Thermal expansion coefficient	α^{π}	m/K	1.5×10^{-5}	1.5×10^{-5}	4.0×10^{-4}	Ideal gas equation
Molar mass	M^{π}	kg/mol	–	0.124	0.018	0.016
Gas constant	R	J/mol/K	–	–	–	8.314

Note π is an indicator for the solid, liquid, or gas phase or hydrates

model. The values of other pertinent parameters are listed in Table 5.11.

Figure 5.31 compares the simulated and measured temperatures, gas pressures, and gas production amounts (Ye et al. 2023). Evidently, the simulated results derived from the coupled model are consistent with the experimental results reported by Masuda et al. (1999) and the numerical simulation outcomes documented by Sun et al. (2019).

(2) Comparison with hypergravity tests

To restore the stress fields and temperature and pressure environments of in-situ hydrate reservoirs, Zhejiang University developed a hypergravity experimental apparatus simulating the hydrate exploitation process. This apparatus, capable of servo-controlling gas–liquid pressures and temperatures under hypergravity conditions, reproduces actual hydrate exploitation conditions, as depicted in Fig. 5.32 (Zhu et al. 2020, 2021; Wang et al. 2024a). It can be installed on a geotechnical centrifuge. By superimposing the hypergravity field produced by the centrifuge with the high-pressure and low-temperature environments of NGH reservoirs, this apparatus replicates the evolution process of NGH exploitation, offering effective means to simulate its evolution and disaster processes on a model scale (Wang et al. 2023, 2024a).

As depicted in Fig. 5.33a, the high-pressure vessel of this apparatus features an internal cylindrical cavity with a diameter of 0.23 m and a height of 0.35 m. The side walls of this high-pressure vessel are connected to a constant temperature water bath, facilitating servo-controlled temperature regulation. Within the vessel, four columns of data collection points are arranged at different heights, enabling three-dimensional monitoring of temperatures and pore pressures within the sediment. During experiments, real-

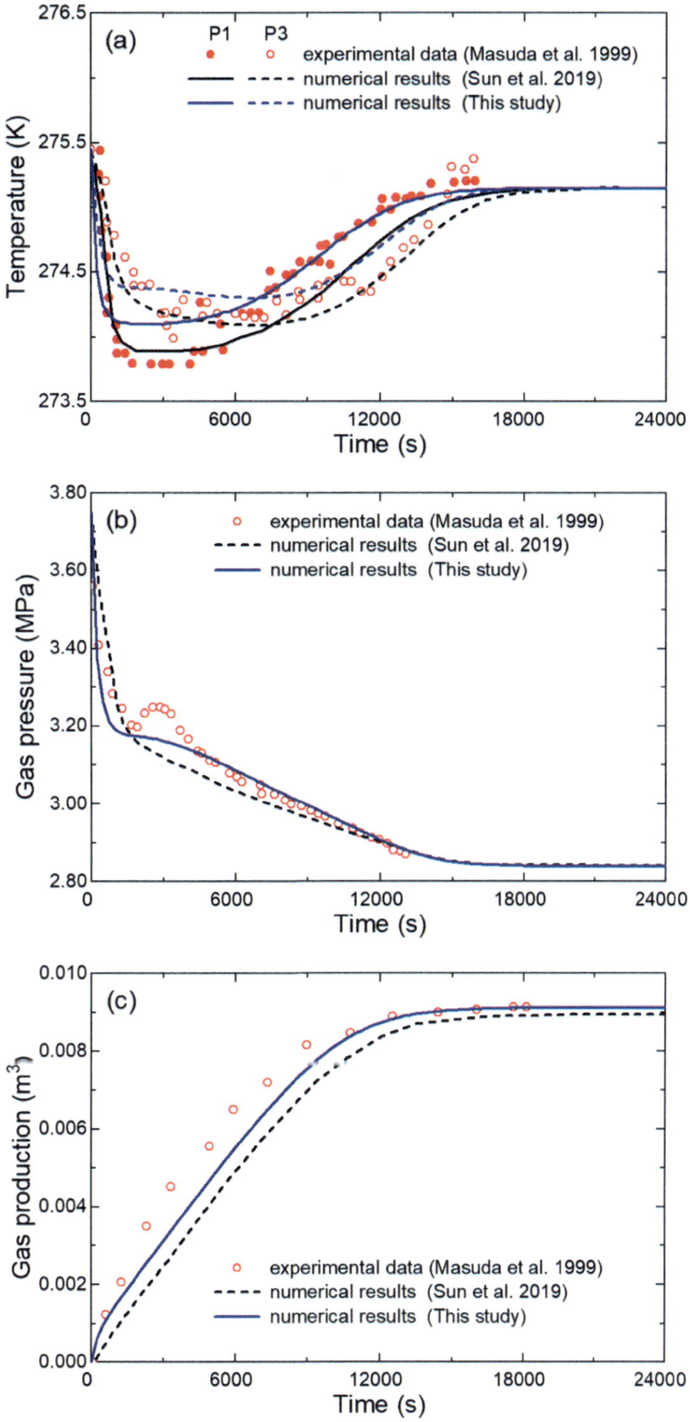

Fig. 5.31 Comparison of the temporal evolution curves of physical parameters during hydrate decomposition: **a** temperatures at various monitoring points, **b** gas pressure along the right boundary, and **c** gas production (Ye et al. 2023)

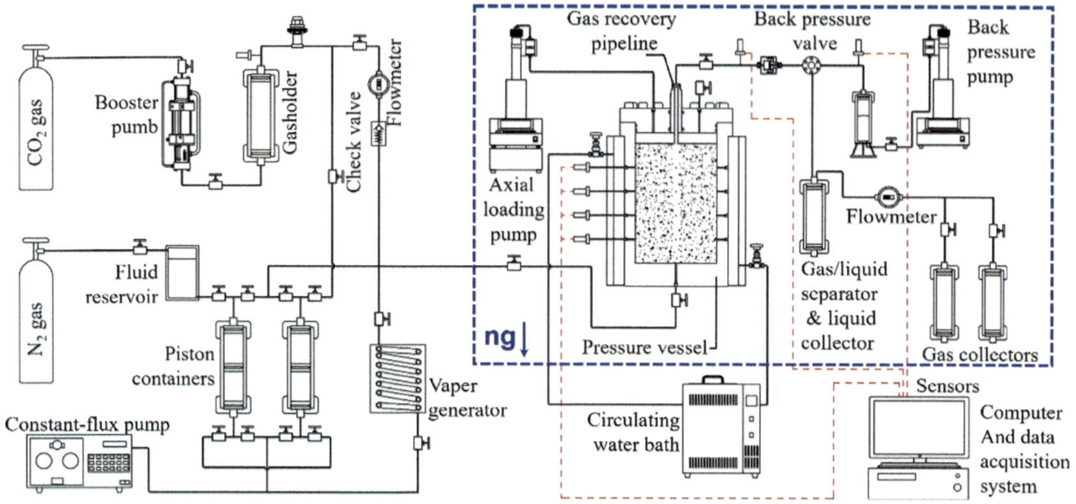

Fig. 5.32 Schematic of the hypergravitational pressure servo control apparatus (reprinted from Wang et al. 2024a, with permission from the Chinese Journal of Geotechnical Engineering)

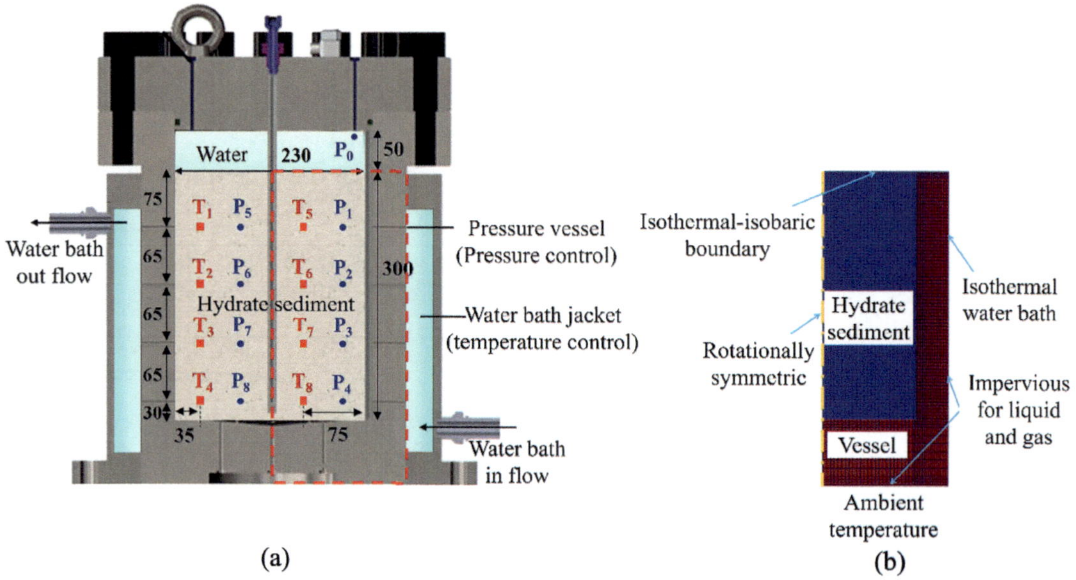

Fig. 5.33 Schematic of the high-pressure vessel: **a** distribution of temperature and pore pressure measuring points, **b** model mesh and boundary conditions (**a** is reprinted from Wang et al. 2024a, with permission from the Chinese Journal of Geotechnical Engineering; **b** is sourced from Ye 2023)

time measurements of gas production are achieved using a gas–liquid separator and mass flow meter (Zhu et al. 2021; Wang et al. 2024a).

Using this apparatus, depressurization–decomposition experiments of hydrates under 100 g hypergravity conditions was conducted. To prevent interference from settlement caused by hypergravity, the model was preconsolidated before the experiment. Detailed information on model preparation can be found in Wang et al. (2024a), Wang et al. (2022) and Wang (2022). During hydrate exploitation, the pore pressure

Table 5.12 Parameter model and values (Anderson 2003; Carroll et al. 1991; White et al. 2020)

Parameter model	Equation	Value
SWRC	$p_c = p_0\left((s_E)^{-\frac{1}{n_{pc}}} - 1\right)^{(1-n_{pc})}$; $s_E = \frac{s_L - s_{Lr}}{1 - s_{Lr}}$	$p_0 = 12.5$ kPa; $n_{pc} = 0.45$; $s_{Lr} = 0.11$
Relative permeability coefficient	$k_{relL} = \left(\frac{s_L - s_{Lr}}{1 - s_{Lr}}\right)^{n_L}$; $k_{relG} = \left(\frac{s_G - s_{Gr}}{1 - s_{Lr}}\right)^{n_G}$	$n_L = 3.0$; $n_G = 3.0$; $s_{Gr} = 0.0$; $s_{Lr} = 0.1$
Intrinsic permeability (m^2)	$K = K_0(1 - s_H)^{N_p}$	$K_0 = 1 \times 10^{-12}$; $N_p = 6$
Diffusion coefficient (m/s)	–	$D = 2.55 \times 10^{-5}$
Henry's law constant (mol/Pa/m^3)	–	$H^a = 3.4 \times 10^{-5}$
Porosity	–	$\varphi = 0.4226$
Enthalpy (J/mol)	$\Delta H = C_0 + C_1 T$	$C_0 = 207,699.5$; $C_1 = -530$
Thermal conducting (W/m/K)	$k_T = k_{Tdry} + (\sqrt{s_L} + \sqrt{s_H})(k_{Tsat} - k_{Tdry})$	$k_{Tdry} = 1.0$; $k_{Tsat} = 2.0$

Table 5.13 Model parameters for the hypergravity experiment (Ye 2023)

Parameter	Symbol	Unit	Solid-phase	Hydrates	Liquid-phase	Gas-phase
Density	ρ^π	kg/m^3	2,640	1,120	1000	–
Thermal conductivity	k^π	W/m/K	2.50	0.39	0.56	0.01
Specific heat capacity	C^π	J/kg/K	1,000	800	4,200	2,100
Viscosity coefficient	μ^π	Pa·s	–	–	1×10^{-3}	7×10^{-6}
Thermal expansion coefficient	α^π	m/K	1.5×10^{-5}	1.5×10^{-5}	2.1×10^{-4}	–
Molar mass	M^π	kg/mol	–	0.1475	0.018	0.044

was reduced from 3.5 MPa to 1.5 MPa at a rate of 0.05 MPa/s to achieve hydrate decomposition.

Based on the degradation–consolidation process during hydrate decomposition, a numerical model corresponding to the hypergravity depressurization–decomposition experiment was established. As depicted in Fig. 5.33b, the gravitational constant in the numerical model was set to 981 m/s^2 to simulate the 100 g hypergravity centrifugal acceleration. The boundary conditions of the model were set based on the outlet pressure (P_0) and ambient temperature. Accordingly, a numerical model for hypergravity experiments was established (Ye 2023). Parameter values of the hydrate decomposition rate, SWRC, relative permeability coefficient, thermal conductivity, and other relevant parameters are specified in Tables 5.12 and 5.13 (Anderson 2003; Carroll et al. 1991; White et al. 2020). Furthermore, the constitutive relationship of the sediment follows the modified elastic model as indicated in Eq. (3.43). The specific model parameters include: $E = 300$ MPa, $c_H = 1$, $E_H = 1,350$ MPa, and $\mu = 0.35$.

Figure 5.34a displays the time evolution curves of pore pressure during the hydrate exploitation stage of the hypergravity experiment, comparing measured values with numerically simulated results. Notably, under hypergravity conditions, pore pressure increases with depth (from P_5 to P_8). The numerical simulation results agree well with the measured values.

Figure 5.34b depicts the temperature–time curves at different measurement points during the hydrate exploitation stage. From 0 h to 0.6 h, the

Fig. 5.34 Results of the hypergravity experiment: **a** pore pressures at P5–P8, **b** temperatures at P5–P8, **c** pore pressure–temperature evolution, **d** accumulative gas production and gas production rate, and **e** pore pressure profiles (Ye 2023)

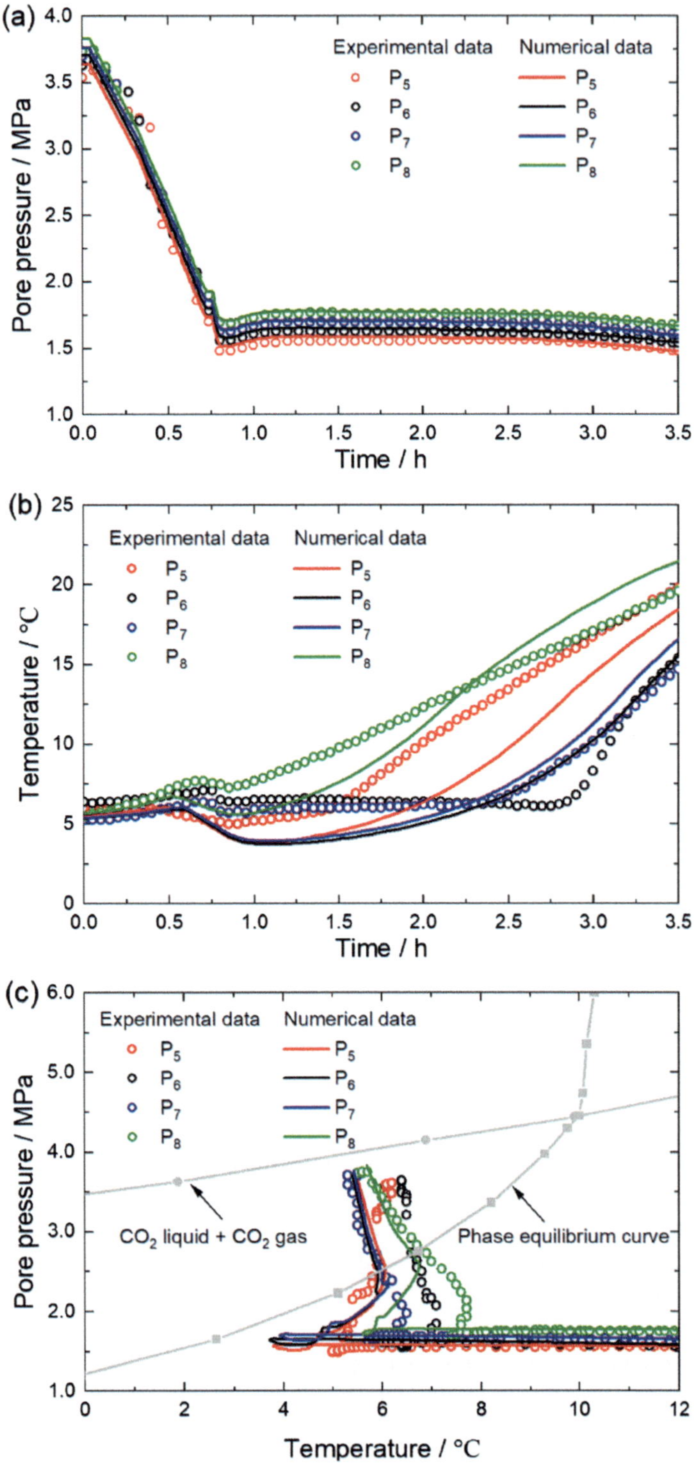

Fig. 5.34 (continued)

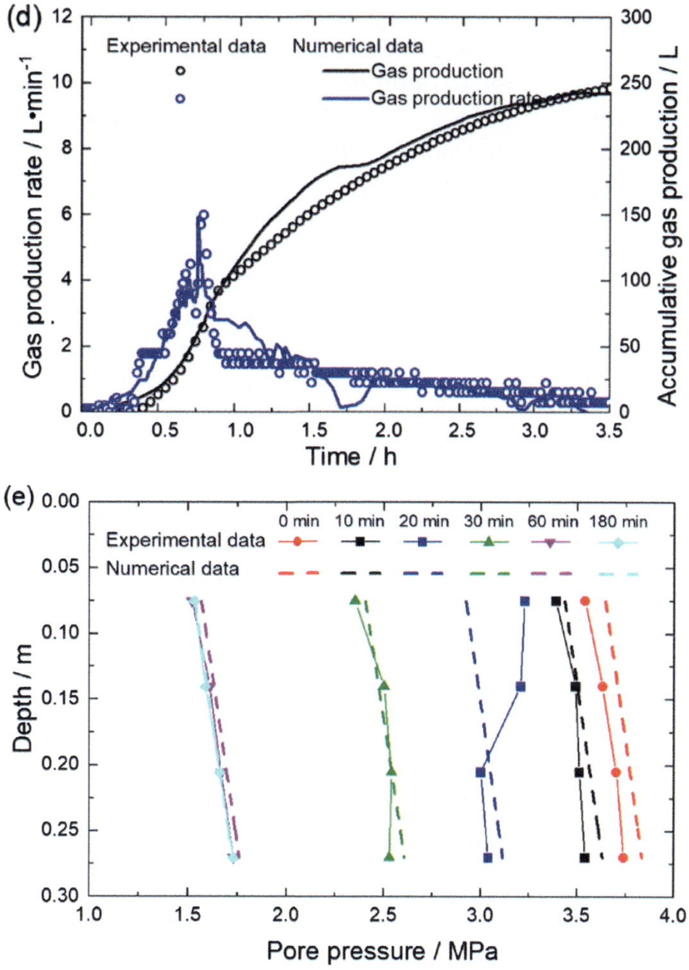

temperature at each measurement point increases gradually owing to the rise in control temperature with ambient temperature. At this stage, hydrate decomposition is yet to commence. Subsequently, as the pressure approaches the equilibrium pressure and hydrate decomposition ensues, the temperature at each measurement point begins to drop. Given the proximity of points P_5 and P_8 to the top and bottom of the model, respectively, they are more affected by the boundary temperature; hence, their temperatures rise again after a period of decline. Meanwhile, points P_6 and P_7, being farther from the boundary, experience prolonged periods of nearly constant temperatures before the temperature increases again. The numerical simulation results reflect the gradual temperature increase before decomposition and the local drop during decomposition at each measurement point. The simulation results also correctly capture the quicker and slower temperature recoveries of points P_5 and P_8 and points P_6 and P_7, respectively.

Figure 5.34c illustrates the temperature–pressure evolution curves of points P_5 to P_8 during the hydrate exploitation stage. Initially, certain differences are apparent between the pressures and temperatures across different points owing to the linear distribution of pore pressure with depth caused by hypergravity. Notably, as the temperature and pressure conditions at each point surpass the phase equilibrium curve, hydrate decomposition ensues, and the conditions approach the phase equilibrium curve.

Ultimately, as the pressure approaches the preset pressure, hydrate decomposition proceeds gradually, leading to a subsequent rise in the temperature.

Figure 5.34d presents the temporal variation trends of the gas production amount and gas production rate during the hydrate exploitation stage. During the initial 0 h to 0.5 h, the gas production rate is almost zero. This is because the temperature and pressure conditions at this stage are within the phase equilibrium curve and hydrate decomposition is yet to commence. In the period between 0.5 h to 1.75 h, hydrate decomposition occurs at an accelerated rate. At this stage, the gas production rate peaks at approximately 8 L/min, and the cumulative gas production rapidly increases to approximately 175 L. Subsequently, as the decomposition rate decreases, fluctuating between 0 and 0.1 L/min from 1.75 h to 3.5 h, the cumulative gas production gradually increases to approximately 250 L.

As depicted in Fig. 5.34e, initial values of the pore pressure are high. However, as depressurization progresses, these values gradually decrease. Given that the profiles at 60 min and 180 min are not significantly distinct, the system can be inferred to stabilize by 60 min.

Figure 5.35 compares the evolution curves of EPP during hydrate decomposition in 1 g and 100 g experiments. Notably, the gas produced by hydrate decomposition first increases the EPP, particularly in closed and partially closed pores. The EPP value rises to a certain peak value before rapidly decreasing. This rapid EPP dissipation over time is primarily attributed to the formation of gas-induced fractures and seepage channels. Furthermore, this phenomenon resembles the pore structure damage and seepage characteristic variations resulting from hydrate decomposition and gas production, as observed in micro-scale CT scans (Wang et al. 2018). Comparing the conditions of point P_7 in Fig. 5.35a and b reveals that under 100 g, the initial effective stress of the skeleton is approximately 200 kPa, with an EPP peak of 300 kPa. Meanwhile, under 1 g, the initial effective stress is 2 kPa, with an EPP peak of 100 kPa. Similar phenomena are evident at measurement point P_6 in Fig. 5.35a. In the 1 g experiment, the critical hydrate decomposition degree for the formation of dominant seepage channels is 0.035–0.045, whereas in the 100 g experiment, this value reaches 0.05–0.168. Notably, the peak value of EPP accumulation in the above experiments is significantly influenced by the initial effective stress. This suggests that greater burial depths of reservoirs lead to greater EPP tolerance of sediments, requiring a greater degree of hydrate decomposition for fracture formation.

Figure 5.36 displays the temporal evolution of gas production rate during hydrate decomposition under different stress conditions. Particularly, in the 100 g experiment, a double peak in the gas production rate is evident. When the upper boundary pressure drops to 1.5 MPa, the pressure differential for hydrates peaks, causing the gas production rate to increase rapidly and reach its first peak. At 45 min, as EPP dissipation ensues, the gas production rate increases rapidly again, forming a second peak. This indicates that fracture formation near the measurement point leads to an instantaneous release of gas accumulated during hydrate decomposition. The temporal variation of the hydrate decomposition degree exhibits significant fluctuations at the peaks. Specifically, by 100 min, some hydrates remain undecomposed. As illustrated in Fig. 5.36c, gas production characteristics in the 100 g experiment resemble those observed in the second offshore test within the Nankai Trough in Japan, with an initial peak followed by reduction and stabilization (Yamamoto et al. 2019).

5.6.2 Decomposition and Consolidation Behaviors of Hydrate Reservoirs

Leveraging the degradation–consolidation model for NGHs, Ye et al. (2022) established a two-dimensional axisymmetric hydrate reservoir model using OGS to analyze the decomposition and consolidation behaviors of hydrate reservoirs. The model dimensions included length H

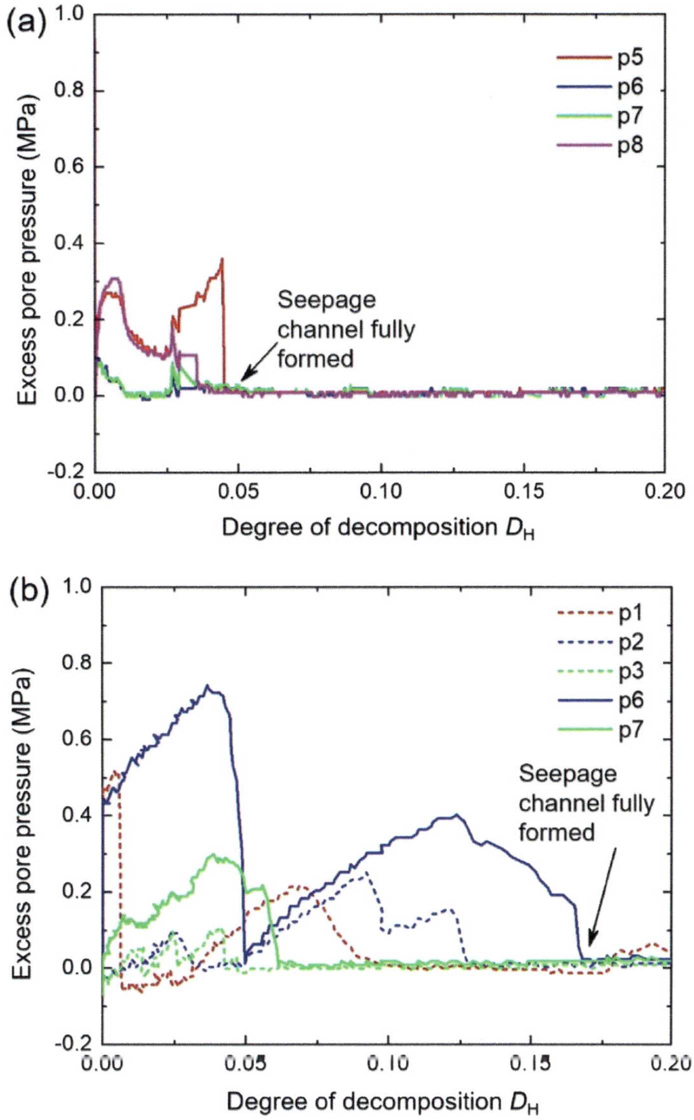

Fig. 5.35 EPP variations with hydrate decomposition: **a** 1 g experiment and **b** 100 g experiment (Wang et al. 2024b)

and radius R of 10 m and 200 m, respectively. A uniformly distributed load of 2 MPa was applied to the upper boundary to simulate the pressure exerted by the overlying soil layer, as depicted in Fig. 5.37. The parameters values for hydrates and the solid, liquid, and gas phases are listed in Table 5.11. The constitutive relationship of the sediment followed the modified elastic model as indicated in Eq. (3.43). The model parameters included: $E = 300$ MPa, $c_H = 1$, $E_H = 1{,}350$ MPa, and $\mu = 0.35$. Furthermore, the length of the production well was designated as H_{well}. Notably, water and gas pressures were reduced from 3.75 MPa and 2.84 MPa to 2.34 MPa and 1.71 MPa, respectively, to facilitate hydrate extraction.

Figure 5.38 illustrates the distribution of gas saturation increments at 10^6 s and 10^7 s, depicting an increase in gas saturation near the bottom of the extraction well but a sharp decrease on either side of the production well. A prominent hydrate dissociation front is also evident (Selim and Sloan 1990; Yousif et al. 1991). This dissociation interface propagates from near to far

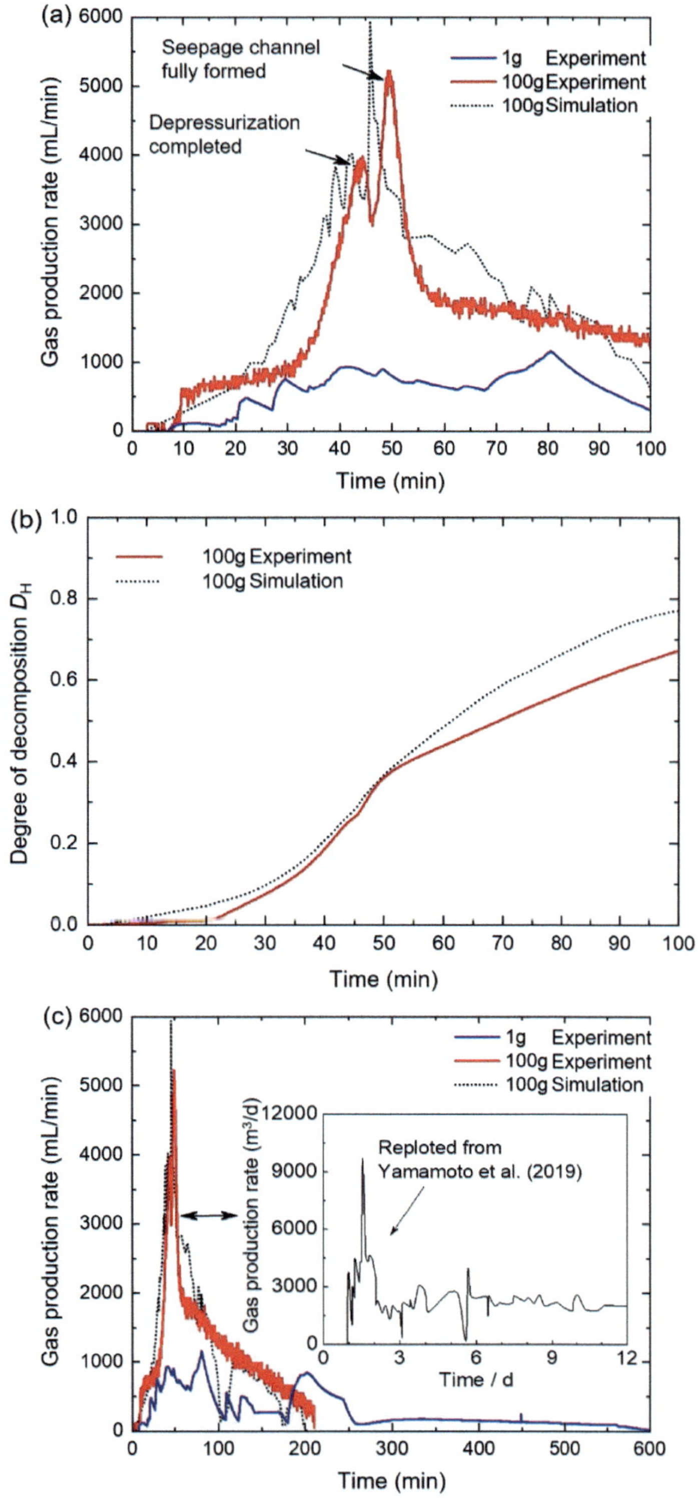

Fig. 5.36 Comparison of the gas production behavior between simulations and experiments: **a** gas production rate in the first 100 min, **b** degree of decomposition in the first 100 min, and **c** gas production rate over complete tests (Wang et al. 2024b)

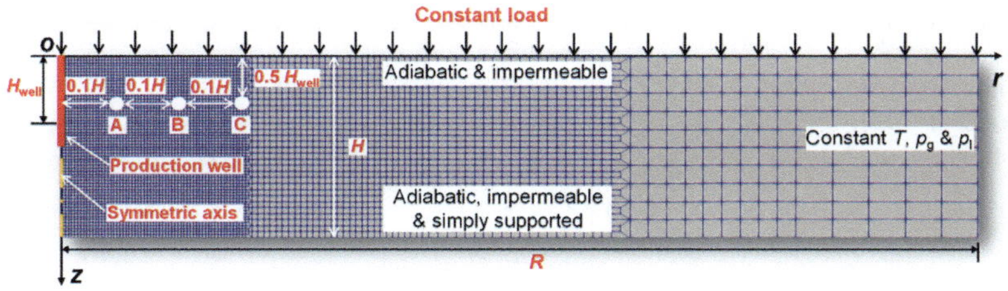

Fig. 5.37 Model domain, mesh, and initial and boundary conditions (reprinted from Ye et al. 2022, with permission from Elsevier)

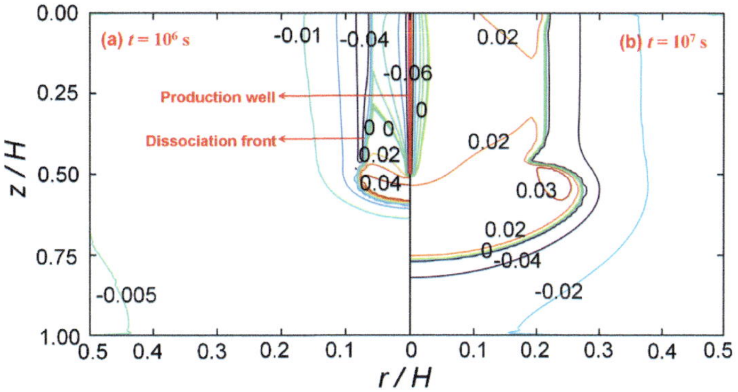

Fig. 5.38 Distributions of gas saturation at 10^6 and 10^7 s (reprinted from Ye et al. 2022, with permission from Elsevier)

over time. The hydrate sediment is segmented into dissociation and non-dissociated zones by this dissociation front. Complete hydrate decomposition occurs within the dissociation front, whereas minimal amount of hydrate decomposition occurs outside the dissociation front.

As illustrated in Fig. 5.39a, hydrate saturation markedly decreases near the production well but exhibits slight reduction in the outer area of the dissociation front, attributed to the reduction in gas pressure induced by gas phase dissipation. During hydrate dissociation, a substantial amount of heat is absorbed, resulting in an inward depression of the dissociation front at a depth of $z/H = 0.4$. As depicted in Fig. 5.39b, the dissociation zone gradually expands over time, particularly near the bottom of the well owing to increased heating.

Figure 5.40a illustrates significant compression in the soil layer near the production well owing to the rapid reduction in water and gas

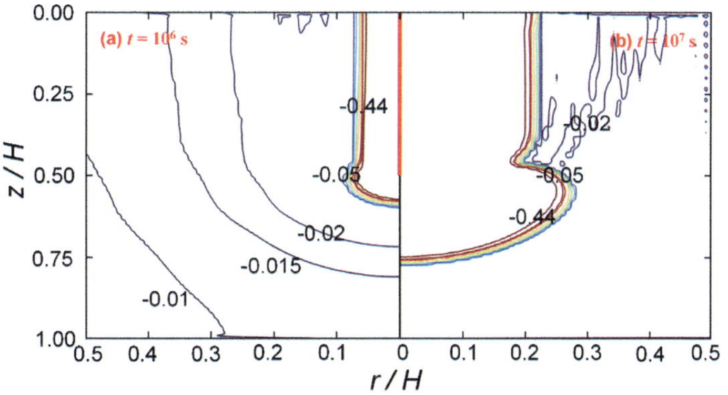

Fig. 5.39 Distributions of hydrate saturation at 10^6 and 10^7 s (reprinted from Ye et al. 2022, with permission from Elsevier)

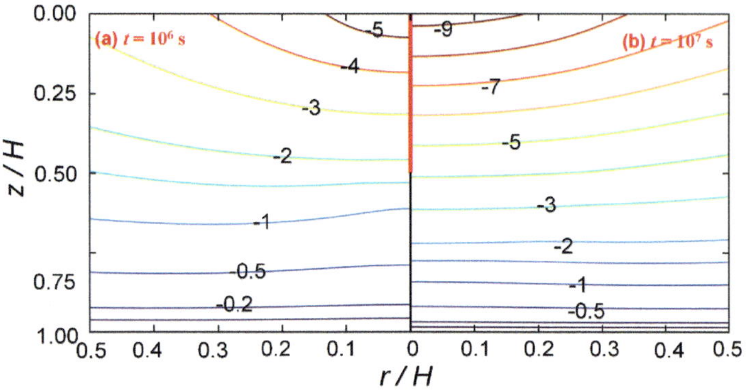

Fig. 5.40 Distributions of vertical displacement at 10^6 and 10^7 s (unit: mm) (reprinted from Ye et al. 2022, with permission from Elsevier)

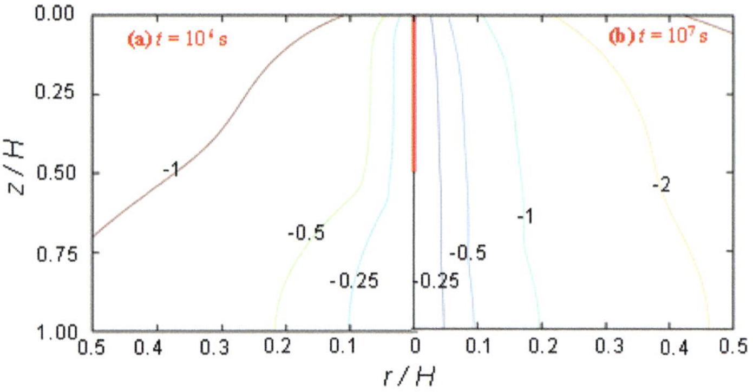

Fig. 5.41 Distributions of horizontal displacement at 10^6 and 10^7 s (unit: mm)

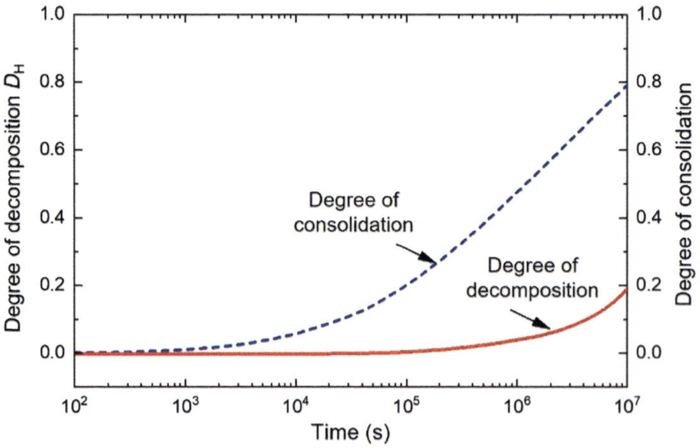

Fig. 5.42 Variations in the decomposition degree of hydrates and the consolidation degree at the center of the production well

pressures, alongside the decrease in sediment skeleton stiffness, with hydrate decomposition. Furthermore, the magnitude of compression increases over time, as illustrated in Fig. 5.40b.

Similar to the previous scenario, Fig. 5.41a depicts significant horizontal compression in the soil layer near the production well. Furthermore, the magnitude of compression increases over time, as indicated in Fig. 5.41b.

As depicted in Fig. 5.42, during hydrate exploitation through depressurization, the decomposition degree of hydrates within the reservoir and consolidation degree at the center of the production well gradually increase over time. Owing to environmental heat replenishment and the limited scope of depressurization, a substantial portion of hydrates in the reservoir remains stable; hence, the decomposition degree of hydrates is far below one at 10^7 s. Conversely, at the center of the exploitation well, where depressurization stabilizes quickly, the consolidation degree approaches 0.8 at 10^7 s.

References

Anderson GK (2003) Enthalpy of dissociation and hydration number of carbon dioxide hydrate from the Clapeyron equation. J Chem Thermodyn 35(7):1171–1183

Carroll JJ, Slupsky JD, Mather AE (1991) The solubility of carbon dioxide in water at low pressure. J Phys Chem Ref Data 20(6):1201–1209

Chen YM (2014) A fundamental theory of environmental geotechnics and its application. Chin J Geotechn Eng 36(1):1–46

Chen YM, Guo RY, Li YC, Liu H, Zhan TL (2016) A degradation model for high kitchen waste content municipal solid waste. Waste Manage 58:376–385

Chen YM, Liu XC, Xu WJ, Li YC, Lan JW, Zhan LT, Li H, Li CM (2019) Analysis on stabilization characteristics and exploitability of landfilled municipal solid waste: case of a typical landfill in China. Scientia Sinica Technologica 49(2):199–211

Chen YM, Xu WJ, Ling DS, Zhan LT, Gao W (2020) A degradation–consolidation model for the stabilization behavior of landfilled municipal solid waste. Comput Geotech 118:103341

IPCC 2019 (2019) Refinement to the 2006 IPCC guidelines for national greenhouse gas inventories. In: Calvo Buendia E, Tanabe K, Kranjc A, Baasansuren J, Fukuda M, Ngarize S, Osako A, Pyrozhenko Y, Shermanau P, Federici S (eds) IPCC, Switzerland

Kim HC, Bishnoi PR, Heidemann RA, Rizvi SS (1987) Kinetics of methane hydrate decomposition. Chem Eng Sci 42(7):1645–1653

Kolditz O, Bauer S, Bilke L, Böttcher N, Delfs JO, Fischer T, Görke UJ, Kalbacher T, Kosakowski G, McDermott CI, Park CH, Radu F, Rink K, Shao H, Shao HB, Sun F, Sun YY, Singh AK, Taron J, Walther M, Wang W, Watanabe N, Wu Y, Xie M, Xu W, Zehner B (2012) OpenGeoSys: an open-source initiative for numerical simulation of thermo-hydro-mechanical/chemical (THM/C) processes in porous media. Environ Earth Sci 67(2):589–599

Kolditz O, Shao H, Wang W, Bauer S (2015) Thermo-hydro-mechanical chemical processes in fractured porous media: modelling and benchmarking—closed-form solutions. Springer, Berlin

Li K, Chen YM, Xu WJ, Zhan LT, Ling DS, Ke H, Hu J, Li JL (2021) A thermo-hydro-mechanical-biochemical coupled model for landfilled municipal solid waste. Comput Geotech 134:104090

Li K, Xu WJ, Chen YM, Zhan LT, Ke H, Xu H, Xiao DK (2023) Biochemical, hydrological and mechanical behaviors of high food waste content MSW landfill: numerical simulation analysis of a large-scale experiment. Waste Manage 171:557–567

Liu HL (2016) Research on biodegradation-consolidation-solute migration coupled behaviors of municipal solid waste and landfill stabilization. Ph.D. thesis, Zhejiang University, Hangzhou, China (in Chinese)

Liu HL, Zhou JW, Chen YM, Li YC, Zhan LT (2016) Evaluation of municipal solid waste landfill stabilization. J Zhejiang Univ (Eng Sci) 50(12):2336–2342

Ma PC (2022) Study on constitutive models for soils with degradable solid skeleton and their application. Ph.D. thesis, Zhejiang University, Hangzhou, China (in Chinese)

Ma PC, Chen YM, Xu WJ, Hu J (2024) A novel constitutive model for degradable soil: municipal solid waste as a case study. Computers and Geotechnics (under review)

Ma PC, Ke H, Lan JW, Chen YM, He HJ (2019) Field measurement of pore pressures and liquid-gas distribution using drilling and ERT in a high food waste content MSW landfill in Guangzhou, China. Eng Geol 250:21–33

Masuda Y, Fujinaga Y, Naganaw S, Fujita K, Sato K, Hayashi Y (1999) Modeling and experimental studies on dissociation of methane gas hydrates in Berea sandstone cores. In Proceedings of the third international gas hydrate conference, Salt Lake City, USA, vol 7, pp 18–22

McDougall J (2007) A hydro-bio-mechanical model for settlement and other behaviour in landfilled waste. Comput Geotech 34(4):229–246

Merry SM, Kavazanjian Jr E, Fritz WU (2005) Reconnaissance of the July 10 2000 Payatas Landfill Failure. J Perform Constr Facil 19(2):100–107

Selim MS, Sloan ED (1990) Hydrate dissociation in sediment. SPE Reserv Eng 5(2):245–251

Sun X, Li YH, Liu Y, Song YC (2019) The effects of compressibility of natural gas hydrate-bearing sediments on gas production using depressurization. Energy 185:837–846

Wang LJ, Wang P, Zhu B, Wang XP, Yang SQ, Chen YM (2024a) Development and application of in-flight centrifuge apparatus for modelling hydrate exploitation. Chin J Geotechn Eng 46(2):316–324 (in Chinese)

Wang LJ, Wang P, Zhu B, Kong DQ, Wang XB, Chen YM (2024b) Physical modelling of hydrate dissociation in sandy sediment by depressurization under hyper-gravity and normal-gravity conditions. J Geotechn Geoenviron Eng 150(10):04024096

Wang DG, Wang CC, Li CF, Liu CL, Lu HL, Wu NY, Hu GW, Liu LL, Meng QG (2018) Effect of gas hydrate formation and decomposition on flow properties of fine-grained quartz sand sediments using X-ray CT based pore network model simulation. Fuel 226:516–526

Wang P, Zhu B, Wang LJ, Wang XB, Yan BF (2023) Effects of depressurization boundary and hydrate saturation on dissociation characteristics of hydrate-bearing sediments. Chin Civil Eng J 56(S1):125–133

Wang W, Kosakowski G, Kolditz O (2009) A parallel finite element scheme for thermohydro-mechanical (THM) coupled problems in porous media. Comput Geosci 35(8):631–641

Wang WQ, Fischer T, Zehner B, Böttcher N, Görke UJ, Kolditz O (2015) A parallel finite element method for two-phase flow processes in porous media: OpenGeoSys with PETSc. Environ Earth Sci 73(5):2269–2285

Wang XB, Wang LJ, Zhu B, Wang P, Yuan SM, Chen YM (2022) Experimental study of behavior of hydrate-bearing sediments during servo depressurization. Rock Soil Mech 43(9):2360–2370 (in Chinese)

Wang XB (2022) Physical model test of depressurization decomposition of hydrate sediments. Master thesis, Zhejiang University, Hangzhou, China (in Chinese)

White MD, Kneafsey TJ, Seol Y, Waite WF, Uchida S, Lin JS, Myshakin EM, Gai X, Gupta S, Reagan MT, Queiruga AF, Kimoto S, Baker RC, Boswell R, Ciferno J, Collett T, Choi J, Dai S, De La Fuente M, Fu P, Fujii T, Intihar CG, Jang J, Ju X, Kang J, Kim JH, Kim JT, Kim SJ, Koh C, Konno Y, Kumagai K, Lee JY, Lee WS, Lei L, Liu F, Luo H, Moridis GJ, Morris J, Nole M, Otsuki S, Sanchez M, Shang S, Shin C, Shin HS, Soga K, Sun X, Suzuki S, Tenma N, Xu T, Yamamoto K, Yoneda J, Yonkofski CM, Yoon HC, You K, Yuan Y, Zerpa L, Zyrianova M (2020) An international code comparison study on coupled thermal, hydrologic and geomechanical processes of natural gas hydrate-bearing sediments. Mar Pet Geol 120:104566

Xiao DK (2022) Aerobic degradation-stabilization model in municipal solid waste landfill and its application. Ph.D. thesis, Zhejiang University, Hangzhou, China (in Chinese)

Xiao DK, Chen YM, Xu WJ, Zhan LT, Ke H, Li K (2022) Biochemical-thermal-hydro-mechanical coupling model for aerobic degradation of landfilled municipal solid waste. Waste Manage 144:144–152

Xu H (2016) Large-scale experiment on biochemo-hydro-mechanical behaviors of high-food-waste-content MSW and applications. Ph.D. thesis, Zhejiang University, Hangzhou, China (in Chinese)

Xu H, Qiu HL, Zhu G, Zhan LT, Zhang ZY, Xu XB, Chen YM, Wang YZ (2019) Comparison of settlement behaviors of high-food-waste-content (HFWC) and low-food-waste-content (LFWC) MSWs and assessment of their prediction models. Sci China Technol Sci 62(12):2271–2292

Yamamoto K, Wang XX, Tamaki M, Suzuki K (2019) The second offshore production of methane hydrate in the Nankai Trough and gas production behavior from a heterogeneous methane hydrate reservoir. RSC Adv 9(45):25987–26013

Ye ZG (2023) Numerical and experimental study on multifield coupling behavior of soil considering phase change. Ph.D. thesis, Zhejiang University, Hangzhou, China (in Chinese)

Ye ZG, Wang LJ, Zhu B, Shao HB, Xu WJ, Chen YM (2022) A thermo-hydro-chemo-mechanical coupled model for natural gas hydrate-bearing sediments considering gravity effect. J Nat Gas Sci Eng 108:104823

Ye ZG, Wang LJ, Zhu B, Chen YM (2023) Thermo-hydro-mechanical coupling model for natural gas hydrate-bearing sediments with depressurization based on OpenGeoSys. Rock Soil Mech 44(11):3191–3202 (in Chinese)

Yousif MH, Abass HH, Selim MS, Sloan ED (1991) Experimental and theoretical investigation of methane-gas-hydrate dissociation in porous media. SPE Reserv Eng 6(1):69–76

Zhan LT, Xu H, Chen YM, Lan JW, Lin WA, Xu XB, He PJ (2017a) Biochemical, hydrological and mechanical behaviors of high food waste content MSW landfill: liquid-gas interactions observed from a large-scale experiment. Waste Manage 68:307–318

Zhan LT, Xu H, Chen YM, Lü F, Lan JW, Shao LM, Lin WA, He PJ (2017b) Biochemical, hydrological and mechanical behaviors of high food waste content MSW landfill: preliminary findings from a large-scale experiment. Waste Manage 63:27–40

Zhan LT, Xu H, Jiang XM, Lan JW, Chen YM, Zhang ZY (2019) Use of electrical resistivity tomography for detecting the distribution of leachate and gas in a large-scale MSW landfill cell. Environ Sci Pollut Res 26(20):20325–20343

Zhu B, Wang LJ, Yang SQ, Chen YM, Lin WA (2020) Hypergravity experimental system for natural gas hydrate depressurization exploitation. Chinese Patent, CN CN201810178557.2

Zhu B, Wang LJ, Yang SQ, Chen YM, Kong DQ, Lin WA (2021) Pressure-control temperature-control hypergravity experimental device for simulating deep seabed responses. U.S. Patent, US11, 187, 691 B2

6 Production and Drainage of Leachate and Landfill Gas

Abstract

Based on the degradation–consolidation model, this chapter introduces a modified calculation formula accounting for leachate production from waste is introduced. Remarkably, this formula demonstrates better predictive accuracy than the traditional leaching coefficient method. Next, a new calculation approach for LFG production is proposed, considering the distinct degradation properties of rapidly and slowly degrading MSW components. Based on this method, the LFG production behaviors of HFWC and LFWC MSWs are compared. Furthermore, the chapter explores three-dimensional collection and drainage technologies for leachate and LFG, presenting detailed discussions on horizontal well dewatering and vertical well vacuum pressure pumping tests. Additionally, strategies for optimizing LFG collection efficiency are explored, along with an examination of the relationship between collection rates and leachate levels. This analysis is complemented by a case study on the Xiaping Landfill in Shenzhen, demonstrating that the implementation of a three-dimensional collection and drainage system and enhanced temporary cover structure increased LFG collection efficiency from 23% to 90%.

Keywords

Municipal solid waste · Landfill, Leachate · Landfill gas · Three-dimensional technology · Collection rate · Enhanced cover · Horizontal well · Vacuum pressure · Pumping test

6.1 General Introduction

In numerous HFWC MSW landfills, actual leachate production often surpasses designed values. Factors such as expanded landfill scales and inappropriate calculation methods for leachate production contribute toward these discrepancies. For instance, the calculated leachate production of a landfill in Shanghai is 1,500 m^3/d; however, its actual leachate production approaches 3,000 m^3/d. Such notable discrepancies between designed and actual leachate productions place substantial burdens on leachate collection and drainage systems, accelerating clogging and reducing drainage capacities. Insufficient drainage capacities, in turn, lead to elevated leachate levels within the landfill, posing stability and safety risks. Furthermore, the actual leachate production of most landfills surpasses the rated processing capacity of leachate treatment plants.

LFG primarily comprises CH_4 and CO_2, along with odorous gases such as H_2S, NH_3, and sulfur

Y. Chen, *Soil Degradation-Consolidation Theory and Its Applications*,
https://doi.org/10.1007/978-981-97-7985-7_6

dioxide. CH_4 has a high global warming potential, demonstrating a 28-fold stronger greenhouse effect compared to CO_2 over a century. LFG escapes from the waste body into the atmosphere, significantly contributing toward carbon emissions. Additionally, LFG accumulation within landfills creates high pore gas pressures, adversely affecting slope stability. For instance, monitoring boreholes at the Xingfeng Landfill in Guangzhou reported pore gas pressures ranging from 180 kPa to 220 kPa, leading to widespread slope failures. Hence, accurate estimations of LFG production and efficient collection are crucial for mitigating geoenvironmental hazards in landfills.

This chapter begins with an overview of different calculation formulas for leachate production, considering rainfall infiltration and waste self-degradation. This is followed by a comparative analysis of these formulas with the leaching coefficient method, which only considers rainfall infiltration. Subsequently, a modified two-stage calculation model for LFG production suitable for HFWC MSW is presented. In addition, to reveal the differences between the gas production patterns of HFWC and LFWC MSW, a comparison with gas production data and model results for LFWC MSW landfills in Europe and the United States is provided. Finally, the chapter explores the applications of three-dimensional collection and drainage technologies for leachates, along with an efficient collection technology for LFG, demonstrating promising results at the Xiaping Landfill in Shenzhen.

6.2 Leachate Production

The water in MSW exists in different forms from that in soil. Specifically, in addition to MSW containing bound water and free water, it also contains intracellular water, which is present in the cells of organic matter. Thus, organic MSW components with high water contents, such as kitchen waste and fruit peels, hold significant amounts of intracellular water. Typically, in the initial stages, MSW exhibits strong water retention capability. However, with layered compression and degradation, the water retention capacity of MSW diminishes, resulting in the gradual release of intracellular water as leachate.

Figure 6.1 illustrates a water balance analysis model for a typical valley-type landfill. Depending on the operational phase, the top of the landfill can be categorized into three different regions: the closure, intermediate cover, and landfill operation areas. Furthermore, leachate production (L) features two components: rainfall infiltration (I) and MSW self-released (E_X). Among these, rainfall infiltration (I) is calculated by subtracting evapotranspiration (E) and surface runoff (R) from precipitation (P). The magnitude

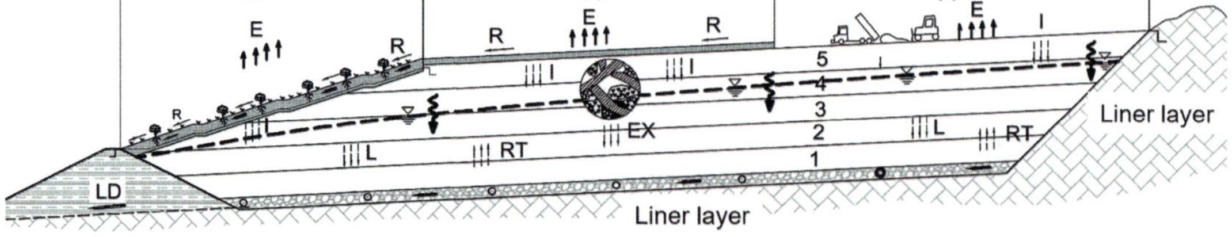

P – precipitation, R – runoff, E – evapotranspiration, I – infiltration, EX – self-released of MSW, L – leachate production, RT – accumulation of leachate in the MSW body due to failure to perform timely drainage, LD – leachate drainage

Fig. 6.1 Leachate production analysis model for staged operation landfills (Lan et al. 2012)

of rainfall infiltration (I) is primarily related to the rainfall patterns of the landfill area and the efficiency of surface runoff management on the landfill site. Meanwhile, MSW self-released (E_X) primarily depends on the initial water content of the landfilled MSW and the field capacities at various stages of landfill degradation.

Based on the water balance relationship illustrated in Fig. 6.1, leachate production (L) can be expressed as follows:

$$L = (P - E - R) + E_X \tag{6.1}$$

Because landfilling occurs in stages, the landfill site can be divided into several operation units. Each operational unit will have its own water retention capacity, and leachate will be generated depending on variations in this capacity. Assuming that the landfilling process comprises m stages, the total leachate generated in the m th stage is obtained as the sum of the leachate generated by rainfall infiltration and the leachate produced by m landfill operation units during that stage:

$$L_m = (P - E - R)_m + \sum\nolimits_{i=1}^{m} E_{X_m}^{i} \tag{6.2}$$

where L_m denotes the leachate production in the m th landfill stage (m^3), P represents the precipitation in the m th landfill stage (m^3), E indicates the evapotranspiration in the m th landfill stage (m^3), R denotes the runoff in the m th landfill stage (m^3), and $E_{X_m}^{i}$ represents the leachate produced (or absorbed) by the i th landfill unit in the m th landfill stage (m^3).

Due to the challenges associated with obtaining actual data on evaporation (E) and runoff (R), the leachate generated from rainfall infiltration, denoted as $(P - E - R)_m$, is difficult to quantify during landfill operations. As an alternative, the leaching coefficient method can be adopted for calculation.

$$(P - E - R)_m = \frac{I_m}{1,000}(C_{L1}A_1 + C_{L2}A_2 + C_{L3}A_3) \tag{6.3}$$

where I_m denotes the rainfall infiltration in the m th landfill stage (mm), C_{L1} represents the leaching coefficient of the landfill operational area (typically ranging from 0.5 to 0.8), A_1 denotes the drainage area of the landfill operation zone (m^2), C_{L2} signifies the leaching coefficient for the intermediate cover area (usually assumed to be $(0.4–0.6)C_{L1}$), A_2 represents the drainage area of the intermediate cover area (m^2), C_{L3} represents the leaching coefficient of the closure area (ranging from 0.1 to 0.2), and A_3 denotes the drainage area of the closure area (m^2).

The amount of self-released leachate by the MSW within the i th unit during the m th landfill stage can be calculated by multiplying the difference between the field capacities of the MSW in the previous and current stages by the mass of MSW within the unit:

$$E_{X_m}^{i} = \frac{M_i \times \left[W_{C(m-i)} - W_{C(m-i+1)}\right]}{\rho_w} \tag{6.4}$$

where $E_{X_m}^{i}$ denotes the leachate produced by the MSW in the i th landfill unit at the m th landfill stage, M_i represents the mass of disposed waste in the i th landfill unit, $W_{C(m-i)}$ indicates the field capacity of the MSW in the i th landfill unit at the $(m - 1)$th landfill stage (at $i = m$, , the resulting value represents the initial water content of the waste), $W_{C(m-i+1)}$ denotes the field capacity of the MSW in the i th landfill unit at the m th landfill stage, and ρ_w represents the density of water.

By combining Eqs. (6.2) to (6.4), leachate production in the m th landfill stage can be obtained as follows:

$$L_m = \frac{I_m}{1,000}(C_{L1}A_1 + C_{L2}A_2 + C_{L3}A_3) + \sum_{i=1}^{m} \frac{M_i \times \left[W_{C(m-i)} - W_{C(m-i+1)}\right]}{\rho_w} \tag{6.5}$$

Figure 6.2 shows the comparison of landfill leachate production amounts obtained from measurements and calculations. Between March

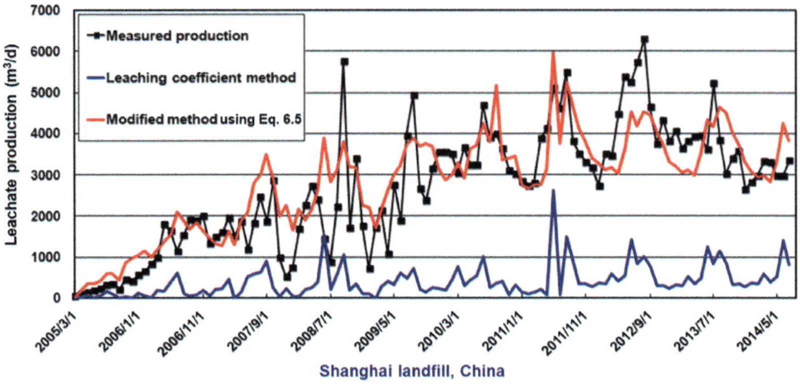

Fig. 6.2 Comparison of landfill leachate production amounts obtained from measurements and calculations of Shanghai Landfill

2005 and May 2014, the average daily leachate production of the landfill was 4,700 m^3/d. The modified formula in Eq. (6.5), which considers gradual leachate production during the layered landfilling process, yielded an average daily leachate production of 4,869 m^3/d, closely matching the measured value. However, the leaching coefficient method that only considers rainfall infiltration would significantly underestimate the actual leachate production by about 5 times.

6.3 Landfill Gas Production

As mentioned in Sect. 2.4.2, the degradation rates of different chemical components in MSW vary greatly and are significantly influenced by substrate concentration. As a result, the quantity of LFG produced by MSW over a certain period under a given set of degradation conditions is mainly influenced by the types and quantities of biodegradable components in MSW. To examine the anaerobic degradation parameters of 28 types of waste components, Zheng et al. (2013) conducted laboratory experiments using the biochemical CH_4 potential test. The CH_4 potential and LFG production rate constant of each waste component were derived by mathematically fitting the experimental data against a first-order kinetic model. Consequently, the average CH_4 potentials for food waste, paper, textiles, and bamboo and wood were determined to be 320.3 L/kg, 233.3 L/kg, 36.2 L/kg, and 129.2 L/kg, with corresponding LFG production rate constants of 18.0 a^{-1}, 14.4 a^{-1}, 0.5 a^{-1}, and 3.2 a^{-1}, respectively. In addition, the experimental data showed that there were significant differences between the LFG production characteristics of different MSW components. For instance, the CH_4 potentials and LFG production rate constants of components such as food waste, toilet paper, and office paper significantly surpassed those of newspaper, textiles, and bamboo and wood. Hence, variations in MSW composition can lead to changes in LFG generation patterns in MSW landfills.

6.3.1 Calculation Method for Landfill Gas Production

It is assumed that the process of LFG production from biodegradable materials under anaerobic conditions follows a first-order kinetic model and that this process in one material is independent of that in another material. Therefore, a model for LFG production in a landfill, considering the differences in the LFG production patterns of different biodegradable materials, can be expressed as follows:

$$Q_n = \sum_{i=1}^{n} M_i L_{R0} \left(\frac{k_R}{12}\right) e^{-\frac{k_R}{12}(t_i - t_{R,lag})} + \sum_{i=1}^{n} M_i L_{S0} \left(\frac{k_S}{12}\right) e^{-\frac{k_S}{12}(t_i - t_{S,lag})} \tag{6.6}$$

where Q_n denotes the theoretical LFG production rate (m^3/a) of the landfill in the n th month following the commencement of its operation; t_i represents the age of the waste in the i th month at the time of calculation (a); M_i denotes the amount of waste in the i th month (t); L_{R0} signifies the LFG potential of rapidly degradable materials per unit mass of waste (m^3/t); L_{S0} denotes the LFG potential of slowly degradable materials per unit mass of waste (m^3/t); k_R denotes the LFG production rate constant of rapidly degradable materials (a^{-1}); k_S represents the LFG production rate constant of slowly degradable materials (a^{-1}); $t_{R,lag}$ denotes the lag time of LFG production from rapidly degradable materials (a), generally assumed to be 0 a to 0.5 a; $t_{S,lag}$ represents the lag time of LFG production from slowly degradable materials (a). Notably, considering the similarity between the LFG production patterns of slowly degradable materials and those of LFWC MSW in western countries, $t_{S,lag}$ can be assumed to have a value of 1 a according to the LANDGEM model proposed by EPA (2005).

There are four main categories of biodegradable components of MSW: food, paper, bamboo and wood, and textile waste. These materials contain various biodegradable chemical substances, primarily including holo-cellulose (a term collectively representing cellulose and hemicellulose), sugars (encompassing carbohydrates such as glucose and starch from food products, which degrade relatively easily), proteins, lipids, and lignin. These biodegradable substances can be grouped into three categories based on their degradation rates: rapidly degradable substances, slowly degradable substances, and recalcitrant substances. Among these, rapidly degradable substances include rapidly degradable cellulose, sugars, lipids, and proteins. Meanwhile, slowly degradable substances include slowly degradable cellulose. Finally, recalcitrant substances include lignin and other inert components. Lignin, in particular, does not typically degrade under anaerobic conditions and requires aerobic environments for biochemical degradation. Table 6.1 provides an overview of the typical chemical components (on a dry basis) of various biodegradable materials found in Chinese MSW (Guo, 2017; Li, 2021).

Based on the LFG generation potential calculation model proposed by the Intergovernmental Panel on Climate Change (IPCC 2006), the LFG generation potential of rapidly and slowly degradable substances per unit mass of MSW can be calculated using the following equations:

$$L_0 = \sum\nolimits_i m_i \cdot (1 - \theta_i) \cdot DOC_{di} \cdot 1{,}867 \quad (6.7)$$

$$L_{R0} = L_0 \cdot \frac{\sum_i m_i^w \cdot (1 - \theta_i) \cdot \varphi_i^R}{\sum_i m_i^w \cdot (1 - \theta_i) \cdot \varphi_i^R + \sum_i m_i^w \cdot (1 - \theta_i) \cdot \varphi_i^S} \quad (6.8)$$

$$L_{S0} = L_0 - L_{R0} \quad (6.9)$$

where L_0 denotes the total LFG generation potential per unit mass of MSW (m^3/a); m_i represents the wet basis content of i th component of fresh MSW (%); θ_i denotes the water content of the i th component (%); and DOC_{di} indicates the degradable organic carbon content of the i th component under anaerobic conditions (%). Notably, typical DOC_{di} values for HFWC MSW in China and LFWC MSW in western countries are listed in Table 6.2. Additionally, in Eq. (6.9), 1,867 represents the volume corresponding to the complete conversion of a unit mass of degradable organic carbon into LFG (L/kg); and φ_i^R and φ_i^S denote the dry basis contents of rapidly and slowly degradable substances in the i th component (%), as indicated in Table 6.1.

The LFG generation rate constants of rapidly and slowly degradable substances specified in Eq. (6.6) are typically determined through onsite LFG extraction tests. The testing procedure for this extraction is as follows:

(1) Selection of test area: The test must be conducted in a flat area of the landfill, where the minimum age of the surface waste is less than 2 years.
(2) Installation of testing and monitoring wells: Testing wells must be equipped with drainage and LFG extraction functionalities.

Table 6.1 Chemical components of biodegradable materials found in Chinese MSW (on a dry basis, %)

Composition	Rapidly degradable substances				Slowly degradable substances	Recalcitrant substances	
	Rapidly degradable cellulose	Noncellulosic sugars	Lipids	Proteins	Slowly degradable cellulose	Lignin	Other inert substances
Food waste	15.00	60.00	5.00	5.00	–	2.00	13.00
Paper	47.40	–	–	–	18.50	14.00	20.10
Bamboo and wood	15.00	–	–	–	38.80	32.60	13.60
Textiles	15.00	–	–	–	38.80	0.00	46.20
Others	0.00	0.00	0.00	0.00	0.00	0.00	100.00

Table 6.2 Degradable organic carbon contents of various components under anaerobic conditions (on a dry basis, %)

Components	DOC_{di} of LFWC MSW	DOC_{di} of HFWC MSW
Food waste	40.70	40.70
Paper	21.90	29.30
Bamboo and wood	24.90	12.70
Textiles	15.00	13.30

Note DOC_{di} of LFWC MSW is based on IPCC (2006), and DOC_{di} of HFWC MSW refers to Guo (2017) and Li (2021)

Monitoring wells must be set up at a distance of 0.75 to 1.25 times the depth of the testing well, with at least three monitoring wells arranged in a straight line.

(3) Conduction of vacuum-pressure LFG extraction tests under different leachate level-lowering conditions: Once the leachate level within the testing well stabilizes, the vacuum pressure of LFG extraction must be gradually increased. Subsequently, when the LFG flow rate stabilizes under each vacuum pressure level, parameters such as the collected LFG volume (C_g), LFG composition, gas pressure, and temperature within the monitoring wells, must be recorded.

(4) Calculation of the influence radius (R_{inf}) of the testing well: Assuming that ΔP denotes the difference between the stable and initial gas pressures within the gas pressure monitoring wells under various levels of vacuum pressure extraction and R represents the separation between the monitoring and extraction wells, the relationship between ΔP and R can be fitted using the following equation:

$$\Delta P = a \ln R + b \tag{6.10}$$

where a and b denote fitting coefficients determined using the least squares method. The difference in pressure change at the influence radius is generally considered to be a vacuum pressure of − 0.25 kPa (Huang and Zeng, 2004), which can be substituted into Eq. (6.10) to compute the influence radius (R_{inf}) of the testing well.

(5) Calculation of the LFG generation rate constant: When the sealing conditions in the test area are relatively good, the LFG collection rate within the influence radius of the extraction well can be assumed to reach 100%. Accordingly, Eq. (6.6) yields the following relation:

$$C_g = \sum_i \pi R_{inf}^2 h_{ti} \rho_i \left[L_{R0} k_R e^{-\frac{k_R}{12}(t-6)} + L_{S0} k_S e^{-\frac{k_S}{12}(t-12)} \right] \tag{6.11}$$

where ρ_i denotes the density of the i th MSW layer within the range of the extraction test (t/m^3), and h_i represents the thickness of the i th landfilled MSW layer within the range of the extraction test (m).

The LFG generation rate constants for rapidly degradable substances are generally two to seven times (an average of four times) higher than those for slowly degradable substances (Li 2021). By incorporating this mathematical relationship into Eq. (6.6) and adopting the least squares method, the LFG generation rate constants k_R and k_S for rapidly and slowly degradable substances, respectively, can be obtained.

Li (2021) compiled and assessed field extraction test data from several landfills, as shown in Table 6.3. This analysis yielded the LFG generation rate constants of both rapidly and slowly degradable substances. The LFG generation rate constants for rapidly degradable substances ranged from approximately 0.40 a^{-1} to 1.90 a^{-1}, while those for slowly degradable substances ranged from approximately 0.09 a^{-1} to 0.55 a^{-1}.

6.3.2 Landfill Gas Production Behavior

Table 6.4 details the LFG generation potentials of various rapidly and slowly degradable substances per unit mass of MSW, derived from Eqs. (6.7) to (6.9), alongside composition and water content data from different countries and regions. Notably, for Chinese MSW, L_0, L_{R0}, and L_{S0} exhibit values of 168.10 L/kg, 149.79 L/kg, and 18.31 L/kg, respectively. These values lead to an L_{S0}/L_{R0} ratio of 0.12, indicating that the LFG generation potentials of rapidly degradable substances (L_{R0}) significantly surpass those of slowly degradable substances (L_{S0}). This suggests that during the initial landfilling stage, a large amount of LFG is produced rapidly, with LFG production reaching a peak within a short time. Subsequently, there is a rapid decline in the LFG generation rate. For American MSW, L_0 values range from 182.43 L/kg to 195.29 L/kg, exceeding that of Chinese MSW. Furthermore, for American MSW, the LFG generation potentials of rapidly and slowly degradable substances per unit mass of waste (L_{R0} and L_{S0}) range from 90.08 L/kg to 93.16 L/kg and 92.35 L/kg to 102.13 L/kg, respectively. Correspondingly, the L_{S0}/L_{R0} ratio ranges from 1.03 to 1.10, indicating that the LFG generation rate of American MSW after landfilling is more gradual

Table 6.3 LFG generation rate constants of rapidly and slowly degradable MSW substances found in Chinese landfills (Li 2021)

Landfill	Stable extraction volume (m^3/h)	Influence radius (m)	L_{R0} (m^3/t)	L_{S0} (m^3/t)	k_R (a^{-1})	k_S (a^{-1})
Yinzhou, Ningbo	63.45–67.33	12.09–13.99	115.4	17.9	0.53–1.66	0.13 ~ 0.42
Tianziling, Hanghzou	57.0–70.7	18.0–30.53	156.8	22.7	0.82–1.75	0.20 ~ 0.44
Jiangcungou, Xi'an	94.0–103.5	15.0–20.0	140.1	22.3	0.48–1.06	0.12 ~ 0.26
Liming, Shanghai	8.9–20.20	11.53–15.52	166.2	17.8	–	0.09 ~ 0.39
Yulongkeng 1, Shenzhen	141.24	28.00	155.7	19.5	1.12	0.28
Yulongkeng 2, Shenzhen	40.0	50.09–62.04	155.7	19.5	–	0.065 ~ 0.100
Shuige, Nanjing	250	29.17	163.6	19.3	1.04	0.26

Table 6.4 Compositions and LFG generation potentials of MSWs sourced from different countries and regions (Li 2021)

Country/Area	Composition (%)				L_{R0} (L/kg)	L_{S0} (L/kg)	L_{S0}/L_{R0} (-)	L_0 (L/kg)
	Food waste	Paper waste	Textile waste	Bamboo and wood waste				
China	57.69	9.71	3.23	1.98	149.79	18.31	0.12	168.10
Slovak, 2005	38.00	13.00	0.00	0.00	104.27	27.53	0.26	131.81
Portugal, 2005	34.00	21.00	0.00	0.00	106.54	43.95	0.41	150.49
UK, 2012	27.10	11.10	3.50	25.90	106.17	68.15	0.64	174.32
Germany, 2017	19.77	24.33	6.60	12.70	94.22	95.58	1.01	189.80
France, 1993	28.60	26.80	5.70	3.30	107.16	76.41	0.71	183.57
Italy, 2017	27.58	21.65	6.16	13.90	108.37	93.05	0.86	201.41
Poland, 2017	22.36	15.22	8.22	12.93	87.23	81.52	0.93	168.75
Russia, 2009	30.00	23.50	8.00	1.00	103.53	69.29	0.67	172.81
U.S. EPA, 2008	18.60	20.70	7.10	15.90	93.16	102.13	1.10	195.29
U.S. EPA, 2012	21.10	14.80	7.35	16.90	90.08	92.35	1.03	182.43
Canada, 2012	27.00	26.00	2.00	2.00	104.33	69.16	0.66	173.49

Note The composition of fresh Chinese MSW is based on average values obtained from more than 20 cities across China

compared to that of Chinese MSW. This, in turn, suggests that American MSW requires longer to achieve its peak LFG generation, following which the rate of LFG generation declines gradually.

Figure 6.3 illustrates the differences in annual landfill amounts, theoretical LFG generation, and measured LFG generation across various landfills: Xi'an Jiangcungou Landfill and Shenzhen Xiaping Landfill in China and the CSEM landfill in Canada. Notably, the theoretical LFG generation amounts for the Xi'an Jiangcungou Landfill and Shenzhen Xiaping Landfill are computed using Eq. (6.6), with corresponding calculation parameters detailed in Table 6.5. Meanwhile, the theoretical LFG generation amount for the CSEM Landfill in Canada is sourced from Lagos (2017). As depicted in Fig. 6.3, the theoretical LFG generation amounts for the two Chinese landfills predicted using Eq. (6.6) align well with the corresponding monitoring results. Furthermore, for both the Jiangcungou Landfill and Xiaping Landfill, LFG generation trends correspond with changes in annual landfill amounts. Particularly, as annual landfill volumes increase, LFG generation rapidly rises to its peak. Notably, the Jiangcungou Landfill and Xiaping Landfill demonstrate peak theoretical LFG generations of 528.30×10^6 m^3/a (Year 25.25) and 427.24×10^6 m^3/a (Year 23.17), respectively. Following a significant decline or termination of landfilling, LFG generation at these landfills also rapidly declines. Approximately 0.25 years and 1.03 years after the termination of landfilling operations, the theoretical LFG generations of the Jiangcungou Landfill and Xiaping Landfill decrease to 50% of their peak values, respectively. Conversely, the CSEM Landfill in Canada demonstrates a gradual, uniform decline after reaching its peak LFG generation, rather than a sharp decrease. Approximately 12 years after the peak LFG generation at the CSEM Landfill, its theoretical LFG generation decreased to 50% of the peak value, indicating that it decreased significantly slower than the theoretical LFG generation of the two landfills in China.

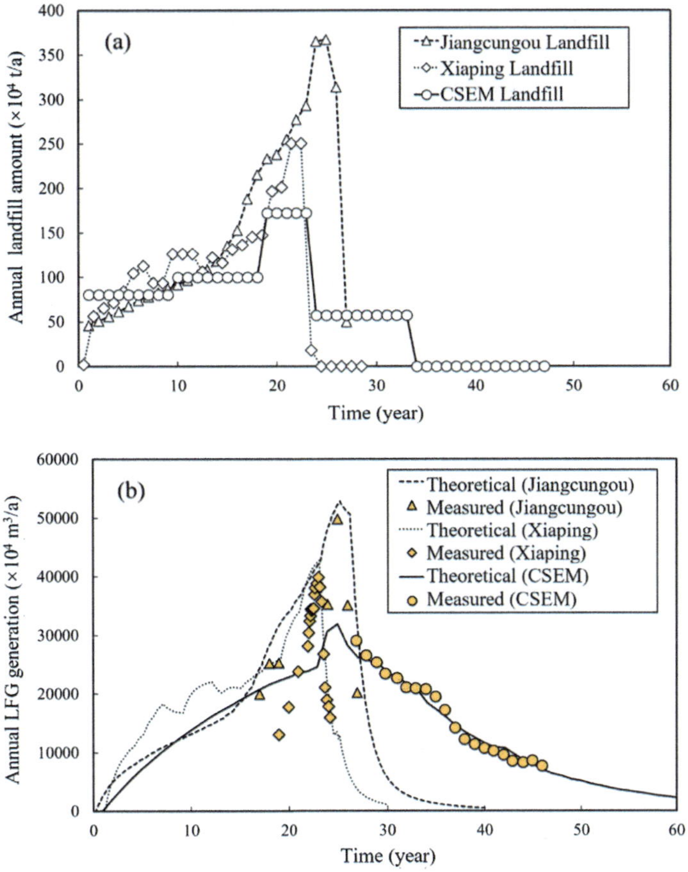

Fig. 6.3 Variations in **a** annual landfill amounts and **b** LFG generations across Chinese and Canadian landfills (Li 2021)

Table 6.5 Parameters used to compute the theoretical LFG generations of the Xi'an Jiangcungou Landfill and Shenzhen Xiaping Landfill (Li 2021)

Parameter	Value	
	Jiangcungou, Xi'an	Xiaping, Shenzhen
L_{R0} (L/kg)	140.1	150.9
L_{S0} (L/kg)	22.3	30.1
k_R (a^{-1})	0.8	1.2
k_S (a^{-1})	0.2	0.3

6.4 Three-Dimensional Drainage Technology for Leachate and Landfill Gas

As described in Sect. 4.7, the leachate level and pore gas pressure in the HFWC landfill are very high. The phenomenon of mutual obstruction in the migration of liquid and gas (as described in Sect. 4.4) makes their discharge extremely difficult (Chen et al. 2018). According to the unsaturated two-phase permeability relationship, the discharge of leachate contributes to the extraction of landfill gas. When the excess pore pressures are encountered, the installation of intermediate leachate collection facilities should be considered, in addition to the installation of leachate collection and drainage systems at the bottom of landfills. Intermediate leachate collection facilities are placed in layers within an

MSW body and effectively lower the main and perched leachate levels in a landfill, while also venting LFG to reduce pore pressure. Furthermore, when high main leachate levels pose instability risks, emergency measures must be promptly undertaken to lower these levels. After such measures have decreased the risk of landfill slope failure, long-term leachate-level control measures should be implemented to ensure that the leachate level remains below the alert level during subsequent operations.

Figure 6.4 illustrates a three-dimensional collection and drainage system. This system includes small-diameter and large-diameter extraction wells, interlayer horizontal drainage trenches, long and short trenches, and horizontal wells. Notably, during the rapid degradation phase, this three-dimensional collection and drainage system prevents liquid and gas migration interference, effectively managing leachate levels and efficiently collecting LFG. Furthermore, during the slow degradation phase, the system facilitates in-situ ventilation and oxygenation, promoting aerobic reactions that expedite landfill stabilization and thus mitigate the environmental impacts of leachates and LFG. Previous studies (Zhan et al. 2017) have conducted independent field experiments on horizontal wells and vacuum pressure extraction wells to refine construction techniques and optimize technical parameters related to pumping performance.

6.4.1 Horizontal Well Dewatering Test

Beaven et al. (2007) performed the first recorded instance of horizontal well installation in an MSW landfill in Rainham landfill, Essex, United Kingdom. After that, Hu et al. (2020) conducted the on-site horizontal well dewatering test at the Tianziling Landfill in Hangzhou, China, which has been operational since 2007. The landfill spans an area of 96 hectares and is equipped with a combination of vertical and horizontal anti-seepage technologies. The landfill features a maximum design elevation of 165 m for the waste body and a total storage capacity of 22.02×10^6 m^3. Extending from east to west, the westernmost part of the landfill functions as the waste dam.

The continuous increase in the amount of MSW disposed per day had resulted in excessive leachate production, which raised the leachate level in the MSW body. The average leachate burial depth is approximately 5 m, posing significant stability risks. To lower leachate levels and hence mitigate stability risks, multiple horizontal and extraction vertical wells have been strategically installed in front of the dam. The layout of the horizontal wells is illustrated in Fig. 6.5. Three horizontal wells—S01, S02, and S03—have been drilled using the horizontal directional drilling method, each featuring a

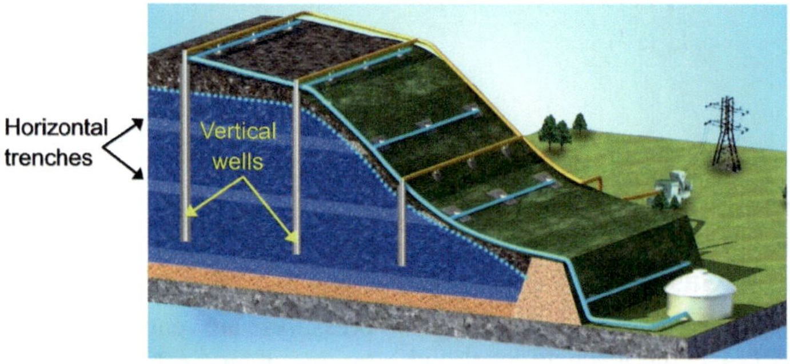

Fig. 6.4 Three-dimensional collection and drainage system for leachate and LFG in landfills (reprinted from Zhan et al. 2017, with permission from Elsevier)

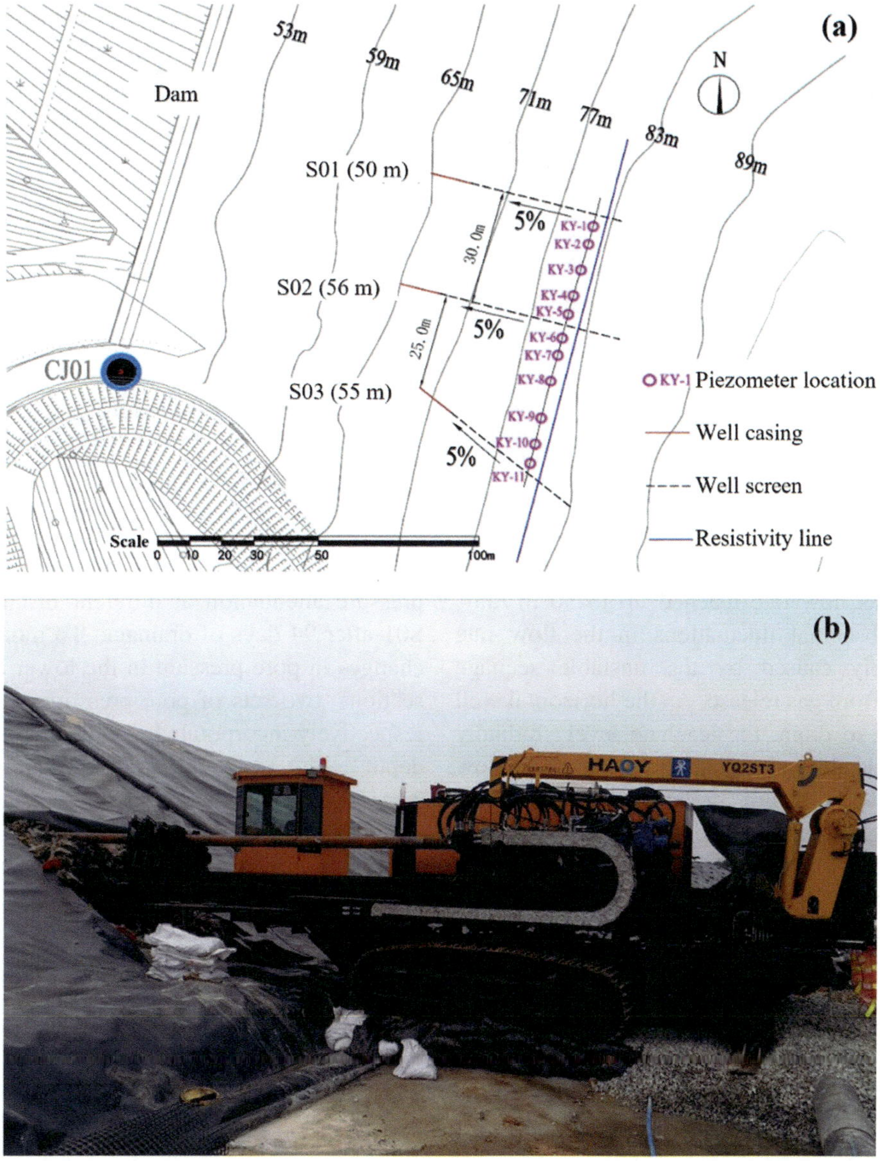

Fig. 6.5 Horizontal wells in landfills: **a** plan view and **b** construction site (reprinted from Hu et al., 2020, with permission from Elsevier)

length of approximately 50 m. Furthermore, these horizontal wells are inclined at an angle of 5% to ensure free leachate flow. Well pipes are made of DN 133 mm galvanized steel with a wall thickness of 6 mm. The upper half of each pipe is perforated with 12-mm-diameter holes, constituting 6% of the total in the screened section. The first 12 m section of each horizontal well is equipped with solid steel pipes, while the remaining portion is perforated. Furthermore, the end of each well is sealed with a plug to prevent waste from entering and blocking the well pipes.

Figure 6.6 illustrates the variation in flow rate with drainage time for horizontal well S01. Initially, on Day 1, the flow rate was high at 75 m^3/day. By Day 3, the flow rate rapidly declined to 21 m^3/day. Moving further, from Day 4 to Day 16, the flow rate fluctuated significantly between

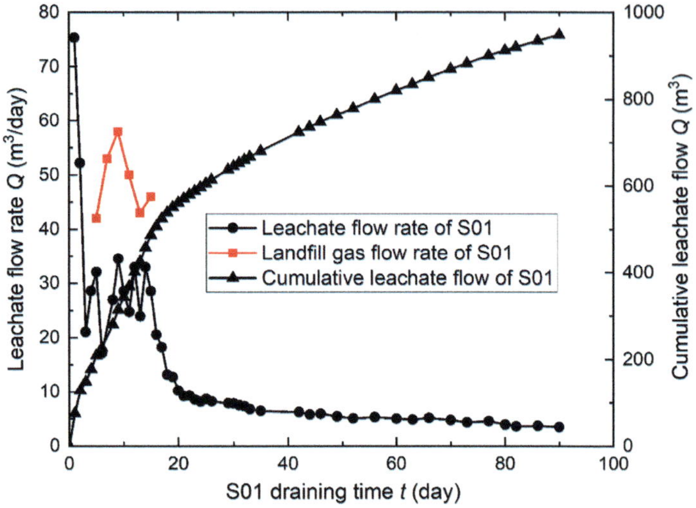

Fig. 6.6 Dewatering performance of well S01 (reprinted from Hu et al. 2020, with permission from Elsevier)

20.5 m^3/day and 34.5 m^3/day, while the concurrent gas flow rate reached up to 50 m^3/day. These substantial fluctuations in the flow rate were likely caused by the unstable seepage resulting from gas release. As the horizontal well continued to drain, the leachate level gradually declined, leading to a gradual decline in its flow rate. By Day 90 of drainage, the flow rate decreased to 3 m^3/day, tending toward gradual termination. The cumulative flow rate for well S01 reached 959 m^3 by Day 93, with an average flow rate of 10.66 m^3/day.

Figure 6.7 depicts the magnitude of pore pressure attenuation at different distances from S01 after 94 days of drainage. To monitor these changes in pore pressure in the lower and upper sections, two sets of pore pressure gauges were respectively incorporated, one buried at the same depth as the horizontal well and another buried 3 m below it. Notably, as the distance from horizontal well S01 increased, the magnitude of pore pressure attenuation decreased. On Day 94, the decrease in pore pressure at 3 m from the horizontal well was 26.95 kPa in the lower

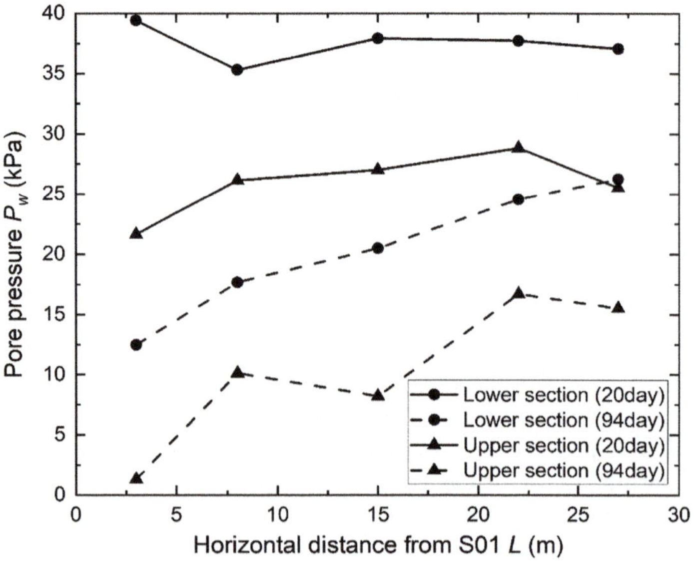

Fig. 6.7 Variations in pore pressure at different distances after drainage from horizontal well S01 (reprinted from Hu et al. 2020, with permission from Elsevier)

section and 20.34 kPa in the upper section. Meanwhile, at 27 m from the horizontal well, the decrease in pore pressure was 10.84 kPa in the lower section and 10.03 kPa in the upper section. Assuming linear attenuation of pore pressure reduction along the horizontal direction, the estimated horizontal influence widths of S01 were 43.1 m for the lower horizon and 50.3 m for the upper horizon.

Traditional methods to assess the effectiveness of different dewatering measures often involve the use of boreholes equipped with pore pressure gauges or monitoring wells to track changes in the leachate level around a site. However, these methods present several drawbacks, including high installation costs, disturbance to geological strata, and discontinuous monitoring data. To address these limitations, Hu et al. (2019) utilized high-density electrical resistivity tomography (ERT) to assess the dewatering effects of horizontal wells.

Figure 6.8 illustrates the changes in the resistivity of MSW around horizontal wells from 30 December 2016 to 28 February 2017, based on the resistivity data from 15 December 2016. On December 30, 2016, and January 15, 2017, noticeable increments in resistivity were observed explicitly near the S01 horizontal well, where drainage was initiated. The area of this increased resistivity expanded over time,

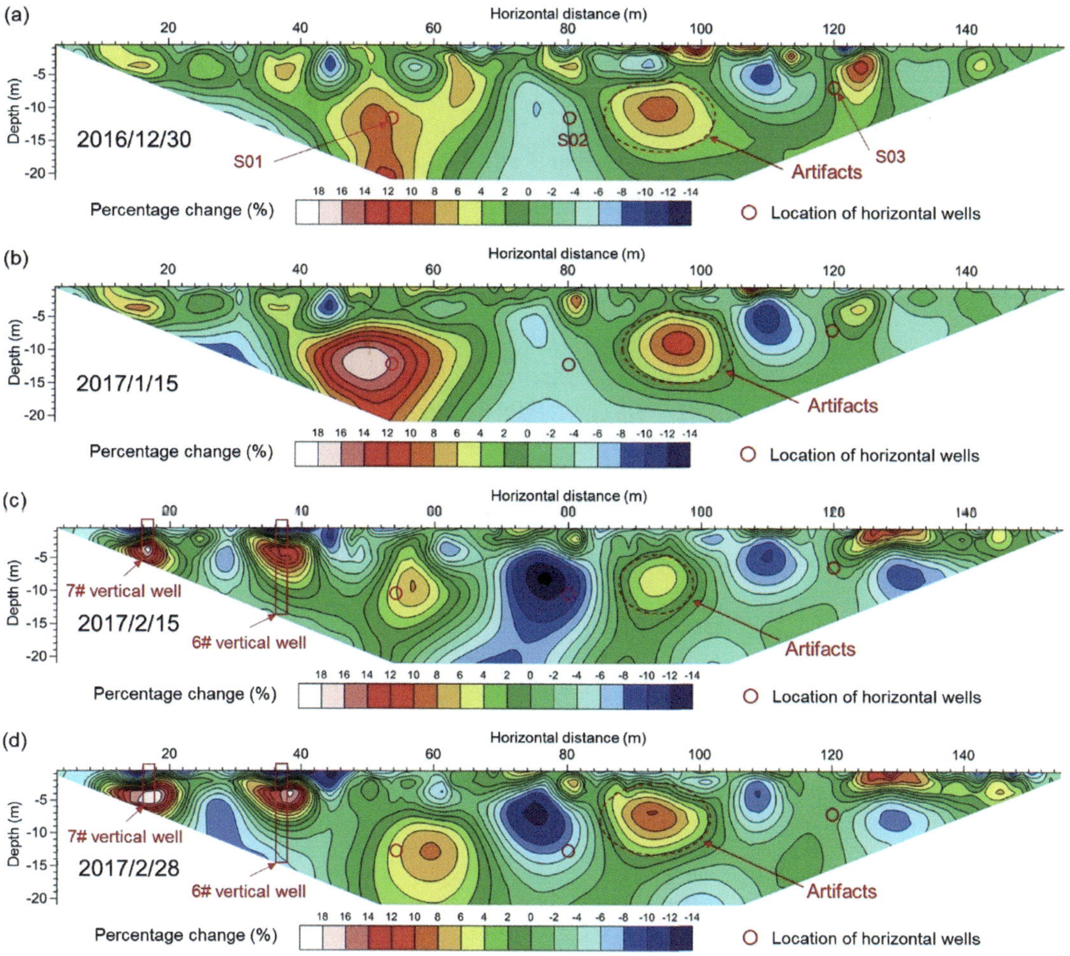

Fig. 6.8 Percentage change in the resistivity of waste around horizontal wells: **a** December 30, 2016, **b** January 15, 2017, **c** February 15, 2017, **d** February 28, 2017 (reprinted from Hu et al. 2019, with permission from Elsevier)

presenting an elliptical distribution on January 15, 2017, thus indicating a gradual expansion of the drainage influence zone. During the above-mentioned period, there was also a pseudo zone of increased electrical resistivity between S02 and S03, a phenomenon that was also reported by Jolly et al. (2011). This phenomenon is attributable to a decrease in the resolution of ERT, resulting from a reduction in current intensity with increasing depth. By February 15, 2017, as drainage commenced from wells 6# and 7# to the left of S01, two elliptical areas with increased resistivity (10–20% increase) appeared, tending toward the resistivity increase area near S01. Additionally, owing to the initiation of drainage from S03, a narrow, elongated area of increased resistivity (10–20% increase) appeared above it, indicating roughly horizontal movement of leachate and reflecting seepage anisotropy. By February 28, 2017, as flow rates from the horizontal wells significantly decreased, no significant changes in resistivity were apparent compared to the conditions on February 15, 2017. However, slight resistivity increments were observed in localized areas. Despite the consistently low flow rates from well S02, the injection of a substantial volume of water during the drilling stage led to a minor decrease in resistivity owing to the slightly elevated water content of the waste near this well. These test results demonstrate that high-density ERT, as an intuitive and operationally feasible geophysical method, can efficiently monitor dewatering effects in landfills and visualize potential leachate migration pathways.

6.4.2 Vertical Well for Vacuum Pressure Pumping Tests

Applying a certain vacuum pressure inside vertical wells can increase the gradient of leachate migration, thereby alleviating the mutual impediment of liquid and gas migration. Consequently, the pumping efficiency of leachate and LFG are enhanced, which effectively reduces the head of leachate and pore pressure within an affected area.

The specific structure of a vacuum pressure well is illustrated in Fig. 6.9. When the leachate was drawn out through submersible pump, the landfill gas was simultaneously extracted through

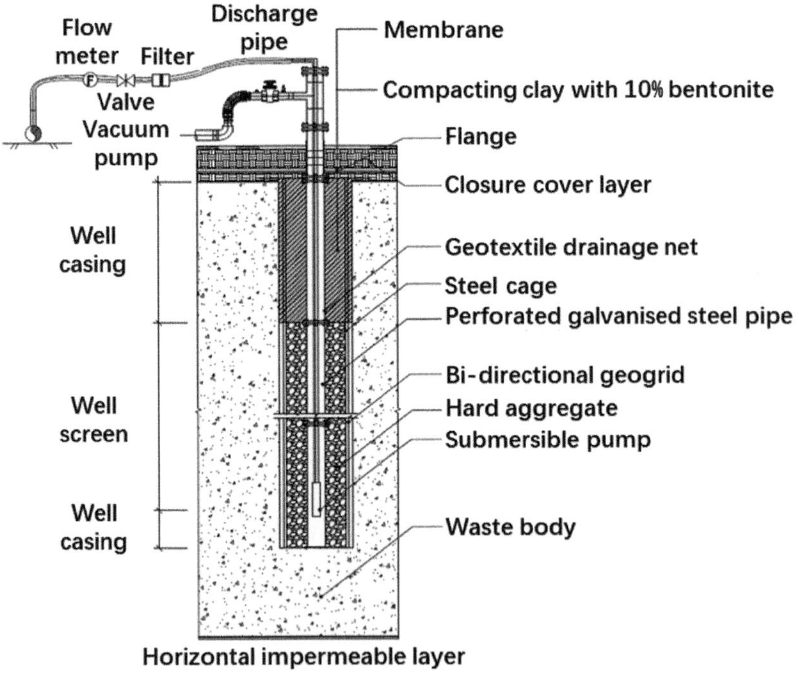

Fig. 6.9 Structure of a large-diameter vacuum pressure vertical well (reprinted from Hu et al. 2021, with permission from Springer Nature)

vacuum pump to form vacuum pressure in the well. The upper section comprises nonperforated galvanized steel pipes, serving as the sealing section for vacuum pressure extraction. This section is filled with compacted clay containing 5–20% bentonite as a sealing material between the pipes and the surrounding waste body, with a sealing section length of at least 8 m. The middle section comprises perforated galvanized steel pipes, acting as the leachate inflow section. Finally, the lower section also comprises non-perforated galvanized steel pipes, functioning as bottom-sealed sand settling section with a length of 1 m. The top of the well pipe is sealed with a blind flange. The area between the middle and lower sections of the well pipe and the surrounding waste body is occupied by various filter materials, arranged from inside to outside as follows: a geotextile drainage net, 25–40 mm of hard aggregate, and a bidirectional geogrid. A submersible pump was installed at the bottom of the leachate inflow section of the well pipe, and one end of the discharge pipe was connected to the submersible pump and the other end was passed through the blind flange. After passing through the blind flange, the discharge pipe is connected to a filter, gate valve, and flow meter, following which it enters the drainage branch. Ventilation openings, provided on the sidewalls of the well pipe protruding from the surface of the waste body, are connected to the vacuum pressure pump unit via piping.

Hu et al. (2021) performed a vacuum pressure well pumping test at the Tianziling Landfill in Hangzhou. Figure 6.10 depicts the layout of the pumping and monitoring wells. In particular, wells 3–6#P, located in a flat region 40 m from the dam, were utilized for the vacuum pressure combined pumping test. These vertical wells were 38 m long and featured a diameter of 300 mm. Monitoring wells 1#W, 2#W, 3#W, and 4#W were placed at 5 m, 10 m, 15 m, and 30 m from wells 3–6#P, respectively, for leachate level monitoring. All monitoring wells had a diameter of 50 mm, and they measured leachate levels using a SET500 leachate level meter. In addition to these, vacuum pressure monitoring wells 1#G and 2#G were installed at 5 m and

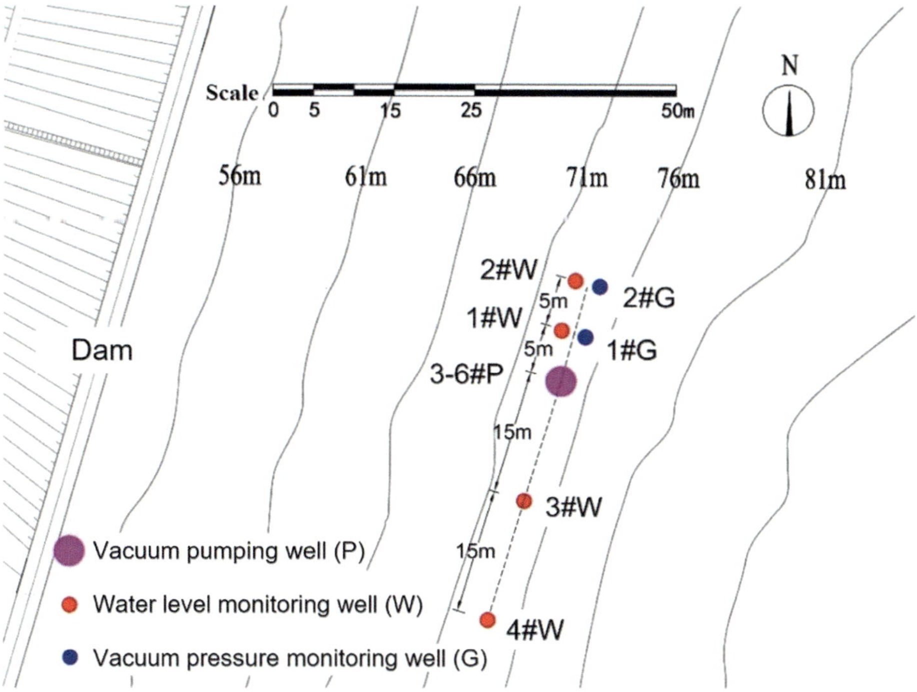

Fig. 6.10 Layout plan of the pumping test (reprinted from Hu et al. 2021, with permission from Springer Nature)

10 m from wells 3–6#P, respectively. Vacuum pressure sensors buried within the boreholes were connected to the surface via vacuum pressure tubes, allowing the pressure readings to be transmitted to vacuum pressure gauges for analysis. Furthermore, the vacuum pressure sensor, comprising perforated polypropylene random copolymer pipes with a diameter of 25 mm and a length of 50 cm, were wrapped in nonwoven geotextile for filtration. They were then buried at depths of 15 m, 20 m, 25 m, and 30 m in wells 1#G and 2#G to examine the distribution of vacuum pressure along the well depth.

Figure 6.11 illustrates the relationship between the transient pumping flow rate and pumping duration under different vacuum pressures during constant drawdown pumping at 20 m for wells 3–6#P. The pumped water initially comprises accumulated wellbore water, resulting in similar initial transient flow rates (Q_w) under different vacuum pressures, all reaching approximately 80 m^3/day. However, as pumping continues, the pumping flow rate under each vacuum pressure gradually decreases. This reduction because the 20 m drawdown in the well lowers the leachate level in the surrounding waste body, thus diminishing the hydraulic gradient and the seepage flow rate. The rate of decrease in the pumping flow rate (Q_w) varies across different vacuum pressures, with the fastest decrease observed when at a vacuum pressure of 0 kPa, reaching 38.0 m^3/day at 44 h. In contrast, the slowest decrease in the pumping flow rate occurs at a vacuum pressure of − 30 kPa, with Q_w reaching 56.1 m^3/day at 44 h. This trend arises because the vacuum pressure within the well extracts LFG from the waste body, increases liquid phase saturation, and consequently increases the value of the liquid phase permeability coefficient (k_w). This increase in k_w partially offset the effects of the reduced hydraulic gradient on the seepage flow rate, thus increasing the pumping flow rate (Q_w). By 44 h, the pumping process approaches a steady-state. As the vacuum pressure within the well gradually increases from 0 kPa to − 10 kPa, − 20 kPa, and − 30 kPa, the steady-state production flow rate correspondingly increases from 38.0 m^3/day to 45.7 m^3/day, 50.7 m^3/day, and 56.1 m^3/day, respectively.

Figure 6.12 illustrates the drawdowns at various distances from wells 3–6#P after a steady pumping state is achieved with a fixed drawdown of 15 m. As the horizontal distance from wells 3–6#P increases, the drawdowns under different well vacuum pressures gradually decrease, reaching extremely low values at a drawdown of 30 m (4#W). Under a vacuum pressure of 0 kPa,

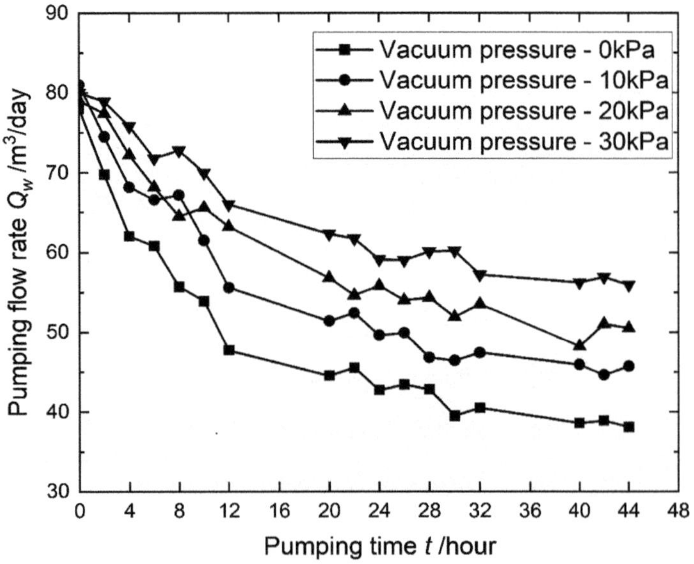

Fig. 6.11 Transient pumping flow rates of well 3–6#P under different vacuum pressures (reprinted from Hu et al. 2021, with permission from Springer Nature)

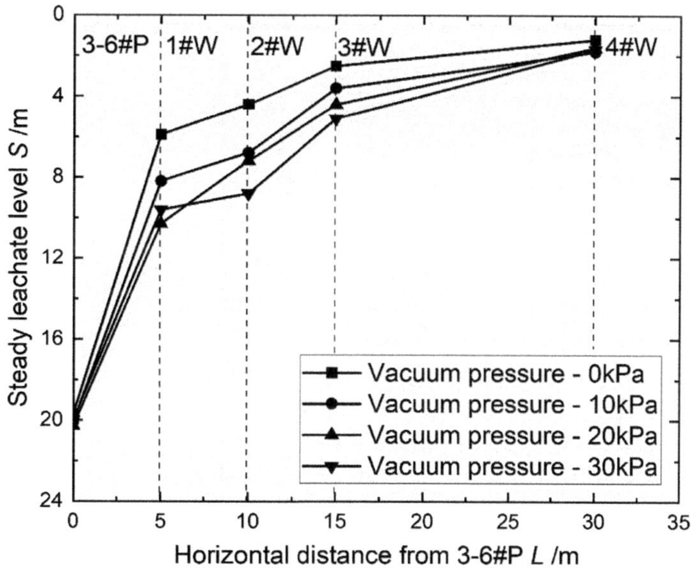

Fig. 6.12 Leachate level drawdowns at different distances from well 3–6#P under different vacuum pressures (reprinted from Hu et al. 2021, with permission from Springer Nature)

the drawdown at a distance of 5 m (monitoring well 1#W) is 5.9 m. As the well vacuum pressure gradually increases to − 10 kPa, − 20 kPa, and − 30 kPa, the drawdown value at 5 m also correspondingly increases to 8.1 m, 9.6 m, and 10.3 m, respectively. Similarly, the drawdowns at 10 m and 15 m gradually increase with increasing vacuum pressures, albeit with a diminishing magnitude of increase over distance. For instance, when the vacuum pressure increases from 0 kPa to − 30 kPa, the drawdowns at 5, 10, and 15 m increase from 5.9 m to 10.3 m, 4.4 m to 8.6 m, and 2.5 m to 5.2 m, respectively. However, at 30 m, the drawdown only increases from 1.0 m to 1.4 m. This attenuation is attributed to the gradual weakening of vacuum pressure as it penetrates the waste body, diminishing its impact on the leachate level. Based on these observations, the range of vacuum pressure that effectively increases the drawdown of the vertical well extends from approximately 15–20 m. Beyond this range, differences in drawdowns under different vacuum pressures are insignificant. This suggests that when designing onsite vacuum pressure-assisted pumping wells for dewatering applications, the separation between individual wells must be maintained within 30–40 m.

Figure 6.13 depicts the variation in steady-state vacuum pressure along the horizontal direction of wells 3–6#P at a depth of 15 m within the waste body. As the horizontal distance from wells 3–6#P increases, the vacuum pressure gradually decreases under different conditions. For instance, under a vacuum pressure of − 30 kPa within the well, an increase in the horizontal distance to 5 m and then to 10 m leads to a decrease in the vacuum pressure from − 30 kPa to − 14.7 kPa and − 6.1 kPa, respectively. Based on the gradient of vacuum pressure attenuation, the vacuum pressure can be roughly estimated to diminish to zero at a horizontal distance of 15 m. This finding is consistent with previous observations indicating that vacuum pressure pumping significantly influences drawdown within a horizontal distance of 15–20 m.

6.5 Efficient Collection of Landfill Gas

The high food waste and water contents of Chinese MSW have led to elevated leachate levels in most landfills across China. Under such highly saturation conditions, the low gas permeability of

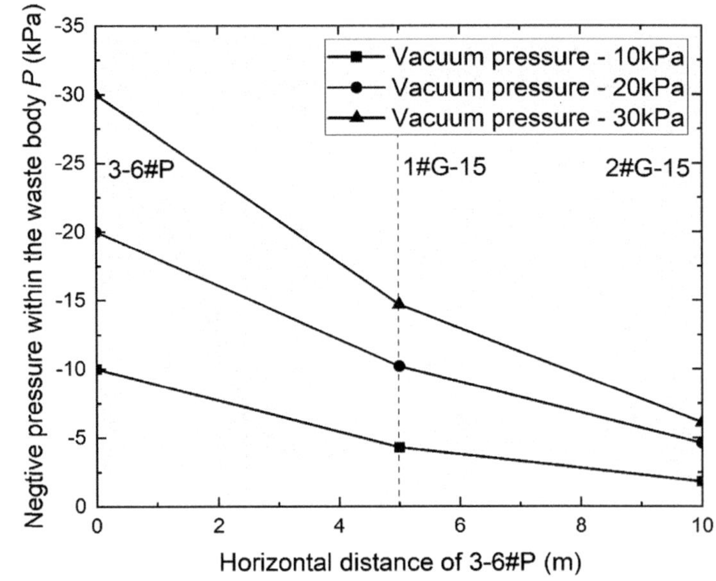

Fig. 6.13 Variations in steady-state vacuum pressure at different horizontal distances from well 3–6#P (reprinted from Hu et al. 2021, with permission from Springer Nature)

MSW hinders efficient LFG collection, resulting in low gas collection rates. A survey on the collection and utilization rates of LFG across 18 landfills in China, as depicted in Fig. 6.14, revealed that the gas collection rates generally range from 25% to 40%, far below the average level of 50% to 95% observed in western countries (Zhan et al. 2015). Among all Chinese landfills, the Xiaping Landfill in Shenzhen stands out with a gas collection rate exceeding 90%, achieved through long-term systematic management efforts.

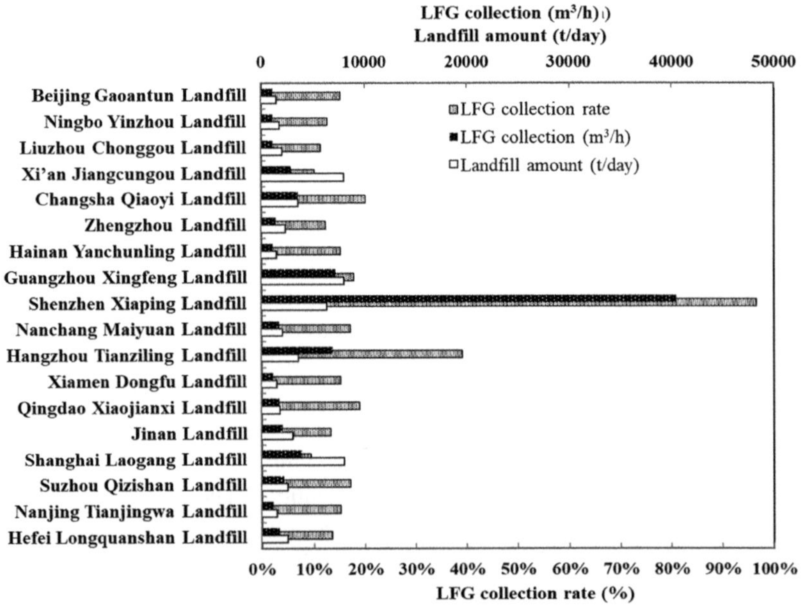

Fig. 6.14 Results of a survey on LFG collection in 18 Chinese landfills (Li 2021)

6.5.1 Relationship Between Landfill Gas Collection Rate and Leachate Level

Zhan et al. (2015) conducted on-site drawdown and gas extraction tests at Liming Landfill in Shanghai to study the influence of leachate levels on LFG collection. Liming Landfill was constructed in 1999 with a designed capacity of 1,500 t/day. It reached full capacity by 2004 and underwent two expansions in 2005 and 2008 owing to delays in constructing a new landfill. Landfilling operations ceased in December 2009, collectively disposing approximately 5.5 Mt of MSW. The experimental area, covering approximately 900 m^2 (30 m × 30 m) of flat terrain, was situated in the northeast section of the old landfill.

The layout of the experimental and monitoring wells during the onsite drawdown and gas extraction tests is depicted in Fig. 6.15. As illustrated, four extraction and pumping wells were installed in the experimental area. Among these, wells 1#, 3#, and 4# formed an equilateral triangle with a side length of 20 m, while well 2# was placed at the center of the triangle at a distance of 11.5 m from the other three wells. To monitor leachate levels and record shallow gas pressures within the waste body, seven leachate level monitoring wells and 20 gas pressure monitoring points were strategically arranged along the lines connecting well 2# with wells 1#, 3#, and 4#. The experimental scheme included gas pressure measurements at different depths within the waste body; hence, two gas pressure monitoring wells were installed at depths of 3 m and 5 m at each monitoring point.

Gas extraction tests were conducted at test well 2# under three different leachate depth reduction conditions: (1) initial leachate level; (2) single-well pumping at test well 2#, with an approximate depth reduction of 11 m within the well; (3) group-well pumping at test wells 1# to 4#, with an approximate depth reduction of 11 m within the wells.

Figure 6.16 depicts the stable leachate levels within the MSW body under these three. Initially, the leachate level within the MSW body appeared at a depth of approximately 4 m. Under single-well pumping at test well 2#, a significant decline in the leachate level decline was observed within a radius of 5 m. The depth of the leachate level at this radius was approximately 5.3 m, and variations in the leachate level beyond a radius of 10 m were minimal. Finally, under group-well pumping, the minimum burial depth of the leachate level between the stabilized test wells 2# and 3# was approximately 7 m.

Table 6.6 details the stable LFG collection rates obtained from the gas extraction tests under different leachate levels. It can be seen that at a given vacuum pressure, the LFG collection

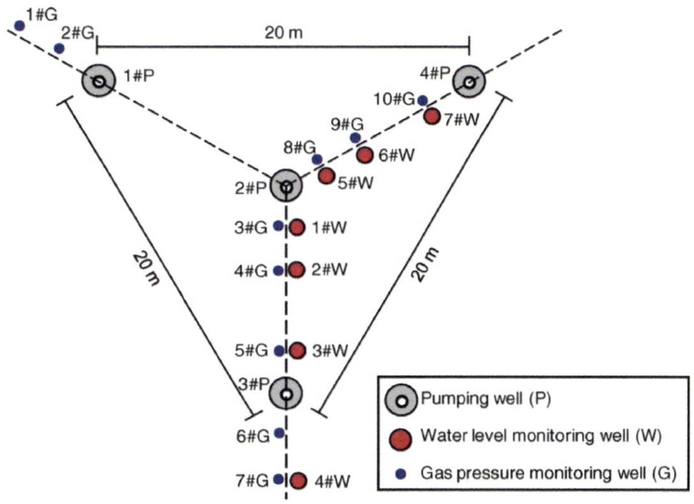

Fig. 6.15 Layout of pumping and monitoring wells (reprinted from Zhan et al. 2015, with permission from the American Society of Civil Engineers)

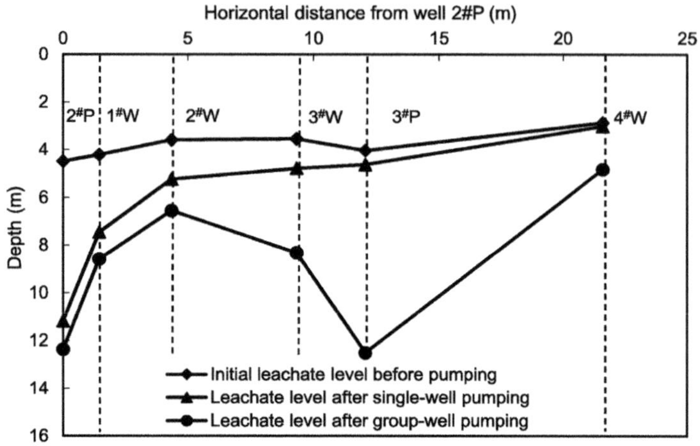

Fig. 6.16 Monitored leachate levels along the 2#P–3#P axis (reprinted from Zhan et al. 2015, with permission from the American Society of Civil Engineers)

Table 6.6 Stable LFG collection rates under different pumping conditions (m^3/h)

Pumping condition	Extraction vacuum pressure			
	−4 kPa	−8 kPa	−11 kPa	−14.3 kPa
Initial leachate level	1.9	7.3	–	–
Leachate level after single-well pumping	–	8.9	10.8	–
Leachate level after group-well pumping	–	15.8	–	20.2

gradually increased as the leachate level lowered. At a vacuum pressure of − 8 kPa in the extraction well, the LFG collection rate during group-well pumping reaches 15.8 m^3/h, approximately 2.2-fold that under the initial leachate level condition. This indicates that reducing leachate levels is an effective means of improving the LFG collection rates of Chinese landfills with high leachate levels.

Zhan et al. (2015) performed numerical simulations to quantitatively analyse the variation in LFG collection rates with leachate levels. The extraction of LFG from vertical wells within landfills can be simplified into a two-dimensional axisymmetric gas transport problem in porous media. The steady-state governing equation for this problem is

$$\frac{K_r}{r}\frac{\partial P}{\partial r} + K_r\frac{\partial^2 P}{\partial^2 r} + K_z\frac{\partial^2 P}{\partial^2 z} = -\frac{2RT\mu}{\omega}a \quad (6.12)$$

where K_r and K_z denote the horizontal and vertical gas intrinsic permeabilities (m^2), respectively; r represents the radial distance from the center of the vertical extraction well (m), z indicates the vertical distance from the top of the waste body (m), P signifies gas pressure (Pa); μ denotes gas viscosity (kg/m/s); ω represents the molar mass of gas (kg/mol); R denotes the ideal gas constant ($kg \cdot m^2/s^2/mol/K$); T represents gas temperature (K); and a denotes gas production rate ($kg/m^3/s$).

Figure 6.17 illustrates the numerical model adopted for vertical well gas extraction analysis. Here, a vacuum pressure of − 5 kPa is selected as the boundary condition for the extraction well. The part below the vertical well on the left side, entire right side, and bottom of the model are considered impermeable boundaries. Meanwhile, atmospheric pressure is selected as the top boundary condition.

In total, two numerical models of different sizes were established, with MSW body thicknesses of 16.5 m (for a medium-scale landfill) and 41.5 m (for a large-scale landfill), respectively. The depths of gas extraction wells in the

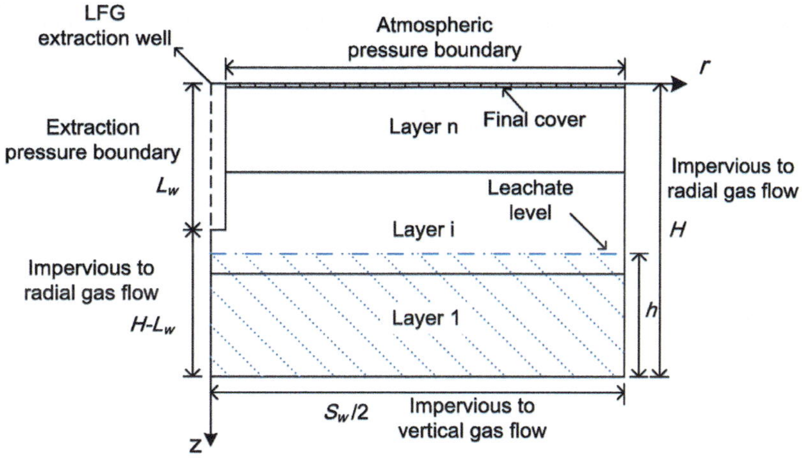

Fig. 6.17 Numerical model of LFG collection from a vertical well within landfills featuring high leachate levels (reprinted from Zhan et al. 2015, with permission from American Society of Civil Engineers)

medium-scale and large-scale landfill models were set to 12 m and 30 m, respectively, with well separations of 25 m and 50 m, respectively. Consequently, the widths of the two-dimensional axisymmetric numerical models for the two landfill sizes were 12.5 m and 25 m (half the well spacing). Both numerical models simulated vertical well gas extraction under four values of the ratio of leachate level height (h) to MSW body thickness (H): 30%, 50%, 70%, and 90%. Based on the numerical simulation results, the LFG collection efficiency was then calculated using the following formula:

$$\eta = \frac{C_n}{Q_n} \tag{6.13}$$

where C_n denotes the collected LFG volume, calculated based on the product of the integral of the radial gas flow velocity along the well length and circumference; and Q_n represents the theoretical gas production of the landfill, calculated according to the burial age and quantity of waste within each layer in the numerical model using Eq. (6.6).

Figure 6.18 illustrates the dependency of the LFG collection efficiency (η) on the relative leachate level (h/H). Notably, for both the large and medium-scale landfills, the influence of leachate level on LFG collection efficiency appears significant, particularly when the relative leachate level (h/H) exceeds 0.3. Furthermore, as

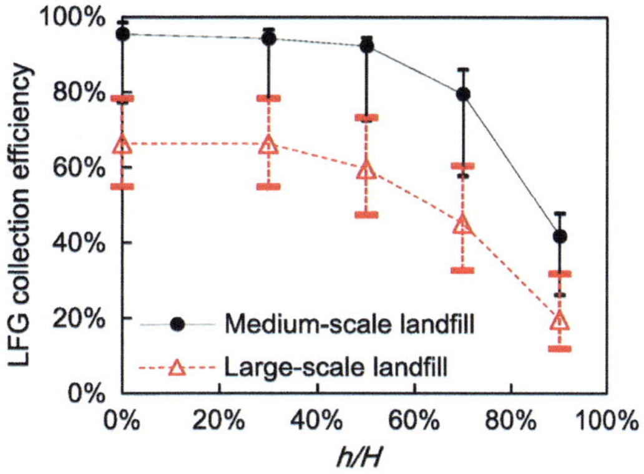

Fig. 6.18 Relationship between the LFG collection efficiency and relative leachate level (h/H) (reprinted from Zhan et al. 2015, with permission from the American Society of Civil Engineers)

h/H increases from 0.3 to 0.9, the LFG collection efficiency (η) decreases by 45%–55% for both landfills. Hence, in practical landfill operations, leachate levels must be maintained below 30% of the MSW body thickness to achieve higher LFG collection efficiencies.

6.5.2 Application in Shenzhen Xiaping Landfill

Xiaping Landfill is located in the Xiaping Valley of Luohu district, Shenzhen, and was one of China's first large-scale sanitary landfills. It plays a critical role in managing MSW from the Luohu and Futian districts. Since its commissioning in 1997, the Xiaping Landfill has accumulated over 20.5 Mt of waste. The first landfilling phase, ongoing until May 2012, witnessed a daily waste intake of nearly 3,000 t and demonstrated a total landfill capacity of approximately 15 Mt. The second landfilling phase was subsequently commenced in May 2012 to accommodate increasing waste volumes from Shenzhen. During the initial stages of the second landfilling phase, the daily waste intake was around 3,500 t, and it reached 5,700 t/day in 2016. However, owing to the implementation of Shenzhen's MSW classification system, the daily waste intake decreased to approximately 500 t/day in 2020. Throughout its operation, Xiaping Landfill has faced significant environmental challenges, particularly with residential areas within 500 m to 5 km from the landfill being affected by odour emissions. To address these challenges, various staged measures have been implemented in the Xiaping Landfill to achieve efficient LFG collection.

In the initial stages of its operation, the Xiaping Landfill predominantly relied on a leachate drainage layer comprising gravel and drainage pipes for leachate drainage from the bottom. However, following three years of operation, the bottom leachate drainage system experienced severe clogging, allowing minimal leachate flow through the main drainage pipes. Consequently, the burial depth of leachate on the slope of the waste body was only approximately 2 m, and leachate flow was frequent along the slope. Meanwhile, in other areas, the burial depth of leachate was generally around 5 m below the landfill surface.

In June 2008, heavy rainfall over three consecutive days caused local failure along the downstream waste slope. Immediate stabilization measures were swiftly implemented to prevent further waste body movement. In particular, 14 small-diameter leachate pumping wells were installed within the unstable sliding areas. Consequently, the leachate level was lowered by 3–5 m within 15 days, reducing the horizontal displacement rate from 75 cm/day to 0.1 cm/day and thus effectively controlling the local failure. Furthermore, to ensure long-term stability and fulfill ongoing landfilling requirements, a three-dimensional collection and drainage system comprising 10 horizontal drainage trenches and 28 wells was constructed. The horizontal drainage trenches were buried approximately 5 m deep and maintained a slope of approximately 2% after settlement. Meanwhile, the 28 wells, each with a diameter of 1 m and a depth of 30 m, were strategically positioned at the top of the slope to perform both leachate drainage and gas extraction. Additionally, the already existing 60 gas collection wells were retrofitted to facilitate both leachate pumping and LFG extraction.

After 6 months of operation of the above-mentioned three-dimensional collection and drainage system, the leachate level had lowered by 10 m, as shown in Fig. 6.19, thereby solving the MSW-slope instability problem and significantly increasing the LFG collection rate. Specifically, from June 2007 to April 2008, the LFG collection rate was approximately 2000 m^3/h, with a collection efficiency of only 10%. Following the construction of the three-dimensional system in June 2008 and the subsequent reduction in leachate level, the LFG collection rate continued to rise. Consequently, by 2011, the LFG collection rate increased to 9,500 m^3/h, with a collection efficiency of approximately 37% (see Fig. 6.20).

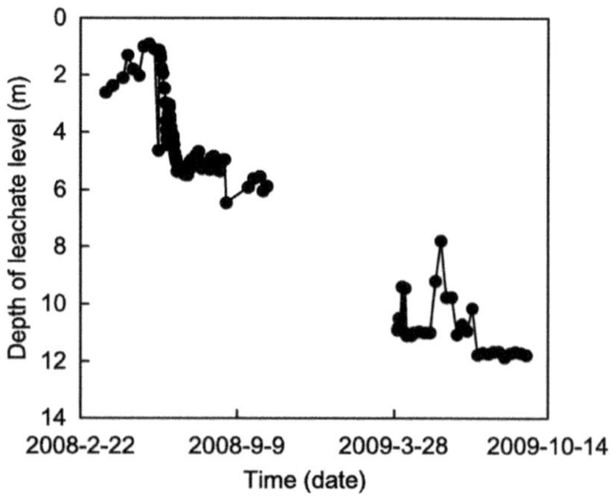

Fig. 6.19 Measured average leachate level in the main filling area of the Xiaping Landfill (reprinted from Zhan et al. 2015, with permission from the American Society of Civil Engineers)

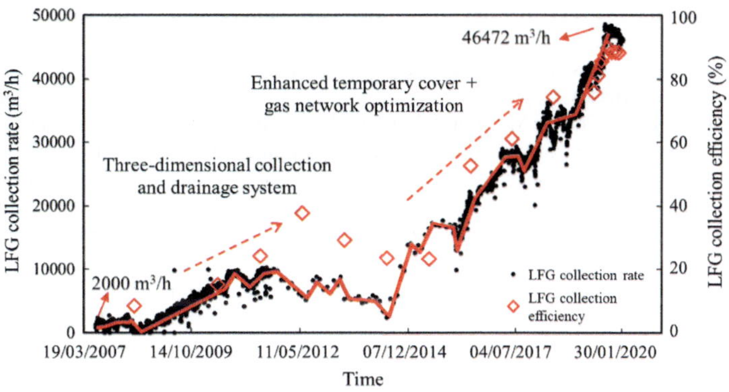

Fig. 6.20 LFG collection rate and efficiency of the Xiaping Landfill

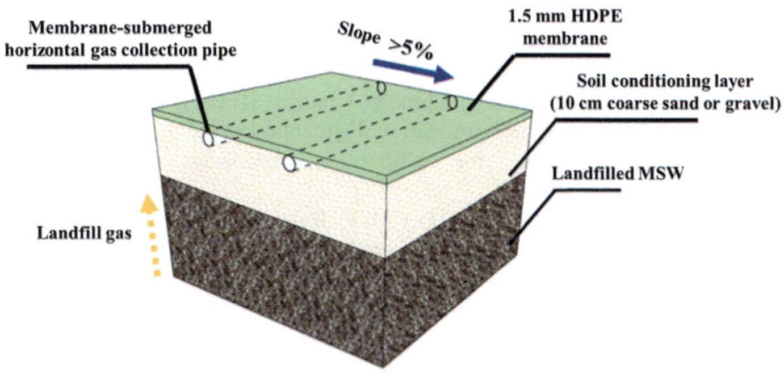

Fig. 6.21 Schematic of the enhanced temporary cover in landfill (Li 2021)

After 2012, with the progress of Phase II landfill operations, horizontal gas wells and drainage trenches were constructed in layers to effectively control the leachate level. These trenches maintained a perched leachate depth of less than 10 m. Additionally, based on the gas production characteristics of HFWC MSW in China, an enhanced temporary cover technology was adopted for the intermediate cover of each landfill unit, as illustrated in Fig. 6.21. Horizontal gas pipes beneath the membrane were spaced 15 m to 20 m apart.

The structure of the enhanced temporary cover is simple, meaning that only a short construction period was required, allowing for quick sealing at the source. Furthermore, given that horizontal gas pipes under the membrane are located at the top of the waste body, their performance is not influenced by the leachate level. LFG generated within the waste body rises to the top and is collected by these horizontal gas pipes. The enhanced temporary cover, combined with the bottom drainage layer, horizontal trenches, and extraction wells, forms a comprehensive collection and drainage system. This system effectively collects LFG during peak gas generation periods, typically observed in the early stages of HFWC MSW landfilling, thus significantly boosting the LFG collection rate. For instance, as depicted in Fig. 6.20, from 2014 to 2019, the LFG collection rate of the Xiaping Landfill increased from 6,413 m^3/h to 45,500 m^3/h, with a corresponding increase in LFG collection efficiency from 23% to over 90%.

References

Beaven RP, Cox SE, Powrie W (2007) Operation and performance of horizontal wells for leachate control in a waste Landfill. J Geotech Geoenvironmental Eng 133(8):1040–1047

Chen YM, Zhan LT, Gao W (2018) Waste mechanics and sustainable landfilling technology: comparison between HFWC and LFWC MSWs. In: The proceedings of the 8th international congress on environmental geotechnics, Hangzhou, China

Guo RY (2017) Research on the biochemical and mechanical behaviors of high food content MSW. Ph.D. thesis, Zhejiang University, Hangzhou, China (in Chinese)

Hu J, Ke H, Chen YM, Zhan LT, Lan JW, Xu WJ (2021) Improvement of vacuum pressure on pumping performance of vertical wells in municipal solid waste landfills. Environ Sci Pollut Res 28:41283–41295

Hu J, Ke H, Zhan LT, Chen ZY, Lan JW, Powrie W, Chen YM (2020) Installation and performance of horizontal wells for dewatering at municipal solid waste landfills in China. Waste Manage 103:159–168

Hu J, Wu XW, Ke H, Xu XB, Lan JW, Zhan LT (2019) Application of electrical resistivity tomography to monitor the dewatering of vertical and horizontal wells in municipal solid waste landfills. Eng Geol 254:1–12

Huang ZH, Zeng S (2004) Field trial of LFG collection and engineering design parameters. China Biogas 22 (1):11–13 (in Chinese)

Inter-governmental Panel on Climate Change (IPCC) (2006) IPCC guidelines for national greenhouse gas inventories. Institute for global environmental strategies.

Jolly JM, Beaven RP, Barker RD (2011) Resolution of electrical imaging of fluid movement in landfills. Proc Inst Civ Eng-Waste Resour Manag 164(2):79–96

Lagos DA, Héroux M, Gosselin R, Cabral AR (2017) Optimization of a landfill gas collection shutdown based on an adapted first-order decay model. Waste Manag 63:238–245

Lan JW, Zhan JW, Li YC, Chen YM (2012) Impacts of initial moisture content of MSW waste on leachate generation and modified formula for predicting leachate generation. Environ Sci 33(4):1389–1396 (in Chinese)

Li H (2021) Degradation-consolidation behaviors and treatment methods for the liquid-gas induced environment disaster of high-food-waste-content MSW landfill. Ph.D. thesis, Zhejiang University, Hangzhou, China (in Chinese)

U.S. EPA (2005) Landfill gas emissions model (LandGEM) version 3.02 user's guide

Wang XM, Nagpure AS, DeCarolis JF, Barlaz MA (2013) Using observed data to improve estimated methane collection from select U.S. landfills. Environ Sci Technol 47(7):3251–3257

Zhan LT, Xu XB, Chen YM, Ma XF, Lan JW (2015) Dependence of gas collection efficiency on leachate level at wet municipal solid waste landfills and its improvement methods in China. J Geotech Geoenvironmental Eng 141(4):04015002

Zhan LT, Xu H, Chen YM, Lan JW, Lin WA, Xu XB, He PJ (2017) Biochemical, hydrological and mechanical behaviors of high food waste content MSW landfill: Liquid-gas interactions observed from a large-scale experiment. Waste Manag 68:307–318

Zheng W, Phoungthong K, Lü F, Shao LM, He PJ (2013) Evaluation of a classification method for biodegradable solid wastes using anaerobic degradation parameters. Waste Manag 33(12):2632–2640

Landfill Settlement and Instability

7

Abstract

This chapter focuses on the application of the degradation–consolidation theory in the context of MSW landfill settlement, capacity, and stability. Typically, the degradation of organic matter within landfills increases leachate levels and gas pressures, significantly impacting effective stress and consequently influencing MSW compressibility and shear strength. These changes ultimately influence landfill settlement and stability. Beginning with an examination of long-term degradation compression tests, this chapter subsequently introduces compression models for MSW, along with methods for landfill settlement and capacity calculations. The discussion then focuses on landfill stability analysis, emphasizing factors such as the shear strength of MSW, instability modes, critical water levels, and critical LFG pressures. Furthermore, centrifugal hypergravity modeling of MSW landfill failures triggered by rising water levels and LFG pressures is presented, proposing critical water level and critical gas pressure values for landfill stability assessments. Strategies for expanding landfill capacity based on stability control are also discussed, including techniques such as dewatering and gas venting for landfill slopes and geogrid reinforcements, along with landfill capacity expansion rate calculations.

Keywords

Municipal solid waste · Degradation compression test · Settlement · Landfill capacity · Landfill stability · Hypergravity modelling · Slope failure modes · Critical water level · Critical LFG pressure · Capacity expansion rate · Geosynthetics

7.1 General Introduction

MSW is highly compressible and hence undergoes vertical compression when deposited in landfills, resulting in a phenomenon known as landfill settlement. This settlement is not simply restricted to the landfilling process but continues even after lsandfill closure. The final settlement of a landfill can reach 25–50% of its initial height. While landfill settlement increases landfill capacity, excessive settlement can compromise the performance and integrity of landfill subsystems, such as slope stability and liner systems. To prevent this, landfill operators employ appropriate settlement prediction models, estimate the actual remaining landfill space, and forecast the landfill lifespan. Accurate estimations of the increase in landfill capacity resulting from waste compression are crucial for efficient landfill disposal planning.

High leachate levels and gas pressures within landfills, combined with external factors such as

Y. Chen, *Soil Degradation-Consolidation Theory and Its Applications*,
https://doi.org/10.1007/978-981-97-7985-7_7

heavy rainfall and earthquakes, considerably reduce the effective stress and shear strength of MSW. This reduction can induce slope instability, resulting in high-energy, large-scale, and high-velocity sliding along potential slip surfaces within the landfill body or soft interface layers at the landfill base. Ensuring landfill stability is critical for the safe and environmentally friendly disposal of MSW as instability and sliding can result in significant leachate and LFG leakage, causing catastrophic environmental pollution. Furthermore, large-scale movements of landfill materials can result in casualties and property damage in downstream areas.

This chapter begins with an introduction to methods for calculating landfill settlement and capacity and subsequently discusses the shear strength of MSW, interface shear strength of geosynthetics, and landfill instability modes. The chapter then delves into centrifugal hypergravity modeling of slope instability induced by rising water levels or gas pressures, proposing critical water levels and gas pressures. Finally, a method for landfill capacity expansion based on stability control is proposed, including measures for stability control and calculations for landfill capacity expansion rates.

7.2 Landfill Settlement and Capacity

7.2.1 Compression Model for Municipal Solid Waste

The compression of MSW typically progresses through distinct stages over time. The first stage involves primary compression, which represents the response of the overall stiffness of the MSW solid-phase framework to the applied overlying stress. Primary compression is generally completed within several days to weeks after the application of overlying stress. Following this, secondary compression ensues, attributed to the weakening of the solid framework due to biochemical degradation and stress creep compression. Given the substantial amounts of degradable components within MSW, biochemical-degradation-induced compression predominates in secondary compression, while the proportion of creep-induced compression is relatively minimal, as described in Sect. 3.6.1. Overall, this biochemical-degradation-induced secondary compression can last for several decades in traditional sanitary landfills.

Xie et al. (2005) and Ke et al. (2010) conducted indoor one-dimensional compression and long-term one-dimensional compression tests on typical Chinese MSW under conditions of inhibited and accelerated degradation. The results of one-dimensional compression tests under different stress levels and accelerated degradation conditions revealed the following: Overlying stress does not significantly influence the LFG production process (Fig. 7.1a). The primary compression of MSW lasts for approximately 7 days, and its secondary-compression-induced strain can reach 8–17% (Fig. 7.1b). Different MSW samples simultaneously subjected to the same stress level exhibit similar total compression levels during compression tests under different stress levels and degradation paths (Fig. 7.1c). Meanwhile, long-term compression tests under inhibited degradation conditions revealed that creep-induced secondary compression is negligible compared to biochemical-degradation-induced secondary compression.

Based on the aforementioned experimental studies, Chen et al. (2010a) proposed a stress–degradation compression model of MSW (Fig. 7.2). This model assumes the following: (1) Creep- and biochemical-degradation-induced secondary compression can be combined, with the latter being predominant and represented by a first-order kinetics model, as indicated in Eq. (7.3). (2) The biochemical degradation rate of MSW is independent of the stress level. (3) The total compression strain of MSW at time t is independent of its stress history and is exclusively related to the stress level and degradation duration (degree of degradation) at time t (Fig. 7.2b).

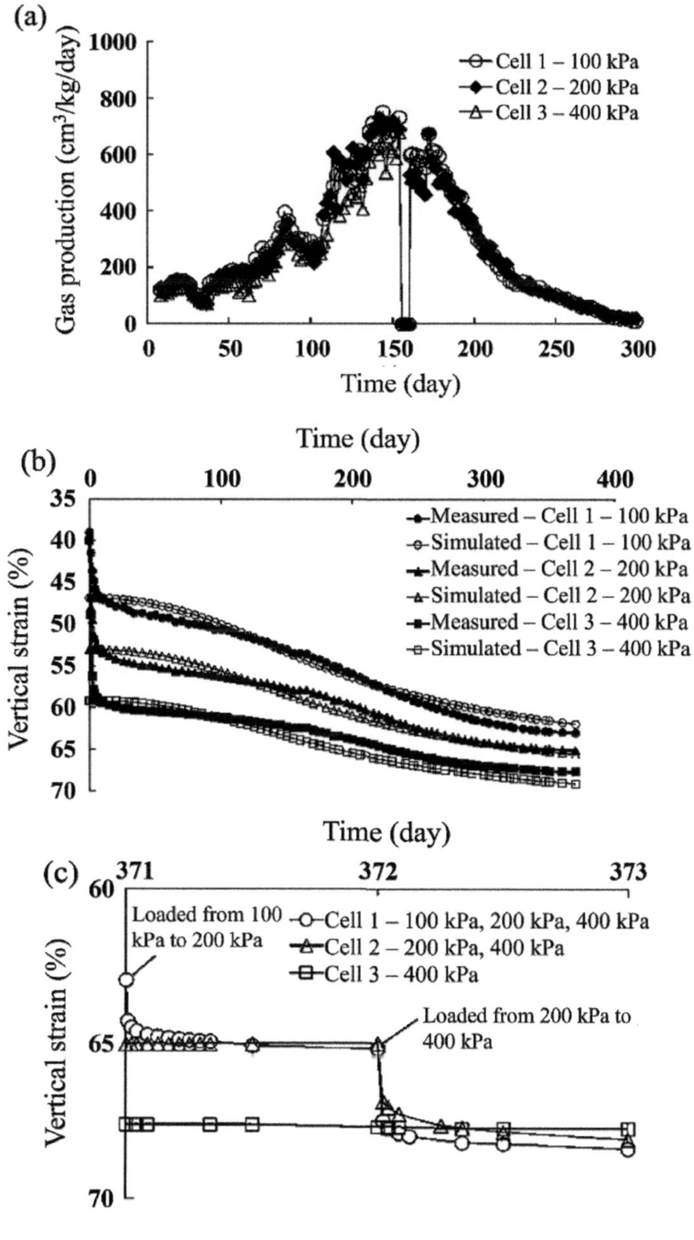

Fig. 7.1 Results of long-term degradation compression tests on MSW under different constant stresses: **a** measured daily gas production from waste, **b** long-term compression under constant stress, and **c** multi-stage compression of decomposed waste (reprinted from Chen 2014, with permission from the Chinese Journal of Geotechnical Engineering)

$$\varepsilon = \varepsilon_p(\sigma) + \varepsilon_s(\sigma, t) \tag{7.1}$$

$$\varepsilon_p(\sigma) = C_c{}'\log\frac{\sigma}{\sigma_0} \tag{7.2}$$

$$\varepsilon_s(\sigma, t) = \varepsilon_{s\infty}(\sigma)\left[1 - \exp^{-c_s(t-t_p)}\right] \tag{7.3}$$

$$\varepsilon_{s\infty}(\sigma) = \varepsilon_{s\infty}(\sigma_0) + (C_\infty{}' - C_c{}')\log\frac{\sigma}{\sigma_0} \tag{7.4}$$

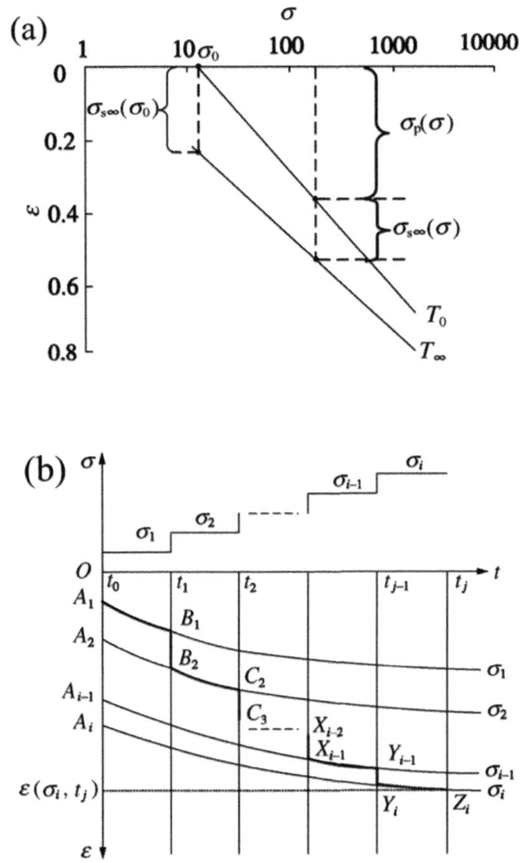

Fig. 7.2 One-dimensional compression model for MSW: **a** determination of parameters of the compression model and **b** working principle of the compression model (reprinted from Chen 2014, with permission from the Chinese Journal of Geotechnical Engineering)

where ε denotes the total compression strain; ε_p represents the primary compression strain; and C_c' and C_∞' denote the modified primary compression coefficients of fresh and completely degraded MSW, respectively, defined as the slopes of the compression curves depicted in Fig. 7.2a. Furthermore, σ_0 denotes the preconsolidation stress (kPa), defined as the stress coordinate intersection of the primary compression curve of fresh MSW with the corresponding zero compression strain; ε_s represents the secondary compression strain; and $\varepsilon_{s\infty}$ denotes the final secondary compression strain, derived from the difference between the primary compression curves of fresh and completely degraded MSW, as illustrated in Fig. 7.2b. Moreover, c_s represents the secondary compression strain rate coefficient (s^{-1}), related to the biochemical degradation rate, which must be calibrated using long-term settlement monitoring data on-site; and t_P denotes the time for primary compression completion (s).

7.2.2 Landfill Settlement Calculation

Landfill settlement is typically calculated using the layer-wise summation method:

$$S = \sum_{i=1}^{n_w} \left(S_{pi} + S_{si}\right) \tag{7.5}$$

where S represents the settlement of the MSW body (m), and n_w denotes the total number of MSW layers, with thicknesses of 2–5 m. Notably, the liquid level within the MSW body must be considered as the interface between these layers. Additionally, S_{pi} denotes the primary compression of the i th MSW layer (m), and S_{si} represents the secondary compression of the i th MSW layer (m).

The primary compression of an MSW layer can be computed as follows:

$$S_{pi} = H_i \frac{C_c}{1 + e_0} log\left(\frac{\sigma_i}{\sigma_0}\right) \tag{7.6}$$

where H_i denotes the initial thickness of the i th MSW layer without considering compression (m); σ_0 represents the preconsolidation stress of the MSW layer (kPa), determined through experiments; and σ_i denotes the effective stress exerted by overlying layers on the i th MSW layer (kPa), representing the effective self-weight stress of the ith MSW layer and above layers. When the effective stress exerted by overlying layers is less than the preconsolidation stress, the primary compression of the given MSW layer is negligible. C_c in Eq. (7.6) denotes the primary compression coefficient, which can be determined through large-scale indoor compression tests on fresh MSW. However, in the absence of experimental data, C_c can be calculated according to the following equation:

$$C_c = \frac{e_0 - e_1}{\log(1000/\sigma_0)} \tag{7.7}$$

where e_1 denotes the void ratio of MSW under a pressure of 1,000 kPa, typically ranging from 0.8 to 1.2. Notably, HFWC MSW tends to exhibit high e_1 values. Furthermore, e_0 represents the initial void ratio of MSW.

Moving further, the secondary compression of an MSW layer can be computed as follows:

$$S_{si} = H_i \varepsilon_{dc}(\sigma_i)(1 - exp^{-C_s t_{wi}}) \tag{7.8}$$

$$\varepsilon_{dc}(\sigma_i) = \begin{cases} \varepsilon_{dc}(\sigma_0) & \text{when } \sigma_i \leq \sigma_0 \\ \varepsilon_{dc}(\sigma_0) - \frac{C_c - C_{c\infty}}{1+e_0} log\left(\frac{\sigma_i}{\sigma_0}\right) & \text{when } \sigma_i > \sigma_0 \end{cases} \tag{7.9}$$

where $\varepsilon_{dc}(\sigma_i)$ denotes the cumulative compression strain of MSW originating from degradation and creep under long-term overlying stress (σ_i), and $\varepsilon_{dc}(\sigma_0)$ represents the cumulative compression strain of MSW resulting from degradation and creep under long-term preconsolidation stress (σ_0). Per recommendations, these values must be determined through indoor compression tests. However, in the absence of experimental data, values of 20% to 30% can be assumed. For HFWC MSW, higher value is recommended. Moreover, t_{wi} in Eq. (7.8) denotes the landfill age of the i th MSW layer (month).

Using the above one-dimensional settlement method, the calculated settlement at Point B (away from the two-dimensional slope surface in Sect. 5.5.1) aligns closely with the vertical displacement from the degradation-consolidation model. However, at Point A (near the two-dimensional slope surface), the settlement calculation differs from the vertical displacement obtained by the degradation-consolidation model, primarily due to the impact of horizontal displacement. This indicates the one-dimensional settlement method is suitable for estimating settlement in landfills.

7.2.3 Landfill Capacity Calculation

Landfill capacity calculations must consider MSW compression during landfilling, which can be computed either using the (1) settlement method or (2) table estimation method. While the latter approach is suitable for landfill capacity estimation and design, the settlement method is preferable for high-accuracy landfill capacity calculations.

The settlement method can be described as follows:

$$W = \sum_{i=1}^{n_s} \left(A_i \sum_{j=1}^{m_i} \gamma_{0ij} H_{ij}\right) \tag{7.10}$$

$$\sum_{j=1}^{m_i} (H_{ij} - S_{ij}) = D_i \tag{7.11}$$

$$D_i = \frac{V_i'}{A_i} \tag{7.12}$$

$$V' = \sum_{i=1}^{n_s} V_i' \tag{7.13}$$

$$Q_m = \frac{W}{V'} \tag{7.14}$$

where W denotes the landfill capacity (t); n_s represents the total number of divided areas in the landfill; A_i denotes the planar area of the i th region (m^2); m_i represents the total number of MSW layers in the i th area; γ_{0ij} denotes the initial unit weight of the j th MSW layer in the i th area (kN/m^3); D_i indicates the average designed effective filling height of the i th area (m); H_{ij} denotes the initial thickness of the j th MSW layer in the i th area disregarding compression (m); S_{ij} signifies the compression of the j th MSW layer in the i th area when filled to height D_i (m); V_i' denotes the effective geometric capacity of the i th area (m^3); V' represents the effective geometric capacity of the landfill (m^3); and Q_m denotes the average unit landfill capacity (t/m^3).

Meanwhile, the table estimation method is described as follows:

$$W = Q_m V' \tag{7.15}$$

where Q_m denotes the average unit landfill capacity, with typical values outlined in Table 7.1.

Zhejiang University developed the LANDFILL software for analyzing landfill settlement and capacity (Chen et al. 2009). Notably, this software discretizes the problem of landfill settlement and capacity calculation into settlement and capacity analysis in landfill columns through spatial and temporal discretization during the landfilling process. As depicted in Fig. 7.3, spatially, the complete MSW body following the landfill cover is segmented into independent landfill column units with a base area of $A = ab$.. Meanwhile, temporally, the construction process of each landfill column unit (initial thickness of each layer and completion time of filling) is guided by the overall landfilling process.

LANDFILL offers various modes for displaying results. Figure 7.4a illustrates the calculation results of landfill capacity and expansion rate. Meanwhile, Fig. 7.4b displays the settlement results of a specific landfill column, including the temporal elevation change curve of the top surface of the landfill column, a bar chart depicting the initial thickness of each MSW layer within the landfill column, curves displaying the temporal elevation change of the top surface of each MSW layer, and curves illustrating the temporal thickness change of each MSW layer within the landfill column. Furthermore, Fig. 7.4c presents the two-dimensional settlement curve of a specific cross-section of the MSW body, while Fig. 7.4d displays the three-

Table 7.1 Average unit landfill capacity considering MSW compression (t/m^3)

Designed effective filling height D_i (m)	20	40	60	80	100
Accelerated degradation conditions	0.95–1.24	1.06–1.44	1.16–1.54	1.21–1.62	1.26–1.70
General conditions	0.82–1.06	0.93–1.15	1.03–1.24	1.12–1.30	1.21–1.35

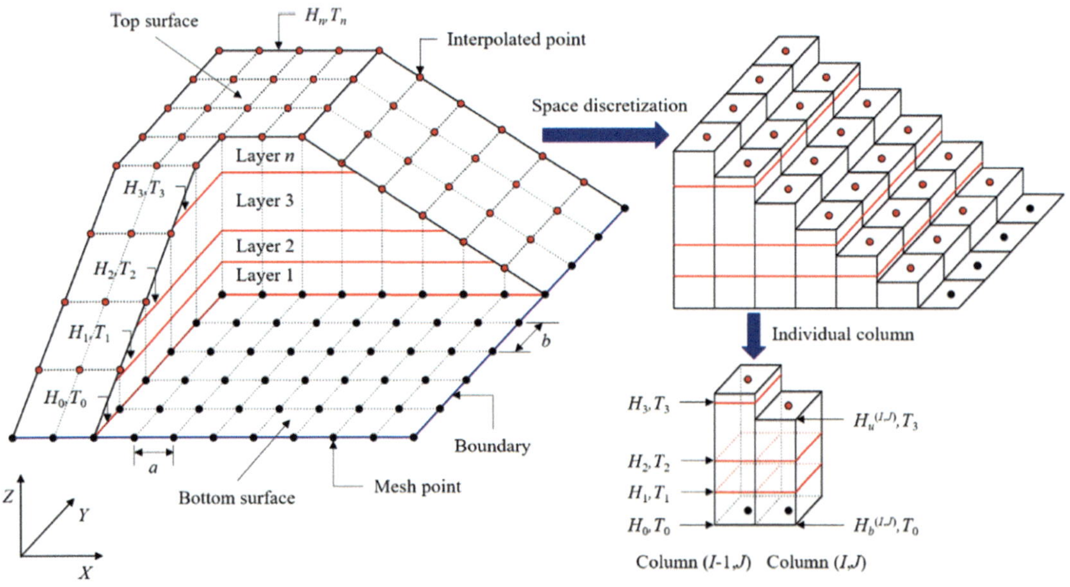

Fig. 7.3 Spatial discretization of the landfilling process (reprinted from Gao et al. 2018, with permission from the American Society of Civil Engineers)

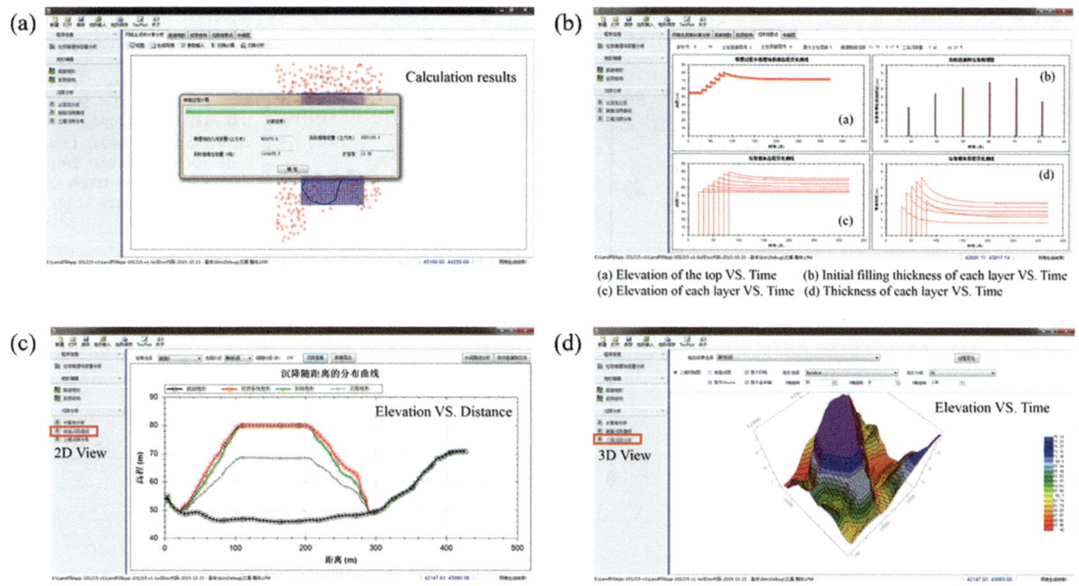

Fig. 7.4 Various result display modes of LANDFILL: **a** landfill capacity and expansion rate, **b** settlement results of a specific landfill column, **c** two-dimensional settlement curve of a specific cross-section of the MSW body, and **d** three-dimensional terrain of the landfill body at various landfilling stages (Gao 2018)

dimensional terrain of the landfill body at various landfill stages, facilitating dynamic observations of the landfilling and terrain deformation processes during settlement progress.

7.3 Landfill Stability

7.3.1 Shear Strength of Municipal Solid Waste

Compared to conventional soils, the triaxial test results and constitutive relationships of MSW (see Sect. 3.6.5) indicate that MSW exhibits strain hardening and aging hardening. Both increasing strain and age significantly impact the shear strength of MSW.

The strain hardening characteristics of MSW are known to vary, complicating the identification of a clear yield strain point. Hence, when conducting stability analysis of landfill bodies, shear strength indices appropriately corresponding to the strain are often selected. Figure 7.5 illustrates the relationship between different shear strength indices of MSW and strain. As evident, with increasing strain, the cohesion (c) and friction angle (φ) of HFWC MSW tend to increase. This trend is however less apparent for LFWC MSW. Corroborating the above results, Singh et al. (2009) reported that increasing strain improves the shear strength indices of MSW. Furthermore, Zhan et al. (2008) demonstrated that the strength indices obtained using 10% strain as the failure criterion are suitable for the static stability analysis of landfills, as verified by projects such as the Suzhou Qizishan Landfill and Shenzhen Xiaping Landfill. Hence, for static stability verification, shear strength indices corresponding to 10% strain are recommended. Conversely, for seismic stability verification, shear strength indices corresponding to 15% strain are recommended (Li 2018).

As MSW undergoes biochemical degradation, its composition and material properties change, affecting its shear strength. Figure 7.6 depicts the relationship between the shear strength indices of MSW and corresponding fill age. Evidently, with increasing fill age, the friction angle of HFWC MSW increases, whereas its cohesion decreases. However, the shear strength indices of

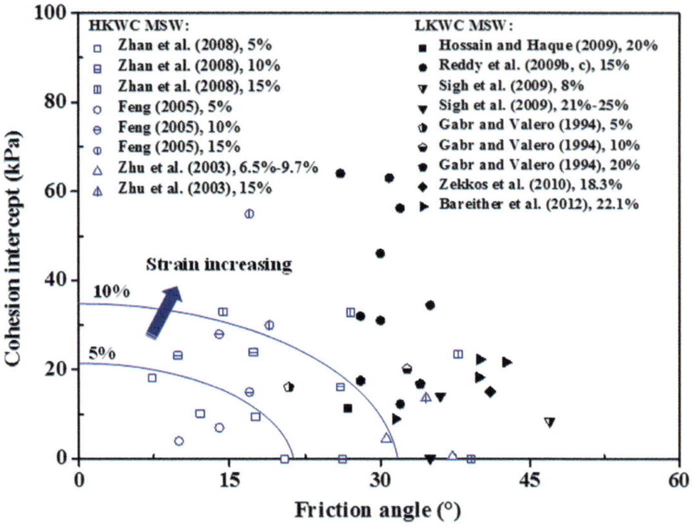

Fig. 7.5 Variations in the shear strength indices of MSW with strain (Gao et al. 2015)

LFWC MSW do not follow this trend. For instance, Reddy et al. (2009) conducted undrained consolidation tests on synthetic MSW. Their results revealed that with biochemical degradation, the cohesion of synthetic MSW increased slightly, whereas the friction angle decreased slightly. Hence, the influence of biochemical degradation on the shear strength of MSW varies across different MSW types.

Based on the test data in Figs. 7.5 and 7.6, reference values for the shear strength indices of MSW are proposed, as listed in Table 7.2, and are adopted by the "Technical Code for Geotechnical Engineering of Municipal Solid Waste Sanitary Landfill" (CJJ 176–2012).

7.3.2 Instability Modes

The intricate forms of leachate and LFG within landfill sites, coupled with the high pressures and multiple weak interfaces at bottom barriers, contribute toward landfill instability. Field monitoring and seepage analysis reveal three primary

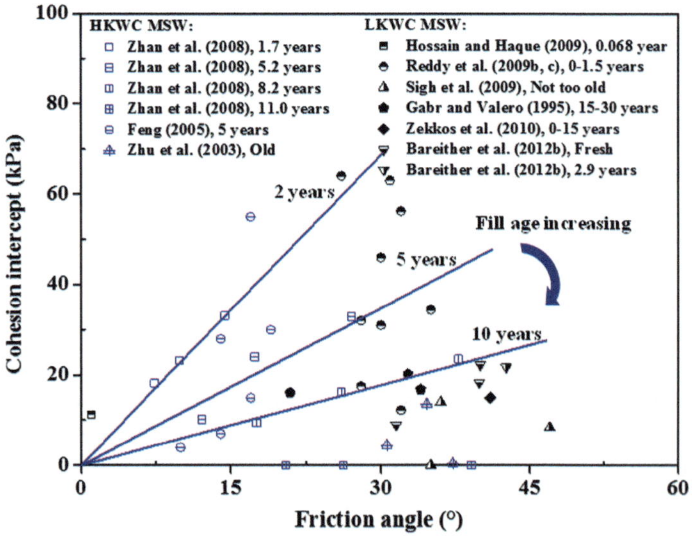

Fig. 7.6 Variation in the shear strength indices of MSW with fill age (Gao et al. 2015)

Table 7.2 Reference values for the shear strength indices of MSW

MSW	Effective friction angle (°)	Effective cohesion (kPa)
Shallow MSW (burial depth is less than 10 m)	12–25	15–30
Deep MSW (burial depth is more than 10 m)	25–33	0–10

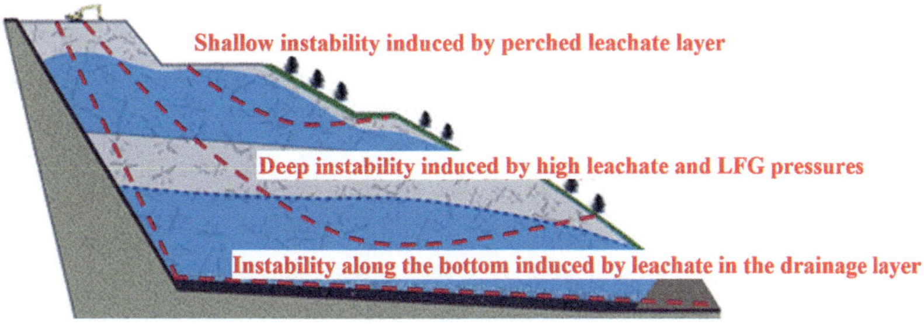

Fig. 7.7 Three primary modes of landfill instability

forms of leachate in landfills featuring high leachate levels: main leachate layer, local perched leachate layer, and leachate within the drainage layer. Considering the strength properties of MSW and the softening characteristics of barrier interfaces, unsaturated seepage simulations and slope stability analyses reveal three primary modes of landfill instability triggered by leachates (as illustrated in Fig. 7.7): (1) Elevated leachate levels and LFG pressures triggering deep global instability of the MSW body, (2) the perched leachate layer inducing shallow local instability, and (3) interactions between leachate in the drainage layer and the main leachate layer prompting global instability along the barrier interface.

7.3.3 Critical Leachate Level

Extensive analyses of engineering accidents have revealed that an increase in the main leachate level leads to a notable decline in landfill slope stability. A stability analysis model for a specific landfill is illustrated in Fig. 7.8. Here, h denotes the height difference between the main leachate level and the crest of the landfill dam, H represents the height difference between the highest point on the landfill slope and the crest of the landfill dam, and h/H denotes the relative position of the main leachate level.

For instance, an H value of 60 m corresponds to a landfill slope with an effective cohesion of 5 kPa, an effective internal friction angle of 28°, and a slope gradient of 1:3.5 (vertical versus horizontal). Figure 7.9 illustrates the impact of rising main leachate levels on landfill stability safety factors. Notably, as the relative position of the main leachate level (h/H) rises, the safety factor significantly decreases, reaching a value of 0.6. This value corresponds to the stability safety factor of 1.35 under normal operating conditions, marking a critical level. Notably, this critical level varies across different landfill strengths, slope heights, and slope inclinations.

Shallow layer slipping is predominantly influenced by the perched leachate level, while deep layer slipping is primarily affected by the main leachate level. As landfill height increases, the depth of the critical level for perched leachate remains relatively constant, whereas that for the main leachate varies significantly.

The use of cohesive soil as an intermediate cover layer in landfills often results in the formation of perched leachate levels. For instance, perched leachate levels have been observed in diverse landfills such as the Suzhou Qizishan Landfill. Incidents of shallow local instability

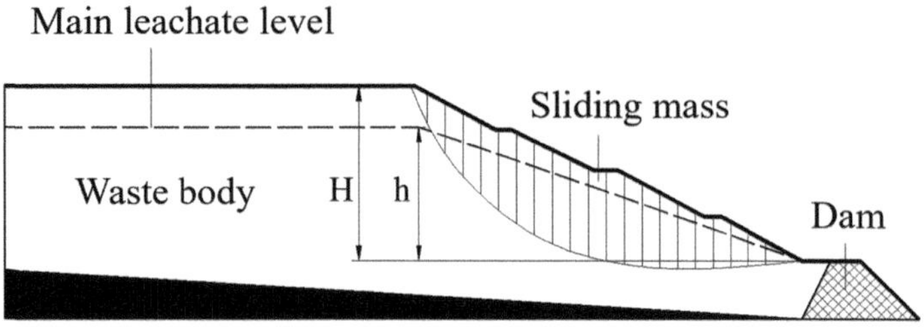

Fig. 7.8 Analytical model diagram of the main leachate level along a landfill slope

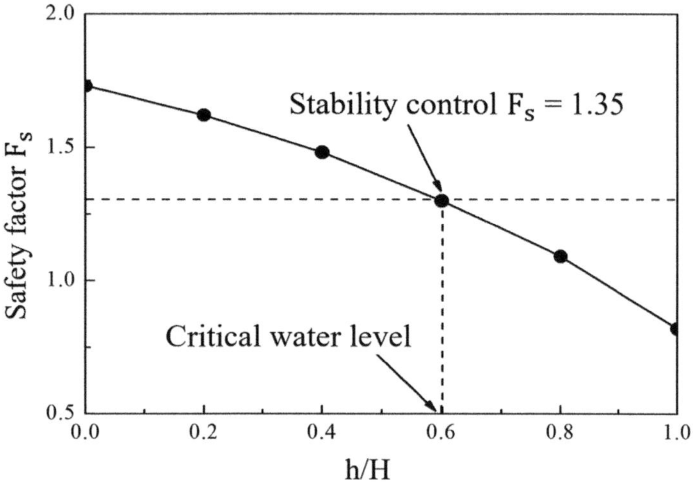

Fig. 7.9 Impact of rising main leachate level on the landfill slope safety factor

triggered by perched leachate levels take place frequently. The relevant analysis model diagram is depicted in Fig. 7.10.

7.3.4 Critical Landfill Gas Pressure

Theoretical analysis in Sect. 5.5 indicates that high gas pressure in the waste body is also a significant factor leading to landfill instability.

Owing to the generally high leachate levels within Chinese landfills, LFG accumulated below the leachate level can exert significant LFG pressures, adversely affecting the stability of landfill slopes. Field monitoring and large-scale model tests at the Xingfeng Landfill in Guangzhou have revealed that LFG accumulated below the leachate level exists as closed bubbles isolated from the atmosphere.

In November 2017, a slope instability incident occurred at the Xingfeng Landfill in Guangzhou. Ma et al. (2019) observed that LFG pressures at the sliding surface ranged from 175 kPa to 210 kPa, aligning closely with the self-weight

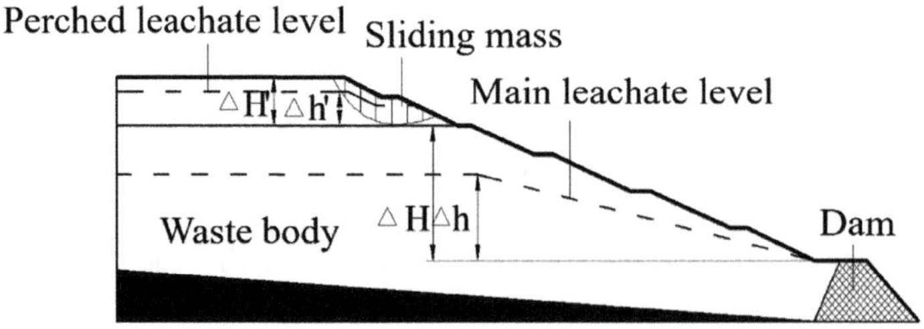

Fig. 7.10 Analysis model diagram for coexisting main and perched leachate levels along a landfill slope

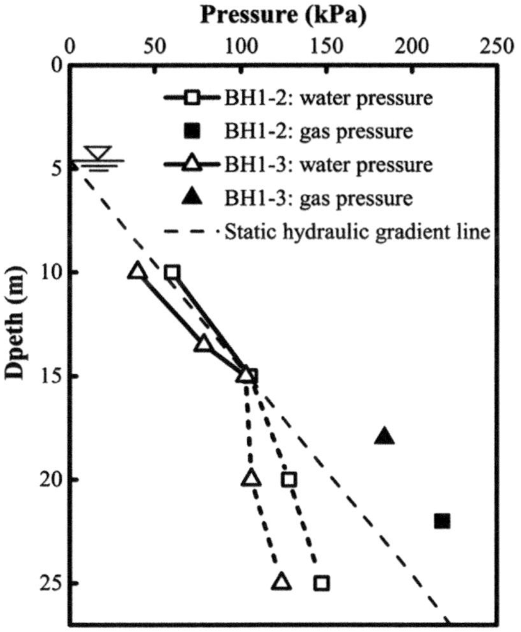

Fig. 7.11 Pore water pressures and pore gas pressures within the waste body at the 135-m platform of Xingfeng Landfill in Guangzhou (reprinted from Ma et al. 2019, with permission from Elsevier)

stresses of 170 kPa to 204 kPa of the overlying waste, as depicted in Fig. 7.11. This suggested that the effective stress at this location approached zero, indicating that the landslide was triggered by high LFG pressures.

To delve deeper into the landslide mechanism, stability analysis was conducted on the central profile of the sliding area using GeoStudio software. In this analysis, the pore pressure field within the waste body was established using the spatial function feature. Furthermore, the unit weight, internal friction angle, and cohesion of MSW were 11 kN/m^3, 14°, and 32 kPa, respectively. In total, two scenarios were considered: Case 1 utilized the measured high gas pressure to establish the pore pressure field, while Case 2 disregarded the influence of high gas pressure and used only the measured pore pressure readings obtained from pore pressure sensors.

The results of this stability analysis, as depicted in Fig. 7.12, indicated that disregarding the influence of high gas pressure (Case 2), the safety factor was 1.151. However, when considering its influence (Case 1), the safety factor decreased to 1.019, and the automatically identified sliding surface closely resembled the actual sliding surface. In this situation, the waste body approached the critical state, rendering it highly susceptible to instability.

Based on the foregoing field monitoring and slope stability analysis results, the corresponding estimated LFG pressures are approximately 69, 74, and 78% of the self-weight stress of overlying waste. Furthermore, to account for monitoring errors and ensure engineering safety, a certain reduction is applied to the above estimates. Specifically, as indicated in Table 7.3, the critical LFG pressures for safety grades I, II, and III landfill slopes are set at 60, 65, and 70% of the self-weight stress of overlying waste, respectively. The LFG collection efficiency must be maximized to effectively reduce pore gas pressures.

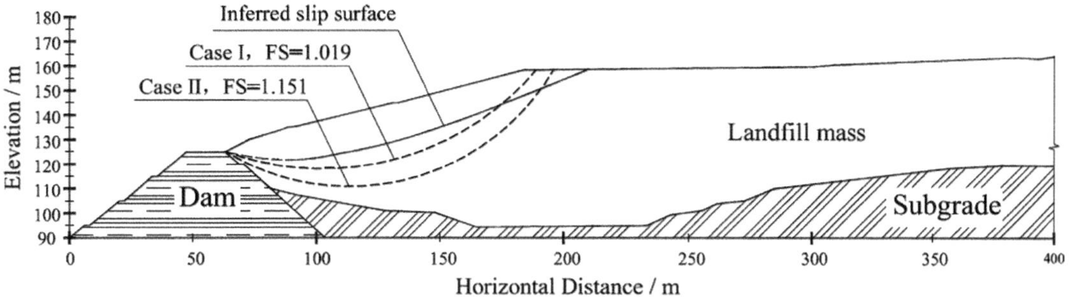

Fig. 7.12 Results of the slope stability analysis conducted on the Xingfeng Landfill in Guangzhou (reprinted from Ma et al. 2019, with permission from Elsevier)

Table 7.3 Landfill gas pressure and collection efficiency requirements

Safety grade	Control requirements
I	LFG pressure < $0.60\rho g h_a$ and LFG collection efficiency > 85%
II	LFG pressure < $0.65\rho g h_a$ and LFG collection efficiency > 80%
III	LFG pressure < $0.70\rho g h_a$ and LFG collection efficiency > 70%

Note h_a denotes the burial depth of the monitoring point (m); ρ represents the wet density of the overlying waste at the monitoring point; and g denotes acceleration due to gravity

7.4 Development of High Gas Pressure in Deep Layer and Its Effect on Slope Stability

As MSW is layered landfilled, the VFAs rapidly generated through the hydrolysis of fresh waste in the shallow layer flow downward with the leachate into deep layers. Then, the VFAs within deep layers undergo methanogenesis, producing a significant amount of LFG, thereby increasing gas pressure in deep layer. Hu et al. (2024a) adopted the degradation-consolidation model proposed in Chapter 5 to simulate this process. Based on the on-site measurement results of the Xingfeng Landfill in Sect. 7.3.5, this section further analyzes the development of high gas pressure in deep layer and its effect on slope stability.

7.4.1 High Gas Pressure Development in Deep Layer During Multi-stage Landfilling

As depicted in Fig. 7.13, after the deposition of the new waste layer, the VFA concentration at the bottom begins to rise. The high concentrations of VFAs in the deep layer are transformed into LFG, thereby increasing gas production in deep layer. Figure 7.14 illustrates the distributions of cumulative gas generation and gas saturation along the landfill height during multi-stage landfilling. Notably, owing to the downward flow of VFAs generated at the top, the cumulative gas generation within deep layers significantly exceeds that within shallow layers. For instance, on Day 150, the cumulative gas generation amounts at heights of 30 m and 0 m

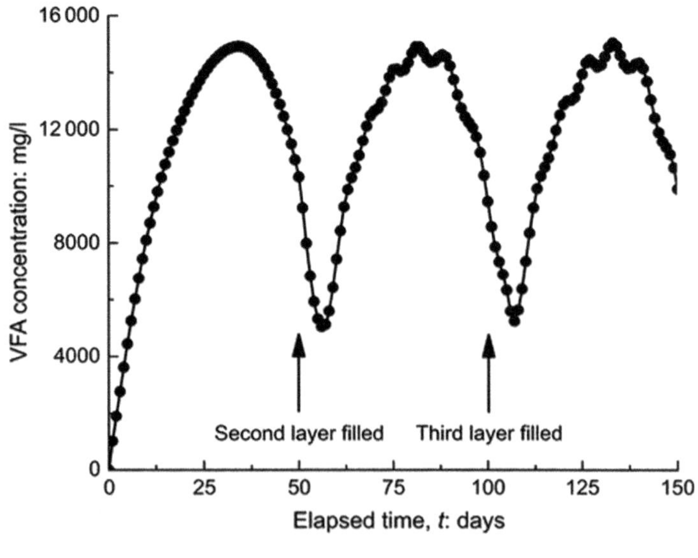

Fig. 7.13 Variation in the VFA concentration within deep layers during multi-stage landfilling (Hu et al. 2024a)

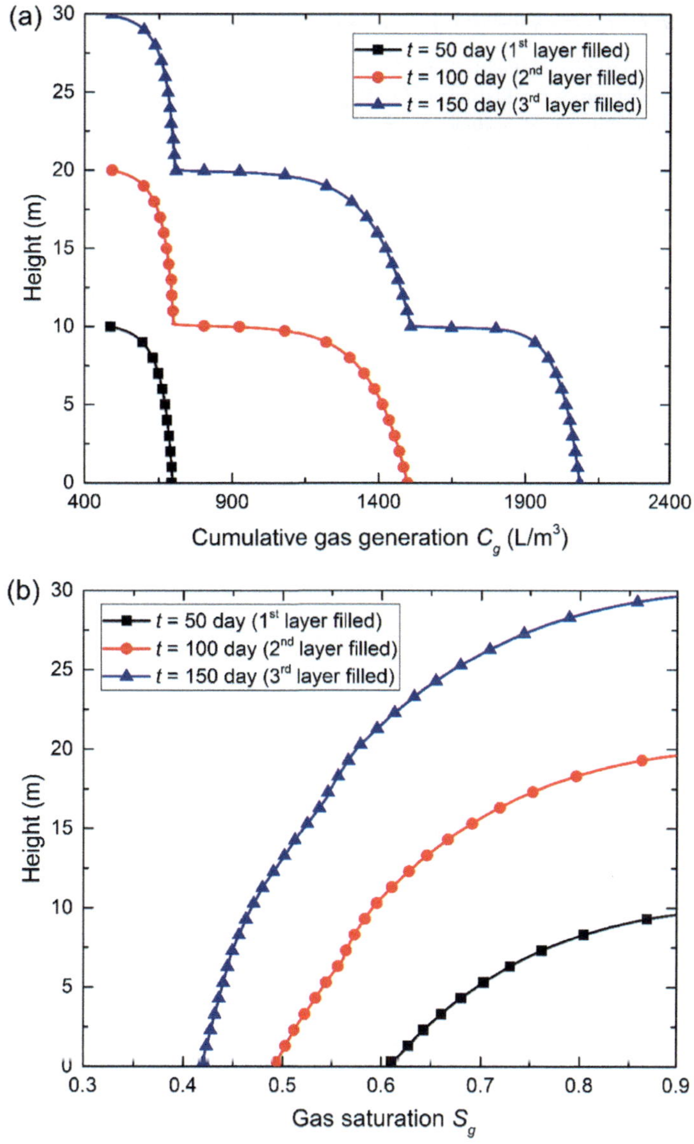

Fig. 7.14 Distributions of **a** cumulative gas generation and **b** gas saturation along landfill height during multi-stage landfilling (Hu et al. 2024a)

are 490 L/m^3 and 2,085 L/m^3, respectively. Meanwhile, the gas saturation at the top is 0.9, and as the landfill height decreases to 0 m, the gas saturation on Day 150 exponentially decreases to 0.42. However, the hydrolysis of newly deposited waste produces a substantial amount of leachate, causing the gas saturation to decrease over time. Accordingly, by Day 150, substantial volume of gas is compressed into a smaller pore space, resulting in the buildup of high gas pressure within deep layers.

Figure 7.15 depicts the two-dimensional distributions of gas pressure during the multi-stage landfilling process. Notably, the first, second, and third layers illustrated in this figure were filled on Days 0, 50, and 100, respectively. At 50, 100, and 150 days, the maximum gas pressures within deep layers reached 37 kPa, 104 kPa, and 185 kPa, respectively, demonstrating a consistent increase over time.

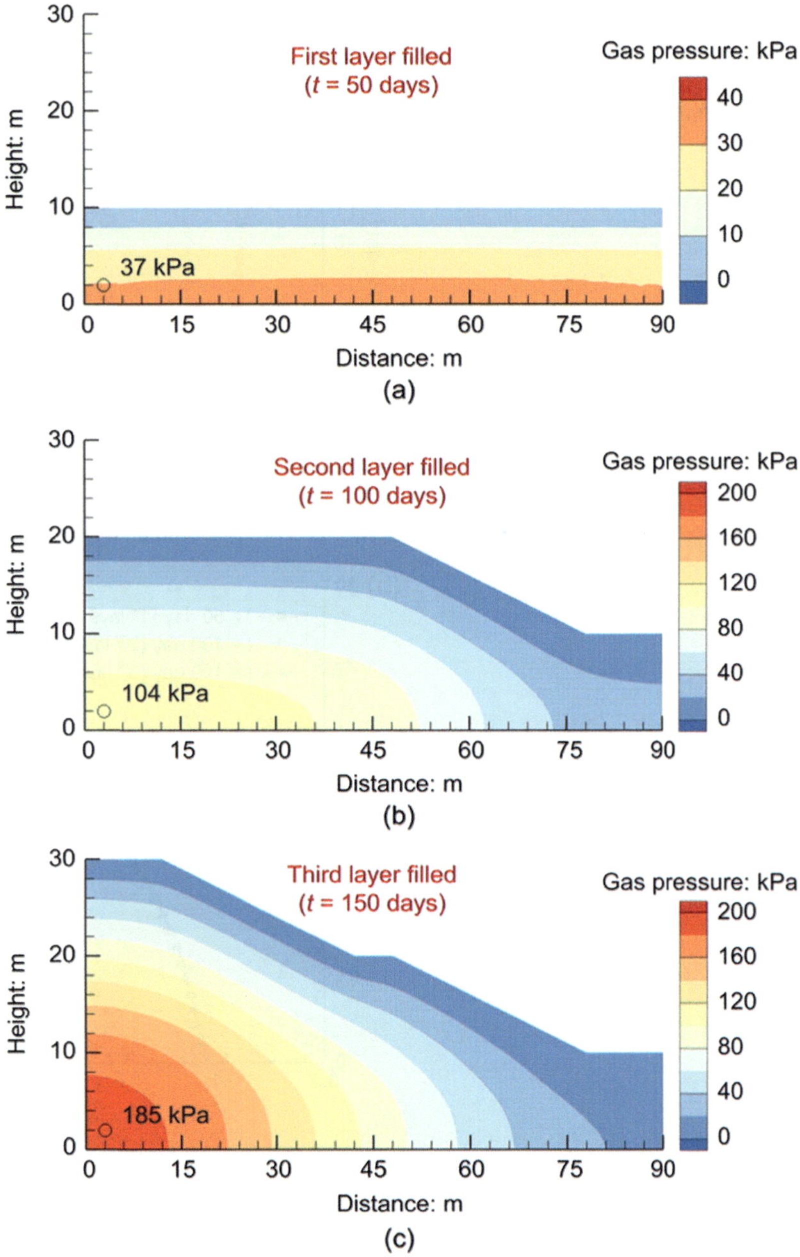

Fig. 7.15 Two-dimensional distributions of gas pressure during multi-stage landfilling: **a** deposition of the first layer, **b** deposition of the second layer, and **c** deposition of the third layer (Hu et al. 2024a)

7.4.2 Variation of Safety of Factor During Multi-Stage Landfilling

The landslide at the Xingfeng Landfill is an instance of a typical unsaturated slope failure triggered by the combined influence of elevated gas and water pressures. As indicated in Eq. (7.16), the shear strength of unsaturated waste can be described using two stress state variables, namely $\sigma - P_g$ and $P_g - P_w$ (Fredlund et al. 1978). The value of the matric suction friction angle (φ^b) is not constant but varies with capillary pressure. Specifically, when the capillary pressure is lower than the entry value, φ^b is approximately equal to φ'. Conversely, when the capillary pressure exceeds the entry value, φ^b decreases with increasing capillary pressure (Vanapalli et al. 1996). The exact relationship between φ^b and matric suction is described in Eq. (7.17).

$$\tau = c' + (\sigma - P_g)\tan\varphi\prime + (P_g - P_w)\tan\varphi^b \quad (7.16)$$

$$\tan\varphi^b = \frac{S_w - S_r}{S_s - S_r}\tan\varphi' \quad (7.17)$$

where P_g denotes the gas pressure (kPa), P_g represents the liquid pressure (kPa), τ denotes the shear strength (kPa), σ signifies the normal stress (kPa), c' represents the effective cohesion (kPa), φ' denotes the effective angle of internal friction (°), φ^b represents the angle related to capillary pressure (°), S_w denotes the saturation of liquid, S_r represents residual liquid saturation, and S_s denotes maximum liquid saturation.

Using Eqs. (7.16) and (7.17), the shear strength of MSW can be calculated, using bulk density, effective friction angle, and effective cohesion values of 12 kN/m^3, 14°, and 32 kPa, respectively. Figure 7.16b depicts the variation in waste shear strength with capillary pressure, where the slope of the curve represents $\tan\varphi^b$. Notably, at a liquid phase saturation of 0.5 within the Xingfeng Landfill, $\tan\varphi^b$ is only 0.19 times $\tan\varphi'$.. Under saturated conditions (i.e., P_c = 0 kPa), the shear strength is 42 kPa. Furthermore, as the capillary pressure increases to 80 kPa, the shear strength only increases to 45 kPa.

The rate of increase in shear strength relative to capillary pressure is described by φ^b. As depicted in Fig. 7.16b, the increase in capillary pressure exerts a weaker effect on the shear strength of MSW compared to that of soil. For instance, when the capillary pressure increases from 0 kPa to 200 kPa, the shear strength of MSW only increases from 42 kPa to 47 kPa, whereas that of soil increases from 11 kPa to 46 kPa. This is because, under a constant capillary pressure, φ^b increases with the increase in entry value. As illustrated in Fig. 7.16a, owing to the presence of smaller pores within soil, its entry value significantly exceeds that of MSW. Hence, the φ^b value of soil is greater, leading to a significant increase in its shear strength with capillary pressure. Conversely, MSW has a lower φ^b value owing to its smaller entry value, resulting in a relatively weaker influence of capillary pressure on shear strength.

Figure 7.17a, b respectively display the development of pore gas and pore water pressures at a landfill depth of 20 m. Notably, as new layers of waste are deposited, both the pore gas and pore water pressures increase, and their calculated values closely align with the measured ones. Upon deriving the pore pressure–time curve, Eqs. (7.16) and (7.17) can be utilized to obtain the shear strength–time curve. Figure 7.17c illustrates the temporal variation of shear strength. For instance, by April 2016, the shear strength reaches 95 kPa, indicating slope stability. Subsequently, owing to the consistent increase in gas pressure, the shear strength gradually decreases over time. By November 2017, the shear strength ultimately drops to a critical value of 45 kPa, triggering the landslide. Figure 7.17d depicts the evolution of the landfill stability safety factor during the landfilling process. During April 2016, the safety factor has a value of 2.075. By December 2017, this value steadily declines to 0.997, indicating the critical sliding state of the landfill site. The calculated sliding time agrees with the actual sliding time.

7.5 Centrifugal Hypergravity Modeling of Landfill Failures Induced by Rising Water Levels

Centrifugal hypergravity modeling has significant advantages in simulating slope stability issues. The University of Cambridge conducted centrifugal hypergravity studies on dry landfill slope instability (Thusyanthan et al. 2006). Chen et al. (2017) performed centrifugal hypergravity tests on the instability of high-water-level landfill slopes, examining the critical water levels for

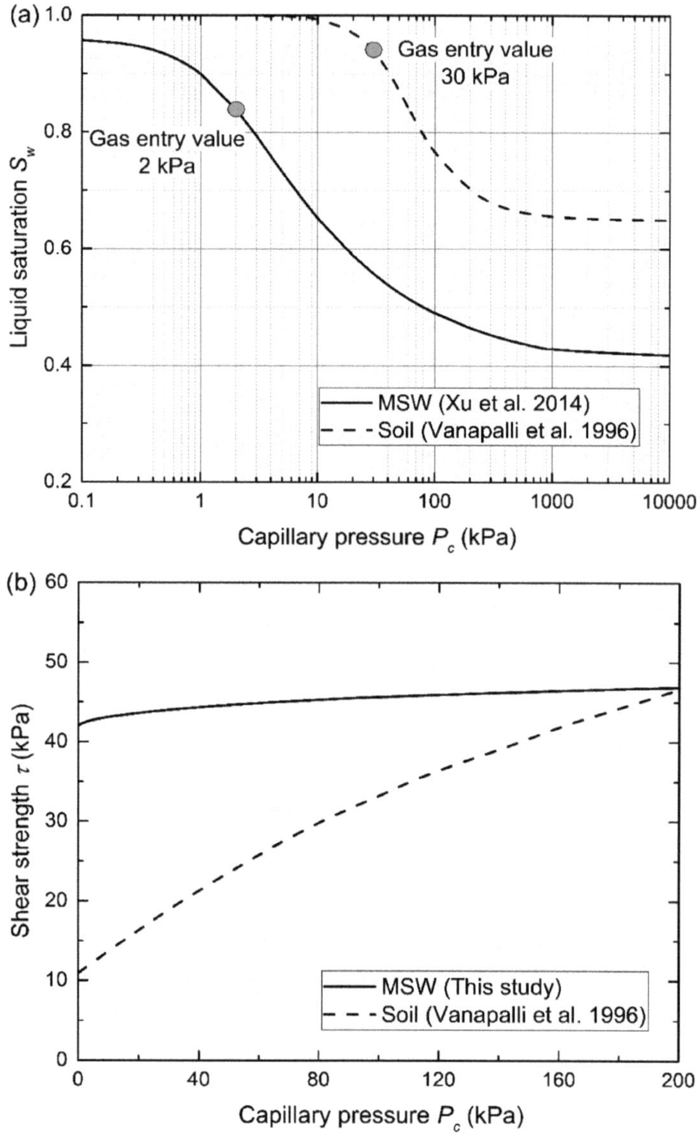

Fig. 7.16 Effect of capillary pressure on the saturation and shear strength of MSW and soil: **a** water retention curve and **b** shear strength curve (Hu et al. 2024a)

landfill instability and analyzing several typical failure modes of landfills with such high water levels. Furthermore, Li et al. (2024) carried out centrifugal hypergravity modeling to explore the seismic failure of landfill slopes with high water levels. Hu et al. (2024b) investigated landfill slope stability under high gas pressure using centrifugal hypergravity tests. This section primarily introduces centrifugal hypergravity modeling on landfill slope instability caused by rising water levels.

7.5.1 Test Setup

Figure 7.18 depicts two typical experimental model landfill bodies, one with an engineered berm and one without. In the model with an engineered berm, a GM covered the surface of the berm. During modeling, synthetic MSW was uniformly spread in 0.05-m-thick layers on the GM, and each MSW layer was compacted to the designed bulk density. The final height of the model MSW landfill body reached 0.3 m.

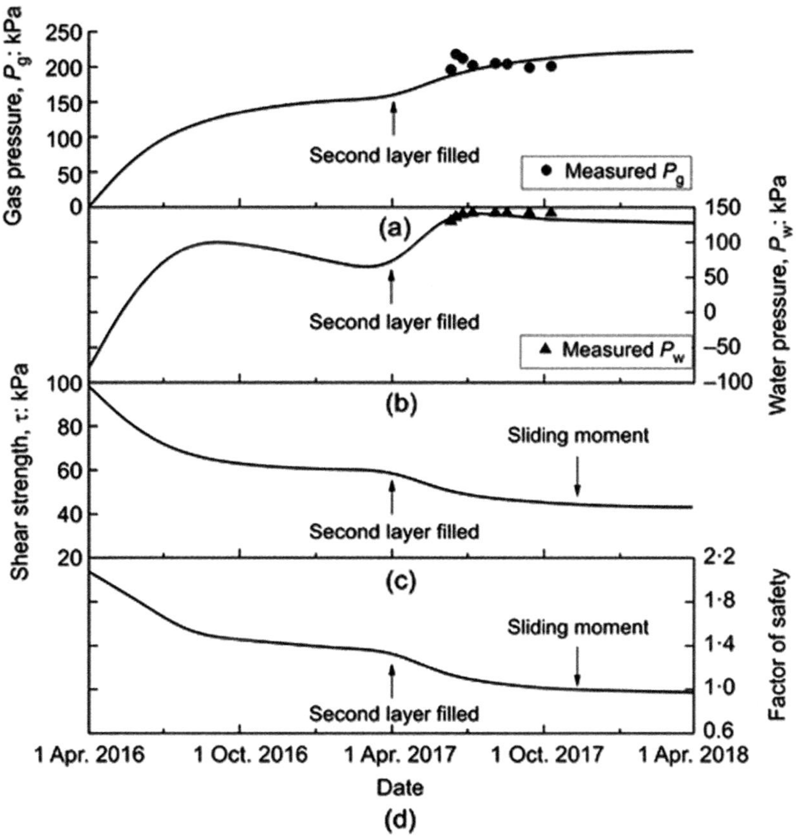

Fig. 7.17 Temporal variation of **a** pore gas pressure, **b** pore water pressure, **c** shear strength, and **d** landfill stability safety factor at a depth of 20 m (Hu et al. 2024a)

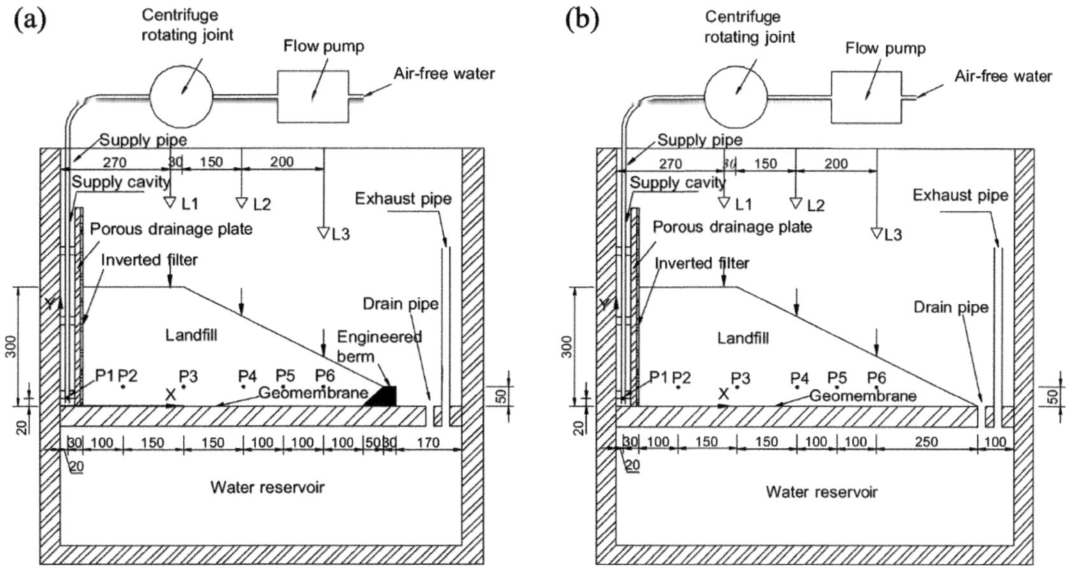

Fig. 7.18 Layout of the test setups for landfill instability (unit: mm): **a** with an engineered berm and **b** without an engineered berm (reprinted from Chen et al., 2017, with permission from Canadian Science Publishing)

Table 7.4 Test setups and critical water levels for the model tests (reprinted from Chen et al. 2017, with permission from Canadian Science Publishing)

No	Slope angle	MSW type	Engineered berm	H_L (cm)	h_{max}/H_L
Test 1	1:1	Fresh	No	24.11	0.75
Test 2	1:1	Partly degraded	Yes	25.28	0.81
Test 3	1:2	Fresh	No	24.01	0.79
Test 4	1:2	Fresh	Yes	24.16	0.80
Test 5	1:2	Partly degraded	Yes	25.11	0.85
Test 6	1:3	Fresh	No	24.18	0.84
Test 7	1:3	Fresh	Yes	24.22	0.87
Test 8	1:3	Partly degraded	Yes	24.86	0.92

Subsequently, the designed slope was meticulously sculpted on the MSW body using the sloping method.

A water level control system, equipped with type BT-300E/153Y flow pump capable of regulating flow in the range of 0.001–2.24 L/min in real time, was designed. The water level within the supply cavity was controlled by the flow pump, and during the experiment, water from the cavity infiltrated the landfill through a porous drainage board. A nonwoven geotextile filter layer (400 g/m^2) prevented local erosion of MSW due to high flow rates, and overflow water from the slope toe was transported to a water reservoir through drainage pipes. Each test setup included six miniature pore water pressure sensors (PDCR81-3G-10), three laser displacement transducers (CP08MHT80), two mini cameras, and one digital camera.

Furthermore, marker points were placed at intervals of 0.045 m on the inner surface of the organic glass side of the model box for image vector analysis. In addition to these, markers were placed on the side of the landfill body adjacent to the glass plate to monitor large deformations and failure modes. Throughout the experiment, high-definition photographs of the landfill model were captured through the organic glass observation window for deformation analysis.

In total, eight model tests were conducted to elucidate the evolution and occurrence mechanism of landfill instability triggered by rising water levels, as summarized in Table 7.4. This analysis focused on the influence of age on landfill stability using synthetic MSW subjected to varying degrees of degradation. Additionally, the effects of the slope gradient and engineered berm on landfill stability were examined.

Once the model was constructed, the centrifuge was set to 66.7 g, and water supply commenced when the settlement of the landfill stabilized. During the experiment, the water level within the supply cavity was raised in four stages—0.05 m, 0.1 m, 0.15 m, and 0.2 m—where the transition between consecutive stages was completed within 1 min at a constant flow rate of 1 L/min.

After reaching the designated level for each stage, the centrifuge continued its operation for an additional 20 min. To maintain the water level within the supply cavity at the designated height during this period, the water flow rates were maintained at 0.06 L/min, 0.1 L/min, 0.14 L/min, and 0.18 L/min, respectively. The simulation outcomes indicated a rise of nearly 3.3 m in the water level over 3 days, followed by a constant level over the next two months. These results aligned with the conditions observed during the heavy rainfall event at the Shenzhen Xiaping Landfill, as reported by Chen et al. (2010b). Based on this simulation, the test replicated the process of landfill slope instability triggered by rapid increments in water level during severe rainfall events. Here, the centrifuge was stopped once significant instability occurred.

7.5.2 Material Selection and Synthesis

7.5.2.1 Synthesis of Synthetic MSWs

According to the synthesis method proposed by Thusyanthan et al. (2006), peat, kaolin clay, and silica sand are selected as the components of synthetic MSW. Notably, the peat, sourced from the Changbai Mountain area of China, is formed through the incomplete decomposition of marsh plants in the presence of abundant water and no air, and it contains substantial amounts of organic matter. Furthermore, its moisture content is approximately 85%, while its specific gravity (D_s) is 1.04. Moreover, it contains fibers and humus of various sizes. Meanwhile, kaolin clay exhibits good plasticity and cohesion, with an average particle size (D_{50}) and specific gravity (D_s) of 0.012 mm and 2.68, respectively. Its plastic and liquid limits are 25.5% and 43.8%, respectively. The silica sand is standard sand from Fujian, China, with an average particle size (D_{50}) of 0.17 mm and a specific gravity (D_s) of 2.622. Its maximum and minimum porosities are 0.952 and 0.607, respectively, while its critical state friction angle is 35°.

During the preparation of synthetic MSW, different sequences were tested for their effect on the uniformity of the synthetic MSW mixture. After repeated attempts, the optimal preparation sequence ensuring both uniformity and stability of characteristics was selected, as depicted in Fig. 7.19. The primary steps included the following: (1) The silica sand and kaolin clay samples were first dried. (2) Subsequently, a certain amount of silica sand was weighed. Following this, its moisture content was adjusted by adding water, and the amount of added water was recorded. Thereafter, the sand mixture was sealed and allowed to mix thoroughly. (3) The moist silica sand was then spread evenly. Subsequently, an equal amount of dried kaolin clay was weighed, uniformly sprinkled over the silica sand, and mixed thoroughly. (4) The moisture

Fig. 7.19 Process of preparing synthetic MSW (Li 2018)

content of the peat sample was measured. Given that the peat used in this study had a low moisture content, preventing it from thoroughly mixing with the silica sand and kaolin clay, its moisture content was adjusted by weighing out the corresponding mass of peat. Following this, the amount of water added was recorded, and large particles were crushed for ease of sampling. (5) The mixture containing silica sand and kaolin clay was spread evenly over the peat and mixed thoroughly. (6) The required amount of water was calculated based on the target moisture content of MSW and the amount of water added during the sampling process. Subsequently, the moisture content of the final mixture was adjusted accordingly, following which the mixture was allowed to stand, ultimately yielding the synthetic MSW required for the experiment.

Considering the influence of the organic fiber content of peat on the properties of synthetic MSW, MSW from the Suzhou Qizishan Landfill was compared with synthetic MSW featuring different composition ratios. In particular, two sand-to-clay-peat ratios—1:0.1:0.4 and 1:0.2:0.8—were selected for the synthetic MSW, resulting in organic fiber contents of 34.3% and 25.75%, respectively. These fiber contents closely matched the organic content of MSW sourced from the Suzhou Qizishan Landfill. Table 7.5 summarizes the basic physical properties of both MSW samples, including their bulk density, moisture content, and void ratio. As indicated, the bulk density of the synthetic MSW aligns with the bulk density distribution of the MSW sourced from the Suzhou Qizishan Landfill. Furthermore, the void ratio of the synthetic MSW approximates the optimal curve of the measured void ratio distribution of the MSW from the Suzhou Qizishan Landfill.

Figure 7.20 compares the saturated permeability coefficients of synthetic MSW and real MSW from the Suzhou Qizishan Landfill, derived from on-site MSW permeability and pumping tests in accordance with the "Technical Code for Geotechnical Engineering of Municipal Solid Waste Sanitary Landfill" (CJJ 176–2012). Notably, the saturated permeability coefficients decrease with increasing landfill age (or depth). Considering the similarity of the permeability coefficients of the two MSW types, reducing the permeability of the synthetic MSW can result in similar seepage flow behaviors within the MSW body under hypergravity conditions. Notably, for MSWs with similar porosity, smaller particle sizes yield smaller permeability coefficients. When the peat used for the synthetic MSW undergoes substantial degradation, the content of organic fibers and larger particles decreases, whereas that of fine clay particles increases, making both fibers and particle sizes finer. Consequently, the permeability coefficient of the prepared synthetic MSW decreases to 2.2×10^{-5} cm/s. Under hypergravity conditions, the similarity ratio of the permeability coefficients of both MSW types is n:1. Therefore, under a gravitational acceleration of 50 g, the permeability coefficient of the prepared synthetic MSW becomes 1.1×10^{-3} cm/s, as depicted in Fig. 7.20, closely matching the optimal curve of the distribution of MSW permeability coefficients. Hence, under a gravitational acceleration of 50 g, the prepared synthetic MSW can accurately replicate the permeability characteristics of real MSW.

Table 7.5 Basic engineering properties of the synthetic and Qizishan Landfill MSW (Li 2018)

MSW	Engineering properties	Fresh	Partly degraded
Qizishan MSW	Age (year)	0	5
	Buried depth (m)	5	15
Synthetic MSW	Water content (%)	55	45
	Bulk density (kN/m^3)	7	9
	Porosity	2.9	1.6
	Compressibility coefficient (MPa^{-1})	3.19	2.62

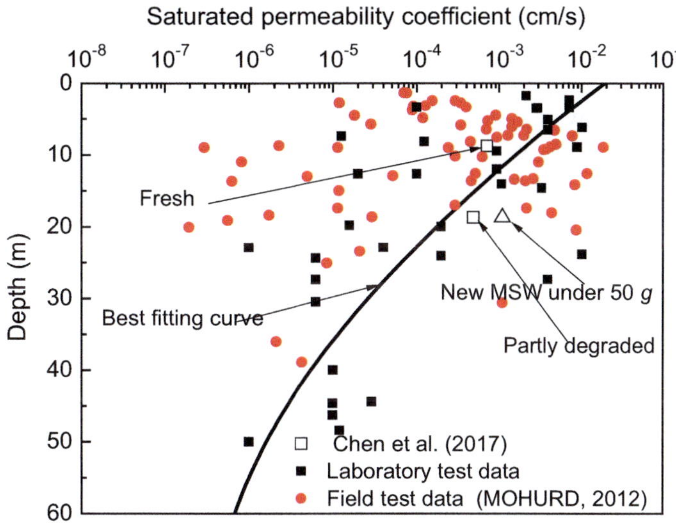

Fig. 7.20 Comparison between the saturated permeability coefficients of synthetic and real MSW (reprinted from Chen et al. 2017, with permission from Canadian Science Publishing)

Figure 7.21 illustrates the stress–strain curves of the synthetic and real MSW samples. Notably, the stress–strain relationship of the synthetic MSW closely resembles that of the real MSW sourced from the Qizishan Landfill in Suzhou, both exhibiting significant strain hardening characteristics. In particular, for fresh MSW, no distinct peak appears in the stress–strain relationship even at 50% axial strain. Conversely, for partly degraded MSW, significant peaks are apparent when the strain reaches 40%–50%. This is primarily attributed to the decrease in organic reinforcing fiber content owing to MSW degradation (Li et al. 2014).

The soil–water characteristic curve of the prepared synthetic MSW with an air-entry suction of 2 kPa, a saturated volumetric water content of 66%, and a residual volumetric water content of 22% is illustrated in Fig. 7.22a. To examine the shear strength characteristics of the unsaturated synthetic MSW samples, two sets of triaxial drained shear tests were conducted before shearing, and the obtained results were compared with those of saturated MSW samples. Notably, the shearing rate was maintained at 0.2 mm/min across both tests. As depicted in Fig. 7.22b, the differences in the stress–strain relationships of the unsaturated and saturated synthetic MSW samples under drained conditions were minimal. This finding suggests that at relatively low shearing rates, the influence of matric suction during shearing is negligible, resulting in minimal buildup of pore water pressure. Hence, as the water level within a landfill rises, the shear strength of synthetic MSW does not significantly change during its saturation.

7.5.2.2 Selection of Geomembranes

In landfill engineering, the poor friction characteristics of liner interfaces make them particularly susceptible to sliding, causing landfill instability. To accurately simulate and analyze the process of global slide instability along the liners of landfills, appropriate GMs were carefully selected for the tests. Typically, when selecting thin film materials for testing under hypergravity conditions, their equivalence is determined based on criteria such as their single-width tensile stiffness or strength and interface friction angle. Table 7.6 summarizes the tensile test results of three HDPE GMs with average thicknesses of 0.1 mm, 0.3 mm, and 0.5 mm, respectively. Notably, the single-width tensile stiffness and strength of the liner system in the Qizishan Landfill in Suzhou range from 33.3 kN/m to 650 kN/m and from 10 kN/m to 65 kN/m, respectively. Remarkably, the single-width tensile stiffness and strength of the 0.1-mm-thick GM closely matched the corresponding parameters of the actual liner system. Hence, this the 0.1-mm-thick GM was deemed suitable for

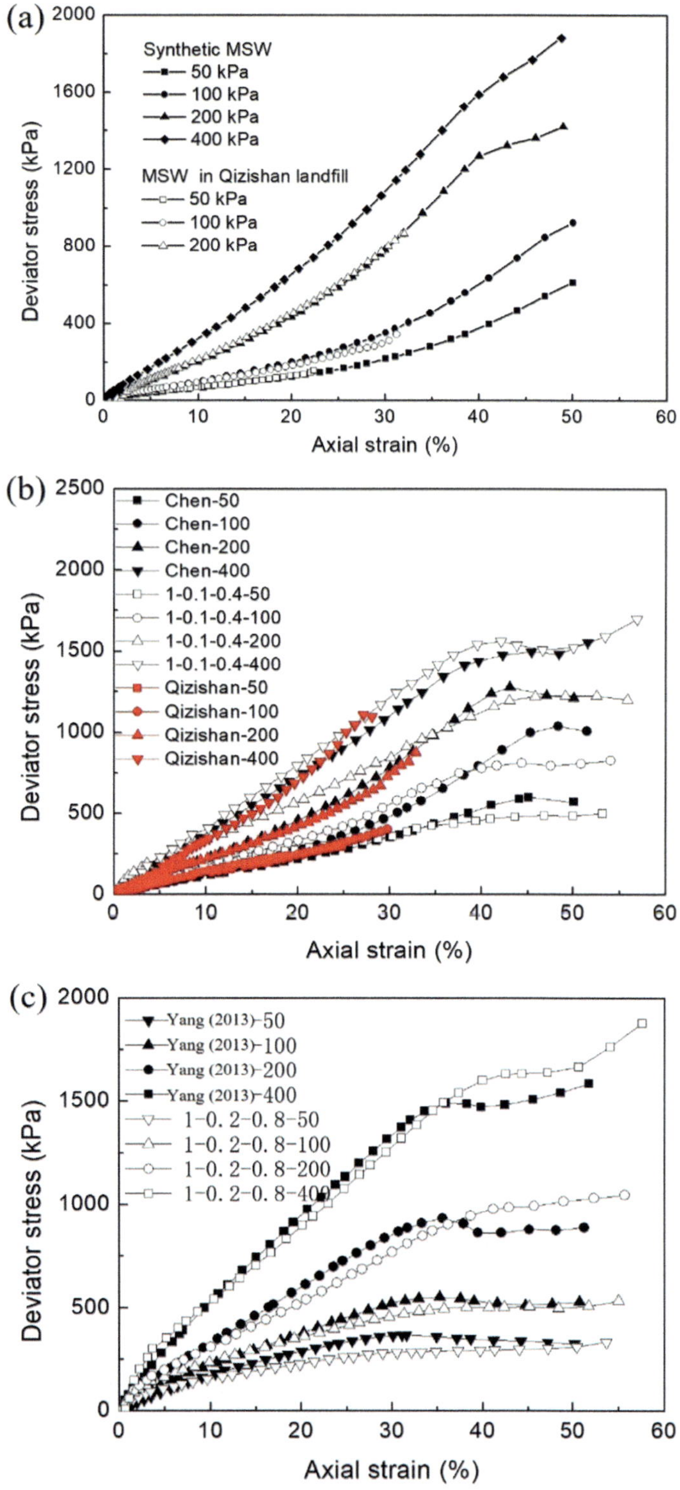

Fig. 7.21 Comparisons between the consolidated drained test results of synthetic and real MSWs: **a** fresh MSW, **b** partly degraded MSW, and **c** completely degraded MSW (reprinted from Chen et al. 2017, with permission from Canadian Science Publishing)

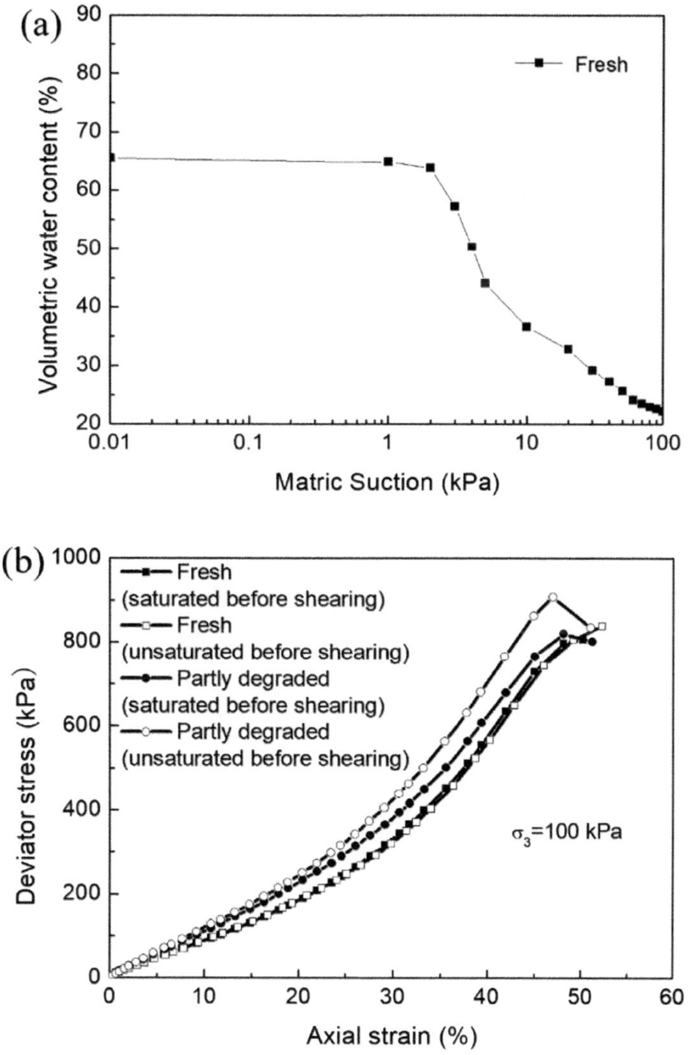

Fig. 7.22 Characteristics of unsaturated synthetic MSW: **a** soil–water characteristic curve of synthetic MSW and **b** comparison between the stress–strain relationships of saturated and unsaturated synthetic MSWs in triaxial consolidated drained tests (reprinted from Chen et al. 2017, with permission from Canadian Science Publishing)

Table 7.6 Parameters of geomembranes (Li 2018)

Thickness of geomembrane (mm)	Single-width tensile stiffness E_t (kN/m)		Single-width tensile strength σ_{tmax} (kN/m)	
	1 g	66.7 g	1 g	66.7 g
0.1	7.74	516	0.96	64
0.3	27.7	1,848	3.06	204
0.5	49.9	3,328	5.3	354

simulating the tensile behavior and performance of actual liner systems within landfills.

Tests were also conducted to evaluate the interface strength between the 0.1-mm-thick HDPE GM and synthetic MSW, and the corresponding results are depicted in Fig. 7.23. Notably, the friction angles corresponding to the peak and residual shear strengths at the interface

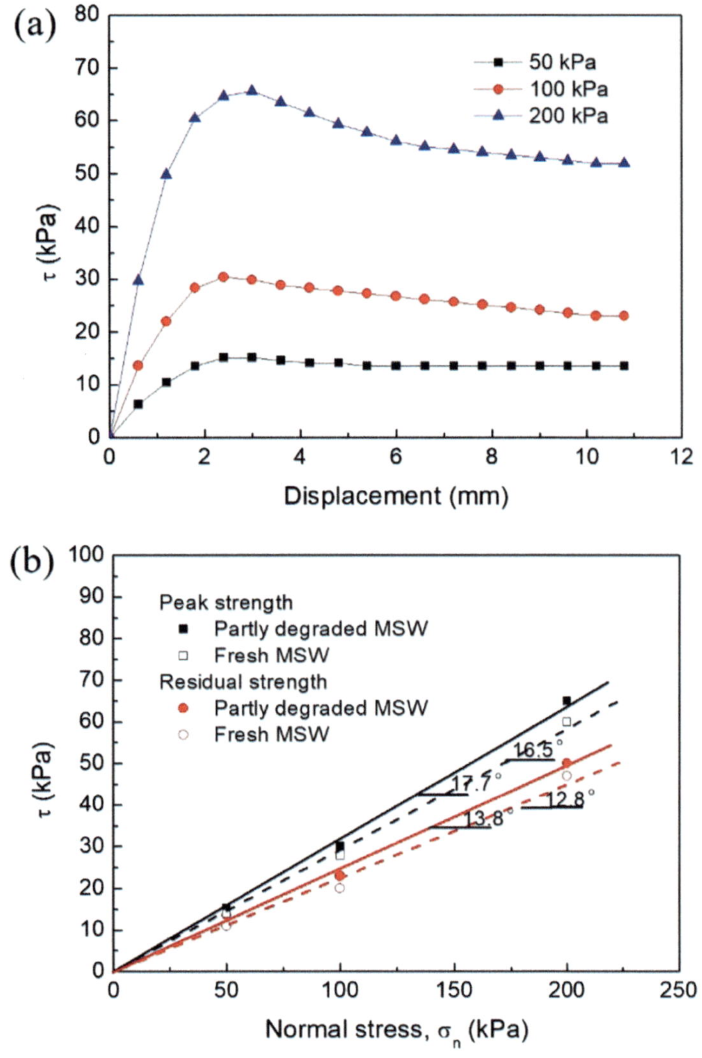

Fig. 7.23 Interface strength between the GM and synthetic MSW: **a** displacement and shear strength and **b** strength envelope (Li 2018)

between the 0.1-mm-thick HDPE GM and fresh synthetic MSW were 16.5° and 12.8°, respectively. Meanwhile, for partly degraded synthetic MSW and the 0.1-mm-thick HDPE GM, the corresponding friction angles were slightly higher at 17.7° and 13.8°, respectively. These values closely aligned with the shear strength characteristics of the liner materials used in the Qizishan Landfill in Suzhou (Chen et al. 2010a), thus confirming the rationale behind using the 0.1-mm-thick HDPE GM to simulate landfill liners in centrifugal hypergravity tests.

7.5.3 Rising of Water Level and Critical Water Level

The centrifugal hypergravity modeling tests were conducted under a centrifugal acceleration of 66.7 g, and water supply commenced once the settlement of the landfill body stabilized. Under the influence of seepage, water within the landfill gradually propagated from the supply cavity toward the slope toe. Figure 7.24 illustrates the progression of water level development within the landfill body. At each stage, once the water

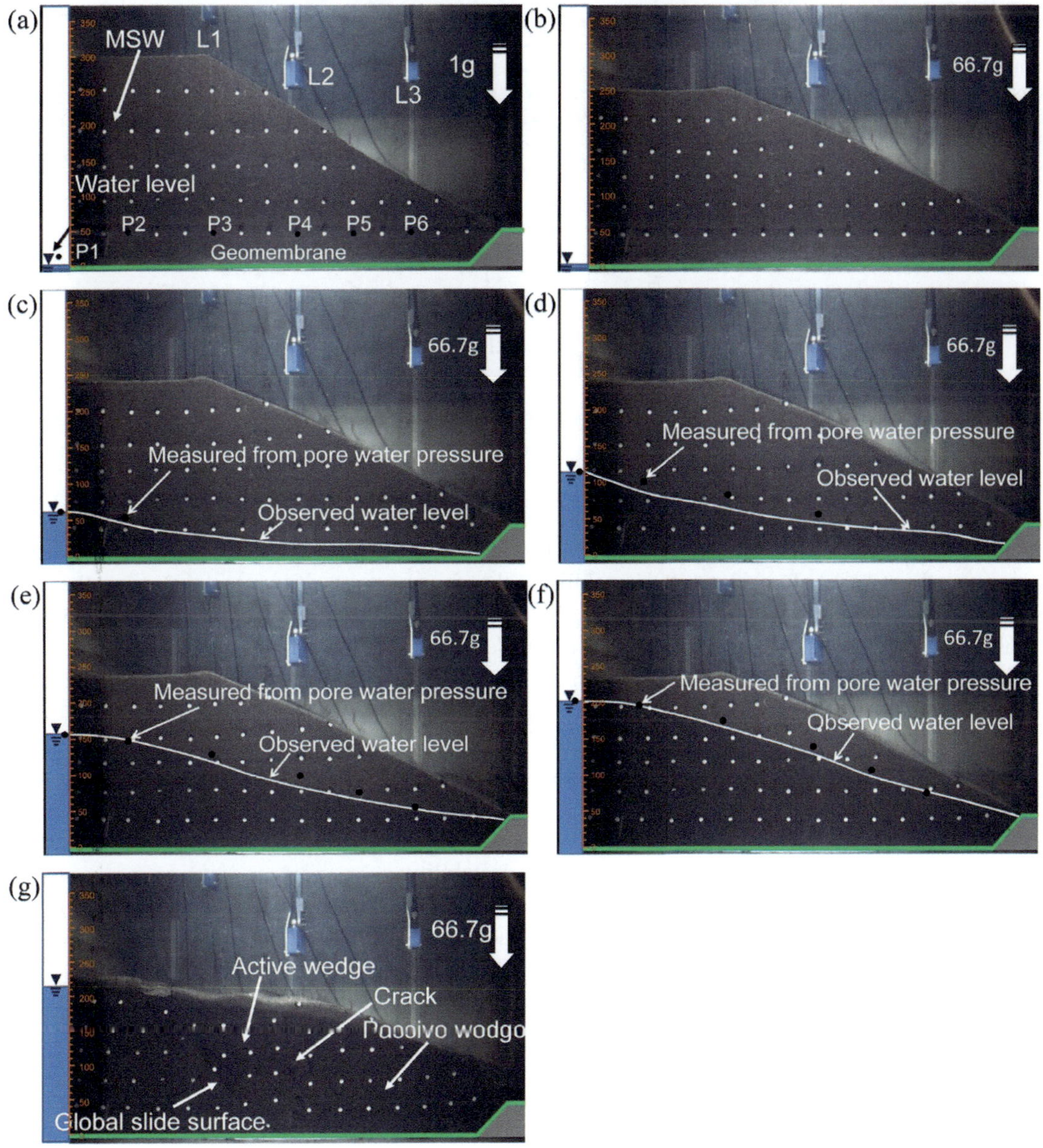

Fig. 7.24 Water level development within the landfill body during the centrifugal hypergravity modeling tests: **a** initial, **b** after settlement at 66.7 g, **c** water level stage 1, **d** water level stage 2, **e** water level stage 3, **f** water level stage 4, and **g** landfill stability failure (reprinted from Chen et al. 2017, with permission from Canadian Science Publishing)

level within the supply cavity reached the desired level, it was maintained stable for 20 min to allow water seepage into the landfill body until the infiltration line within the landfill almost stopped fluctuating. The monitored pore water pressures were converted into static pressure heads and mapped onto images of the modeled landfill. These levels were compared with the observed infiltration lines. Results revealed that the measured static pressure heads closely approximated the observed water levels. Importantly, no excess pore water pressures were

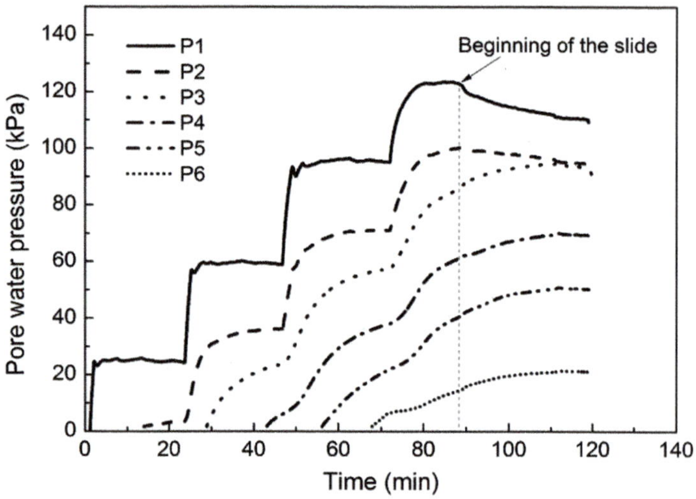

Fig. 7.25 Pore water pressures measured by transducers during the water supply process (reprinted from Chen et al. 2017, with permission from Canadian Science Publishing)

detected within the landfill at the end of each water level stage before landfill instability ensued. Moreover, the difference between the measured static water pressure heads and observed water levels decreased as the water level rose.

Figures 7.25 and 7.26 respectively depict the observed pore water pressures and vertical displacements at various points during the centrifugal hypergravity modeling. During the first three stages of water level rise and the subsequent stabilization stage, a gradual increase was observed in the vertical deformation at the top of the landfill body; however, no significant instability or sliding occurred. However, during the fourth stage of water level rise, the pore pressure at the corresponding crest location decreased abruptly after a brief period of stability. Concurrently, the vertical displacement at the crest increased suddenly, signaling the gradual onset of landfill body sliding. Hence, sudden changes in the water levels within a landfill body and the vertical displacement at the corresponding crest can serve as indicators of slope instability.

Next, the relative water level ratio (h/H_L) was adopted to determine the critical water level for landfill instability, where h denotes the water

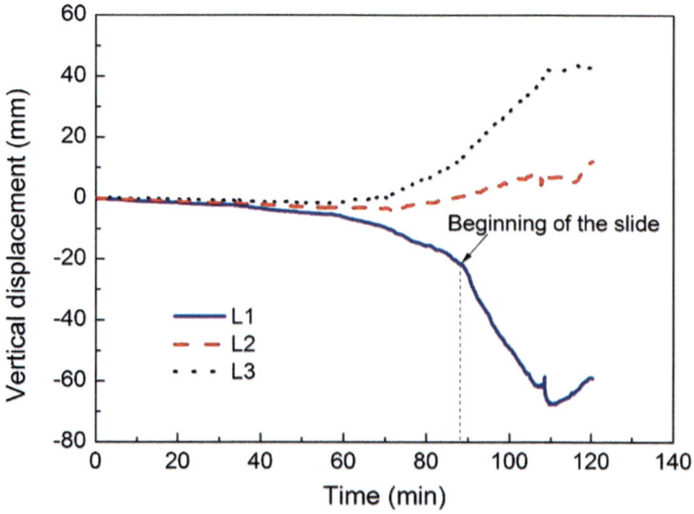

Fig. 7.26 Vertical displacements measured by laser displacement transducers during the water supply process (reprinted from Chen et al. 2017, with permission from Canadian Science Publishing)

level within the supply cavity, and H_L represents the height of the landfill measured by the installed laser displacement transducers before water supply. Leveraging the relationship between the relative water level ratio and the vertical displacement at the crest, Fig. 7.27 displays the critical water level for instability in each test. Initially, as the water level increases, the vertical displacement of the landfill body also gradually increases, albeit with small changes. However, when the water level approaches the critical value, h_{max}, the vertical displacement trend of the landfill body abruptly changes. For instance, under constant water level conditions, the vertical displacement significantly increases. This increase can occur even in conditions with a subsequent decrease in water level. At the point where the sudden change in displacement occurs, h_{max}/H_L serves as the warning water level ratio for landfill instability.

Notably, in each test, the critical water level ratios for landfill slope instability varied from 0.75 to 0.92, primarily owing to differences in the types of MSW used, variations in slope geometry, and presence of engineered berms. In particular, the critical water level ratio for landfill slope instability increased with decreasing slope gradient. Observations revealed that the critical water level increased as the degradation degree of MSW increased. Furthermore, the engineered berm raised the critical water level for landfill slope instability by 1.3% to 9.5%. To validate the results derived from the test involving a slope gradient of 1:3 and an engineered berm, they were compared with the field monitoring results of the Xiaoping Landfill in Shenzhen, featuring an average slope gradient of 1:3.5 and an engineered berm at its base. As mentioned previously, the rate of water level increase observed during the test corresponds to the fastest water level increase experienced by the Xiaoping Landfill in Shenzhen during a continuous heavy rainfall event. Notably, the Xiaoping Landfill in Shenzhen exhibited significant horizontal displacement and settlement at a water level ratio of approximately 0.9 (Chen et al. 2010a). Following this, the top of the landfill experienced a settlement of up to 4 m. Correspondingly, in the test with a similar slope gradient and engineered berm conditions, instability and sliding commenced at a critical water level ratio of 0.92. The landfill settlement in prototype dimensions was approximately 2.7 m, slightly smaller than the value monitored on-site.

7.5.4 Typical Slope Failure Modes

7.5.4.1 Continuous Sliding Caused by Global Instability Along the Liner Interface

Landfills without engineered berms and with slope gradients ranging from 1:1 to 1:3 are susceptible to global instability along the liner interface when the water level within the landfill body reaches the critical value (h_{max}). Once global sliding commences, excessive deformation of the landfill body leads to a decrease in the water level within the landfill body. Depending on the slope gradient, different patterns of subsequent sliding can occur after the initial global slide.

Specifically, for steeper slopes, such as those with a gradient of 1:1 depicted in Fig. 7.28, once the water level rises to a certain height, initial cracking appears at the base of the landfill. This cracking is accompanied by significant horizontal displacement of the landfill body. The resulting cracks progressively develop into sliding planes, resulting in significant global sliding. As the sliding continues, varying deformation within the landfill body gradually forms another significant penetrating crack owing to the combined effects of the resistance of the bottom slope and reinforcement hardening of the MSW. Ultimately, an additional arcuate sliding plane forms at the top of slopes with steep gradients.

Conversely, on gentler slopes, such as those with a slope gradient of 1:2 depicted in Fig. 7.29, once the water level reaches the critical point, global sliding along the liner interface initiates. Subsequently, penetrating cracks develop, and the sliding pattern resembles that of the landfill with a steep slope. However, owing to the gentler slope gradient, the formation of local arcuate sliding planes is typically not observed.

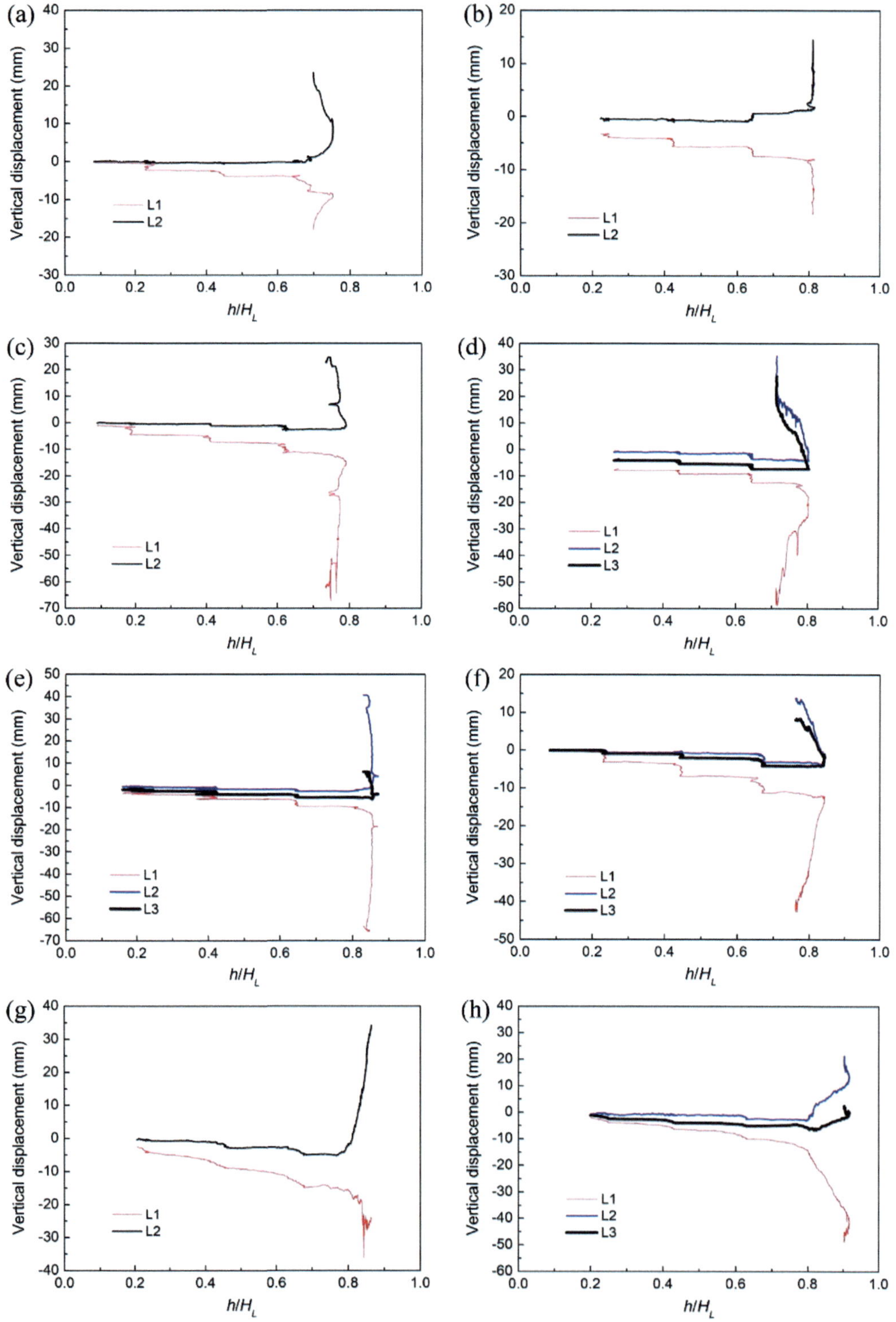

Fig. 7.27 Plots of vertical displacements measured using laser displacement transducers versus water level ratios: **a** test 1, **b** test 2, **c** test 3, **d** test 4, **e** test 5, **f** test 6, **g** test 7, and **h** test 8 (reprinted from Chen et al. 2017, with permission from Canadian Science Publishing)

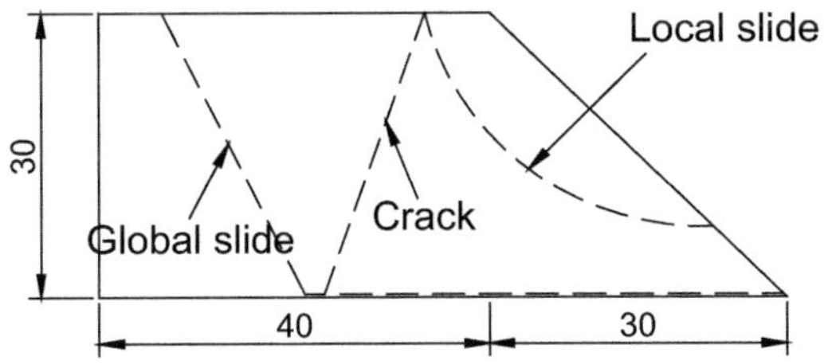

Fig. 7.28 Global sliding along slopes with a gradient of 1:1 (global slide → crack → local slide) (reprinted from Chen et al. 2017, with permission from Canadian Science Publishing)

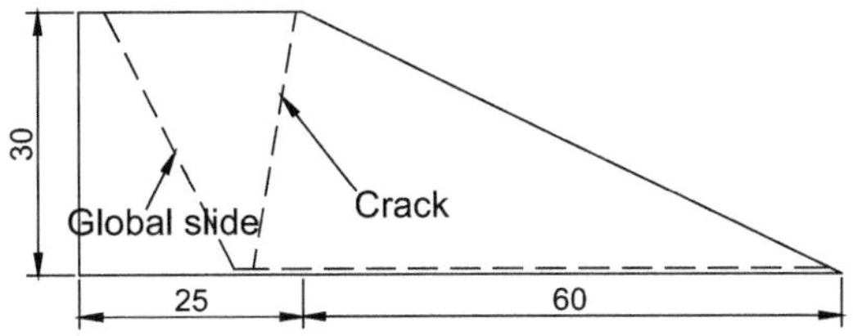

Fig. 7.29 Global sliding along slopes with a gradient of 1:2 (global slide → crack) (reprinted from Chen et al. 2017, with permission from Canadian Science Publishing)

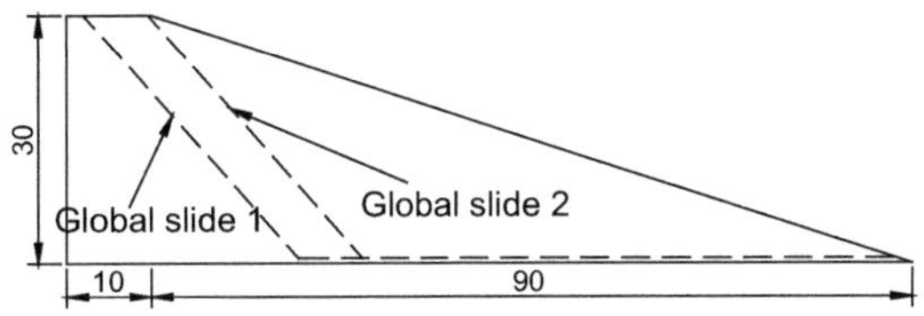

Fig. 7.30 Global slide along slopes with a gradient of 1:3 (global slide 1 → global slide 2) (reprinted from Chen et al. 2017, with permission from Canadian Science Publishing)

For very gentle landfill slope slopes, such as those with a slope gradient of 1:3 illustrated in Fig. 7.30, when global sliding initiates, new global sliding planes immediately form along the sliding body. These new sliding planes are typically parallel to each other, illustrating a multi-stage sliding pattern.

Notably, landfills equipped with engineered berms can also experience global sliding, as depicted in Fig. 7.31. When the MSW within the landfill undergoes significant degradation, the organic reinforcing fibers content within the waste decreases significantly, weakening the strain hardening characteristics of the MSW at high strain values. Under these circumstances, when the landfill body undergoes significant deformation, it may not be capable of providing adequate resistance, leading to global instability and sliding along the liner interface, resembling the behavior observed in landfills without engineered berms. Upon reaching the critical water level, the landfill body first experiences global sliding, followed by the development of penetrating cracks.

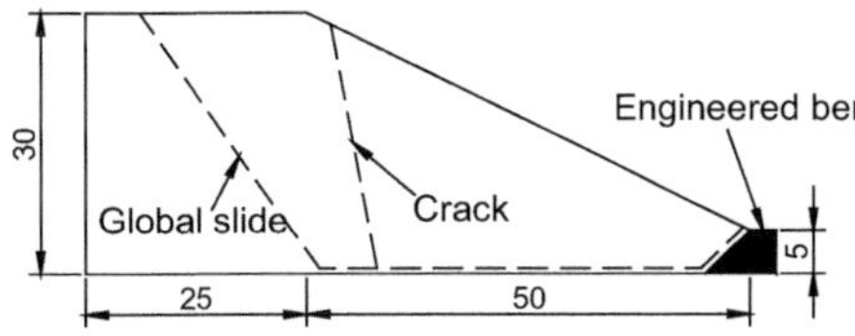

Fig. 7.31 Global sliding of a landfill equipped with an engineered berm (global slide → crack) (reprinted from Chen et al. 2017, with permission from Canadian Science Publishing)

7.5.4.2 Penetration-Crack-Induced Failure

For typical landfills equipped with engineered berms, the reinforcement provided by the MSW enhances the strength of the landfill body, offering support during significant deformations. The presence of an engineered berm also restricts bottom displacement, preventing the occurrence of global sliding along the liner interface when the critical water level is reached. Instead, cracks first appear at the bottom of the landfill. However, these cracks do not evolve into sliding planes within the landfill body owing to the restriction of bottom displacement by the engineered berm. Nevertheless, the displacement of the upper landfill body continues to increase, resulting in the development of penetrating cracks that extend vertically through the top of the landfill. This type of failure is also accompanied by significant horizontal displacement of the landfill body. Figure 7.32 illustrates this phenomenon for a landfill with a slope gradient of 1:2 and an engineered berm, where two distinct penetrating cracks are observed during failure.

In contrast, for landfills with gentle slope gradients, such as the gradient value of 1:3

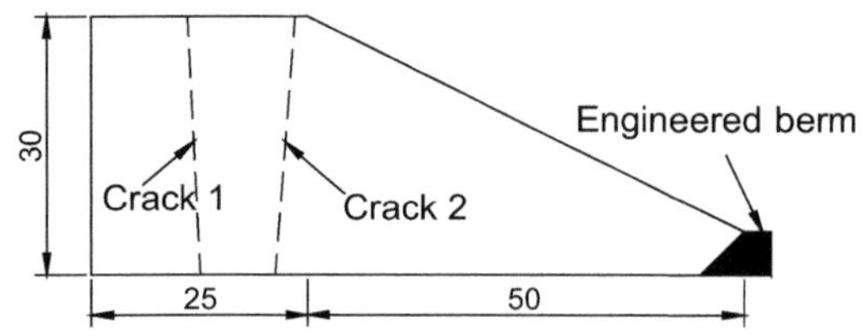

Fig. 7.32 Development of penetration cracks in a landfill with a slope gradient of 1:2 and an engineered berm (crack 1 → crack 2) (reprinted from Chen et al. 2017, with permission from Canadian Science Publishing)

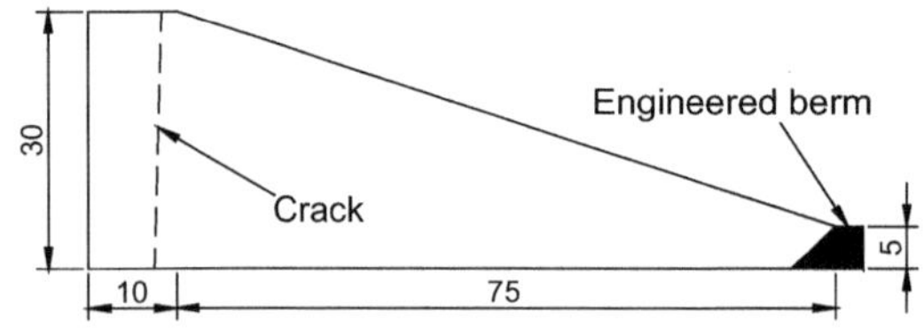

Fig. 7.33 Development of penetration cracks in a landfill with a slope gradient of 1:3 and an engineered berm (crack) (reprinted from Chen et al. 2017, with permission from Canadian Science Publishing)

depicted in Fig. 7.33 and deemed safe by regulations, the failure mode primarily involves the development of penetrating cracks. This is because the gentle slope provides a sufficiently elongated anti-sliding section, complicating the formation of sliding planes within the landfill body.

7.5.4.3 Instability Induced by Toe Erosion

For landfills with steep slopes and significant levels of MSW degradation, a different failure mode can occur, as illustrated in Fig. 7.34. In these landfills, the characteristics of the MSW are more akin to those of sand, possessing weak reinforcement properties and high permeability. Under these circumstances, toe erosion occurs when overflow at the slope toe erodes MSW under the action of seepage, leading to the loss of MSW materials. Several factors contribute to this phenomenon: the high degree of MSW degradation, low organic fiber content, and weak reinforcement hardening effect. As the water level within the landfill body rises, the extent of erosion at the slope toes gradually increases and propagates upward along the slope. When the erosion area develops fully, local sliding initiates from the top of the slope, indicating the critical state of water level. The loss of MSW materials along the slope reduces the resistance to back pressure, ultimately triggering global sliding along the liner interface. Once the global sliding plane forms, the water level no longer rises.

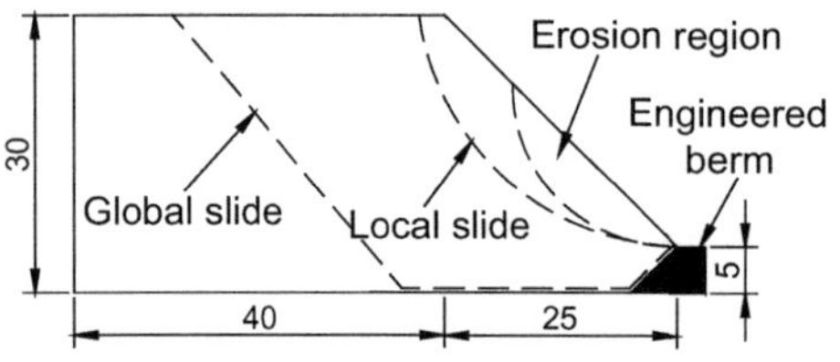

Fig. 7.34 Instability induced by toe erosion (erosion → local slide → global slide) (reprinted from Chen et al., 2017, with permission from Canadian Science Publishing)

7.6 Centrifugal Hypergravity Modeling of Landfill Failures Induced by Rising Landfill Gas Pressures

7.6.1 Test Setup

Hu et al. (2024b) conducted a centrifugal hypergravity test to examine the instability of landfill slopes induced by rising gas pressures. Figure 7.35 illustrates the test design. The landfill model was simplified to represent a homogeneous synthetic MSW landfill with a GM at its bottom, excluding temporary or final cover layers. During testing, synthetic MSW was uniformly compacted in 3-cm-thick layers on the GM, and each MSW layer was compacted to a design unit weight of 9 kN/m^3. Particle image velocimetry (PIV) markers were placed at 3 cm intervals near the sidewalls of each MSW layer. The landfill slope had a final height of 30 cm, with a width of 70 cm and a slope of 1:1.5, which was created using a slope-cutting method.

The centrifugal hypergravity was set to 66.7 g, facilitating the simulation of a real landfill height of 20 m. A 5-cm-wide liquid supply cavity was installed on the left of the model to

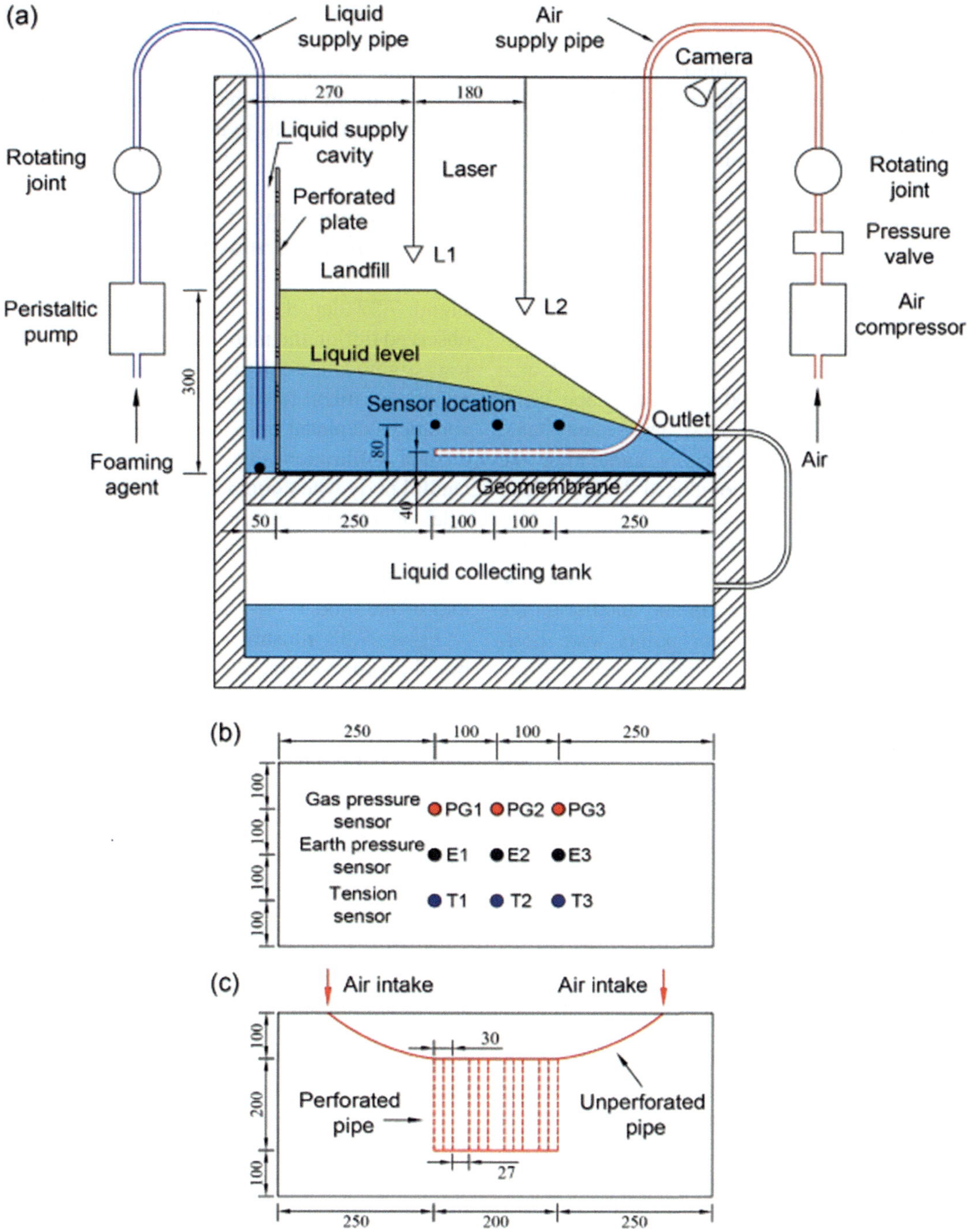

Fig. 7.35 Model setup: **a** cross-section of the landfill, **b** layout of sensors, and **c** layout of the air injection network (unit: mm) (reprinted from Hu et al. 2024b, with permission from Canadian Science Publishing)

inject sodium dodecyl sulfate foaming liquid. Furthermore, a pore pressure sensor was installed at the bottom of the liquid supply cavity to monitor liquid levels during testing. Notably, the liquid level within the cavity was maintained at 15 cm. An overflow outlet, positioned 5 cm above the bottom on the right side of the model, channeled seepage from the slope to a lower liquid collection tank.

An air injection pipe was connected to the bottom of the landfill body. Once the liquid level within the liquid supply cavity stabilized, air was injected into the right side of the model at a constant pressure of 200 kPa until slope instability occurred. The injected air created stable foam with the liquid, thereby simulating the process of liquid–gas coinduced landfill slope instability.

To monitor displacements, two laser displacement transducers were installed above the model surface, one at the top and another on the slope surface. Three sensor points were positioned at the bottom of the model, each housing gas pressure, tension, and earth pressure sensors to monitor pore gas, pore liquid, and earth pressures at different distances, respectively. Cameras were positioned on the side and top surfaces of the model to capture photographs throughout the testing.

7.6.2 Model Consolidation and Liquid–gas Injection Process

The experiment was conducted on the ZJU-400 centrifuge at Zhejiang University. Figure 7.36 depicts the temporal curves of control quantities during the testing process. Once the model was fabricated, it was mounted onto the centrifuge, which was then gradually accelerated from 1 g to 20 g, 40 g, 50 g, and finally to 66.7 g. The centrifuge operated at each acceleration level for approximately 10 min, stabilizing at 66 0.7 g after 35 min. Once the settlement of the landfill body stabilized, liquid supply commenced at the 35th min, progressively increasing until the liquid level within the supply cavity stabilized around 15 cm by the 60th min. Once the liquid level stabilized, air injection commenced at the 65th min, with an injection pressure of 200 kPa, and continued until the 80th min when the injection process was terminated.

7.6.3 Development of Pore Pressure and Deformation

Figure 7.37 depicts side views of the changes observed during the main operational steps of the test. Significant settlement of the landfill body under centrifugal force is evident from the height variations depicted in Fig. 7.37a–c) illustrates the distinct infiltration lines appearing within the landfill body when liquid is supplied from the left side of the model. In Fig. 7.37d, instability induced sliding on the slope crest and the slope side is apparent after air injection from the bottom of the model.

Figure 7.38 illustrates the variation in pore liquid pressure during the liquid and air injection processes. In particular, before liquid injection, pore liquid pressures at the three measurement points T1, T2, and T3 ranged from −3 to −5 kPa, indicating an initial matric suction of 3–5 kPa and an unsaturated condition. After approximately 3 min of liquid injection, the pore liquid pressures at all three measurement points increased to zero, indicating saturation. As liquid injection continued, the liquid within the left side liquid supply cavity gradually moved toward the right side owing to the head difference, thereby increasing the pore liquid pressures at the three measurement points. After 30 min of liquid injection, the pore liquid pressures at points T1, T2, and T3 stabilized at 64.5 kPa, 51.4 kPa, and 41.8 kPa, respectively, suggesting the formation of a stable infiltration line within the slope. Upon the stabilization of liquid phase seepage, air injection commenced at the 65th min, inducing fluctuations in the pore liquid pressure. By the 71st min, continued air injection increased the pore liquid pressure at points T1, T2, and T3 to 75.6 kPa, 61.3 kPa, and 51.8 kPa, respectively, reflecting the effect of raising the liquid level. At this point, the landfill body began to slide, and

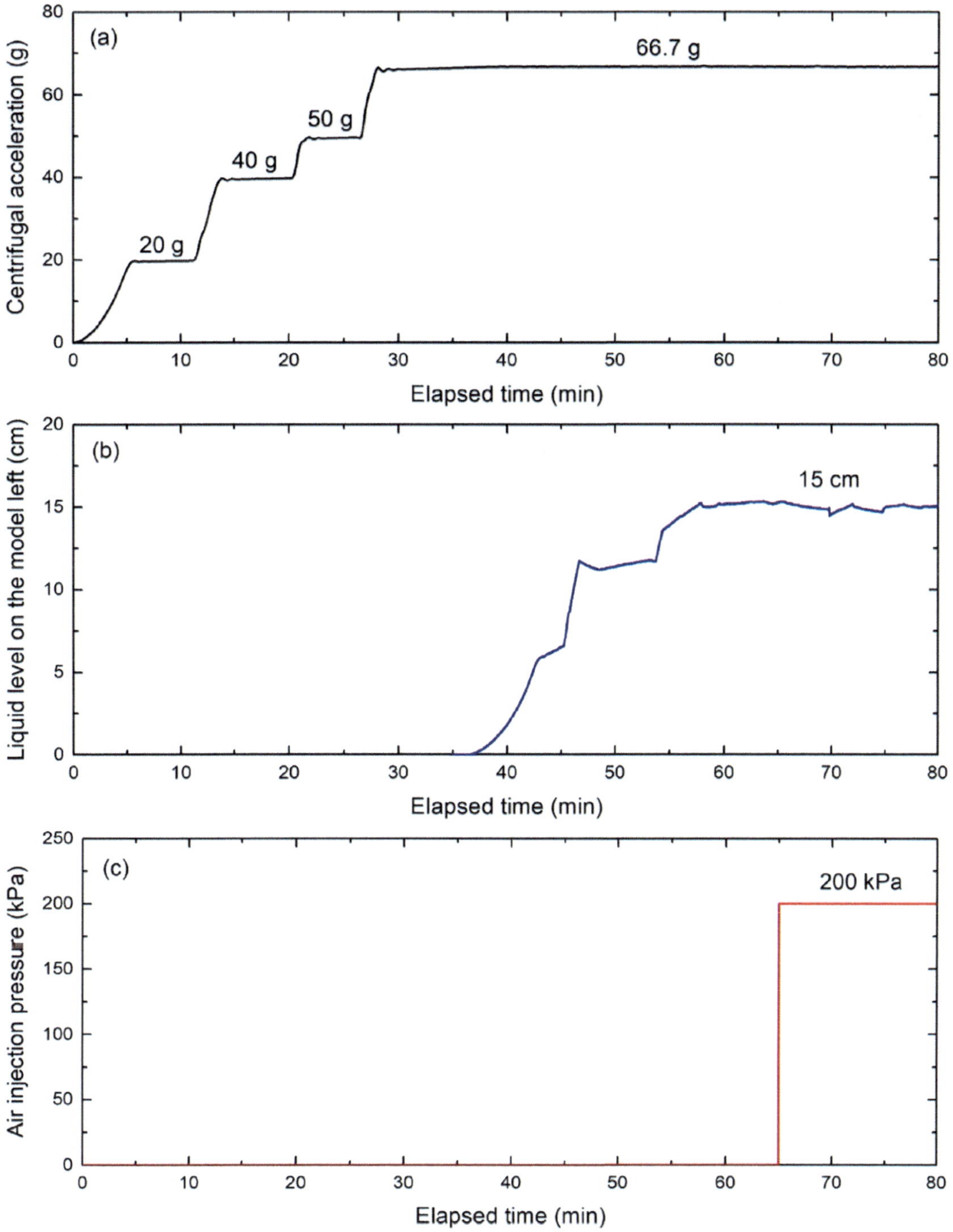

Fig. 7.36 Model testing process: **a** step 1—increasing the acceleration to 66.7 g, **b** step 2—injecting liquid from the left side of the model until reaching a height of 15 cm, and **c** step 3—injecting air at a constant pressure of 200 kPa (reprinted from Hu et al. 2024b, with permission from Canadian Science Publishing)

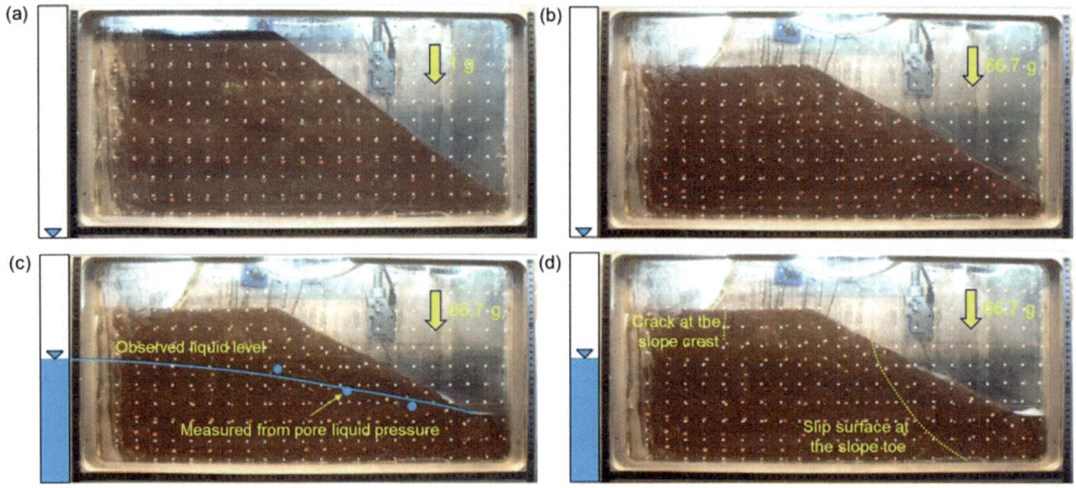

Fig. 7.37 Process of model instability: **a** model mounted on the centrifuge, **b** occurrence of settlement under a centrifugal acceleration of 66.7 g, **c** formation of a stable liquid level within the landfill, and **d** occurrence of slope failure due to air injection (reprinted from Hu et al. 2024b, with permission from Canadian Science Publishing)

Fig. 7.38 Variation in the pore liquid pressure during liquid and air injections (reprinted from Hu et al. 2024b, with permission from Canadian Science Publishing)

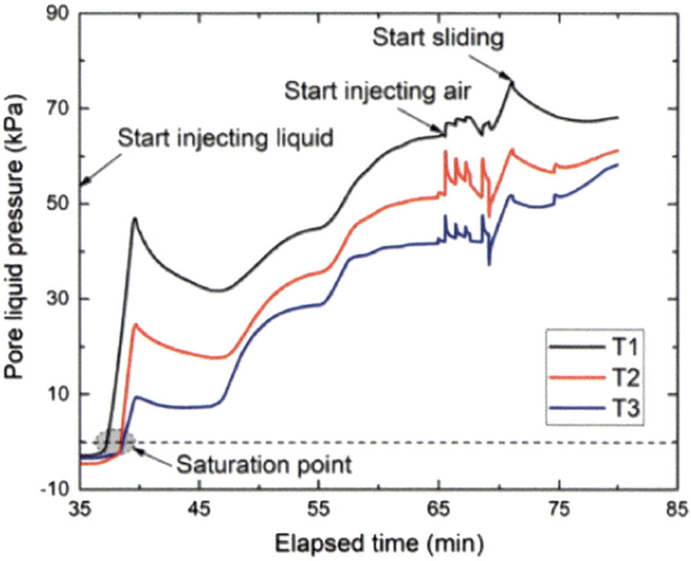

the pore liquid pressure decreased. By the 74.5th min, the pore liquid pressures at points T1, T2, and T3 decreased to 68.6 kPa, 56.5 kPa, and 49.9 kPa, respectively. As the sliding subsequently stabilized, the pore liquid pressures exhibited a slight rebound.

Figure 7.39 illustrates the variation in pore gas pressure during air injection. Initially, the gas pressure was zero. As air injection progressed, pore gas pressure gradually increased. Specifically, by the 71st min, the pore gas pressures at PG1 and PG2 reached peak values of 100.8 kPa and 83.0 kPa, respectively, exceeding the pore liquid pressures at that instant (75.6 kPa and 61.3 kPa, respectively). Following the onset of sliding, the pore gas pressure decreased, stabilizing at values of 85.1 kPa and 79.6 kPa at points PG1 and PG2, respectively. Overall, the pore gas pressure during sliding significantly exceeded the pore liquid pressure, indicating that

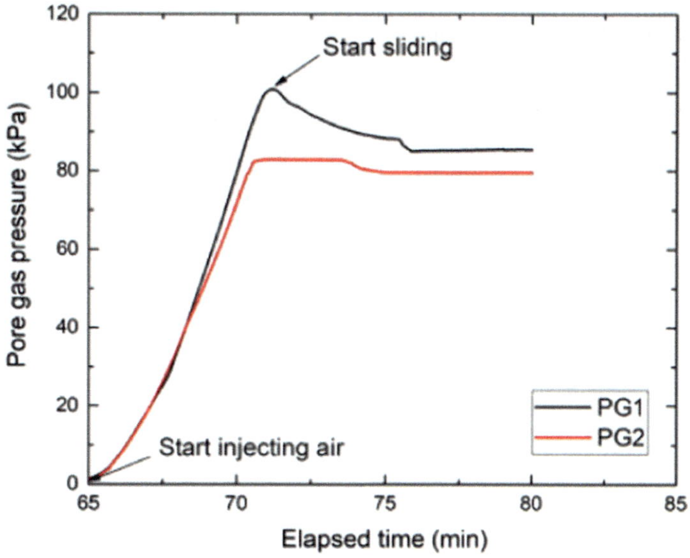

Fig. 7.39 Variation in pore gas pressure during air injection (reprinted from Hu et al. 2024b, with permission from Canadian Science Publishing)

high gas pressure predominantly influenced sliding.

Figure 7.40 depicts the evolution of vertical displacement on slope side L2 during the liquid and air injection processes. In this figure, positive vertical displacements indicate downward settling, whereas negative displacements denote upward bulging. During the liquid injection phase, the vertical displacement was close to zero. However, after 6 min of air injection (at the 71st min), the vertical displacement exhibited significant fluctuations ranging from −21.7 mm to 13.3 mm, indicating the onset of sliding. The abrupt change in displacement at the moment of sliding correlated well with the abrupt change in pore pressure.

Figure 7.41 illustrates the development of vertical displacement at slope crest L1 during the

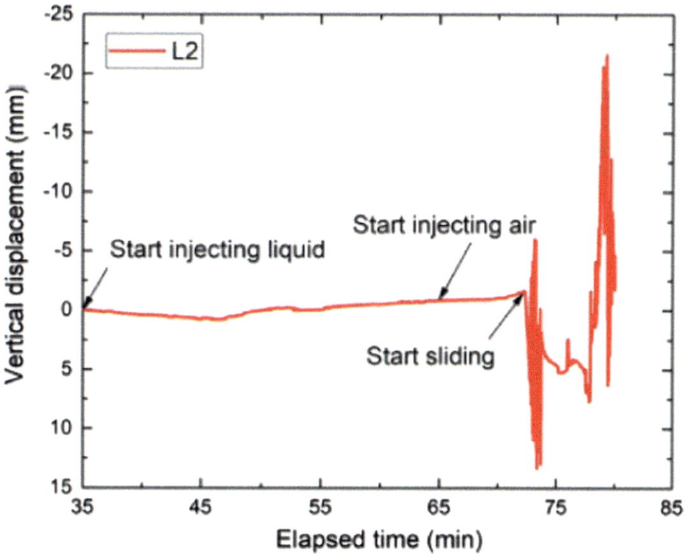

Fig. 7.40 Development of vertical displacement on slope side L2 during the liquid and air injection processes (reprinted from Hu et al. 2024b, with permission from Canadian Science Publishing)

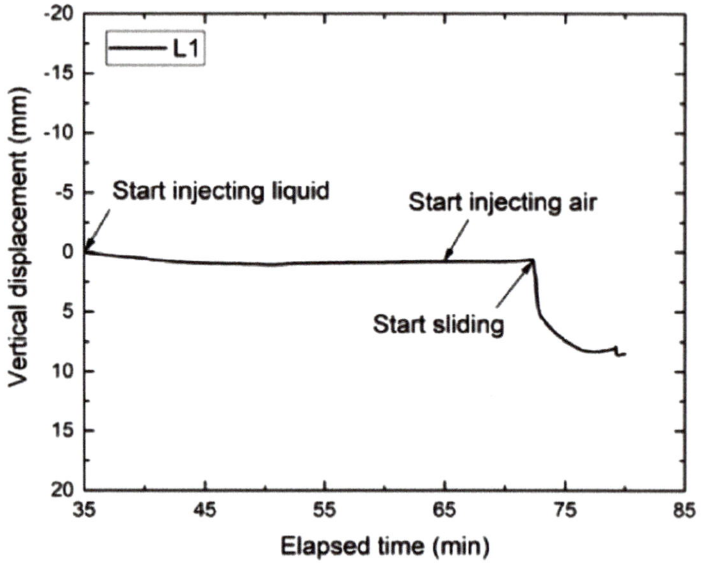

Fig. 7.41 Evolution of the vertical displacement at slope crest L1 during the liquid and air injection processes (reprinted from Hu et al. 2024b, with permission from Canadian Science Publishing)

liquid and air injection processes. Particularly, during the liquid injection stage, the displacement was nearly zero. However, after 6 min of air injection (at the 71st min), the displacement abruptly increased to 8.3 mm. The resulting downward movement of the crest, indicating a settlement of 8.3 mm, was attributed to the traction exerted by slope sliding.

Figure 7.42 illustrates the PIV displacement vector map of the slope side. Notably, a significant downward sliding trend is evident, particularly noticeable at the slope toe, where displacements can exceed 30 mm. The envelope line of the region wherein the displacement vectors at the slope toe exceed 10 mm delineate the observed sliding surface.

7.6.4 Stability Analysis and Critical Landfill Gas Pressure

The stability of landfill slope was assessed using the Slope/W module within the GeoStudio software, employing the Morgenstern–Price method for safety factor calculations. Remarkably, this software automatically identifies the most critical slip surface. In addition to the synthetic MSW, a GM unit was placed at the bottom of the model, characterized by cohesion (c') and friction angle (φ') values of 0 and 18°, respectively.

To accurately compute the stability safety factor, pore pressure conditions across the entire landfill slope had to be appropriately defined. However, only actual pore pressure data at

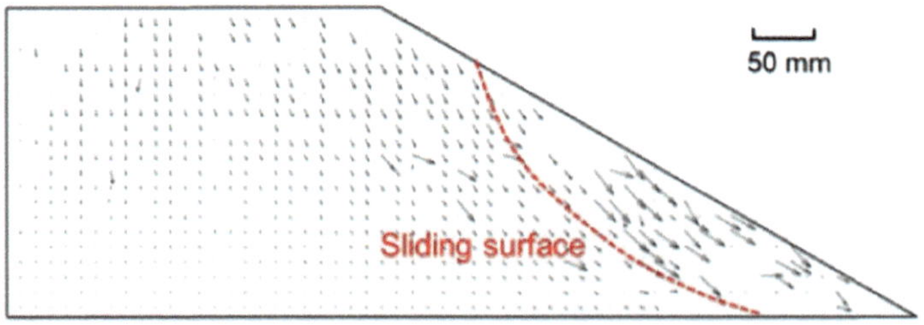

Fig. 7.42 Displacement vectors obtained from the PIV analysis (reprinted from Hu et al. 2024b, with permission from Canadian Science Publishing)

measurement points 1 and 2 were available for this experiment. Therefore, numerical simulations were employed to model the pore pressure distribution across the entire slope, using the data corresponding to measurement points 1 and 2 for calibration. The pore pressure distribution was simulated using the liquid–gas two-phase flow module provided in the COMSOL software.

Figure 7.43 illustrates the simulated pore pressure distributions. As depicted in Fig. 7.43a, beginning from the injection tube, the pore pressure gradually spread outward, creating an elliptical zone of high gas pressure. The calculated gas pressures at measurement points 1 and 2 were 95.6 kPa and 87.4 kPa, respectively, closely matching the measured pressures of 100.8 kPa and 83.0 kPa at these points.

Figure 7.43b depicts that the injection of air increased the liquid level, causing the contours of liquid levels above the air injection pipe to bulge upward. The calculated pore liquid pressures at measurement points 1 and 2 were 72.1 kPa and 66.5 kPa, respectively, also closely matching the corresponding measured pressures of 75.6 kPa and 61.3 kPa. These peak pore pressure distributions were then imported into the GeoStudio software to calculate the safety factor of the slope.

Figure 7.44a, b respectively depict the results of slope safety factors under peak gas pressure and peak liquid pressure conditions. Under peak gas pressure conditions, the factor of safety was 0.991, indicating an unstable state. Meanwhile, the calculated sliding surface closely resembled the observed sliding surface depicted in Fig. 7.42. Under peak liquid pressure conditions, the factor of safety was 1.217, indicating that the slope would not slide. These findings suggest that for MSW landfills subjected to both high gas and liquid pressures, solely considering the effect of pore liquid pressure may overestimate slope stability, potentially leading to unforeseen landslide accidents.

Figure 7.45 illustrates the variation in vertical displacement with the relative gas pressure ratio at test point 1 (away from the slope side). Notably, the relative gas pressure ratio is defined as the ratio of gas pressure to earth pressure at a given point. When the relative gas pressure ratio was less than 0.84, vertical displacement remained nearly zero, indicating no sliding of the landfill slope. However, when the relative gas pressure ratio reached 0.84, the vertical displacements at both the slope side and crest abruptly increased, accompanied by a subsequent decrease in the relative gas pressure ratio,

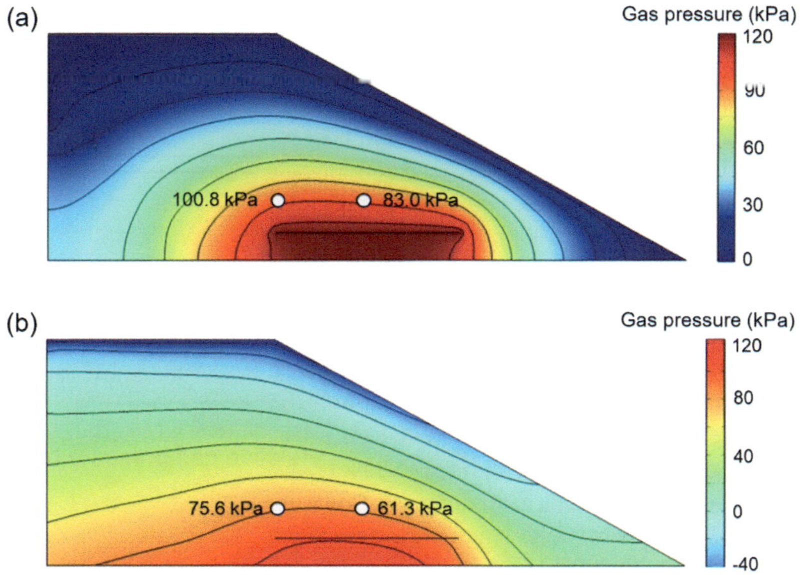

Fig. 7.43 Numerical simulation of pore pressure distributions: **a** peak gas pressure distribution and **b** peak liquid pressure distribution (reprinted from Hu et al. 2024b, with permission from Canadian Science Publishing)

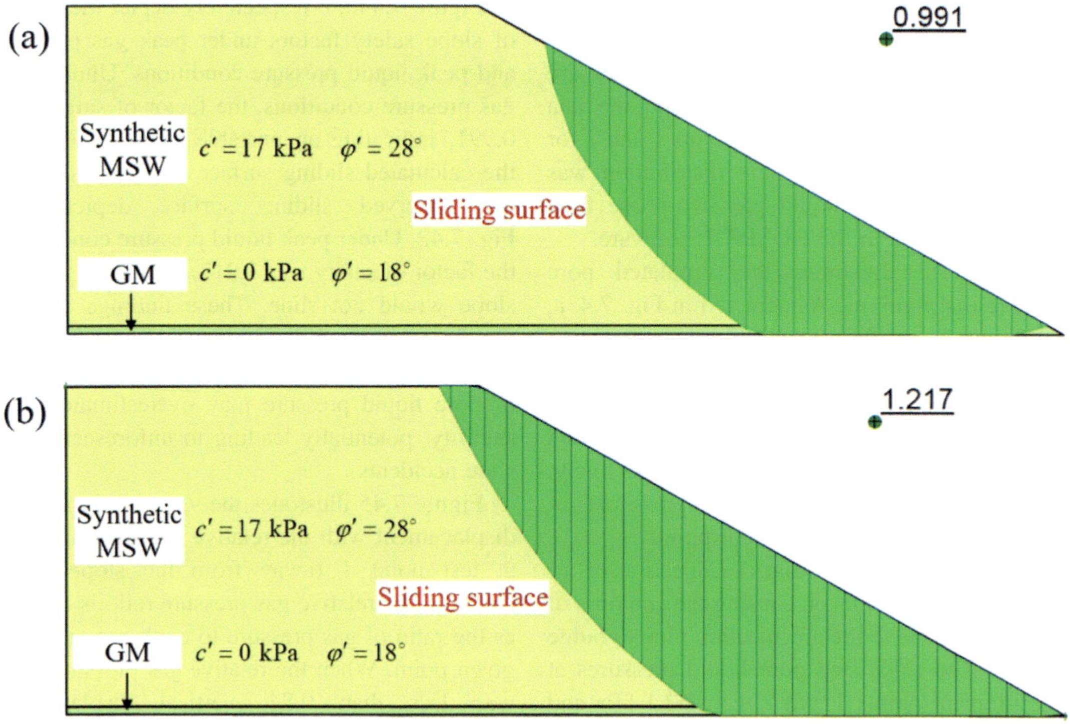

Fig. 7.44 Slope factor of safety: **a** under the combined action of peak gas and liquid pressures and **b** under the action of only peak liquid pressure (reprinted from Hu et al. 2024b, with permission from Canadian Science Publishing)

Fig. 7.45 Variation in vertical displacement with the relative gas pressure ratio at test point 1 (away from the slope side) (reprinted from Hu et al. 2024b, with permission from Canadian Science Publishing)

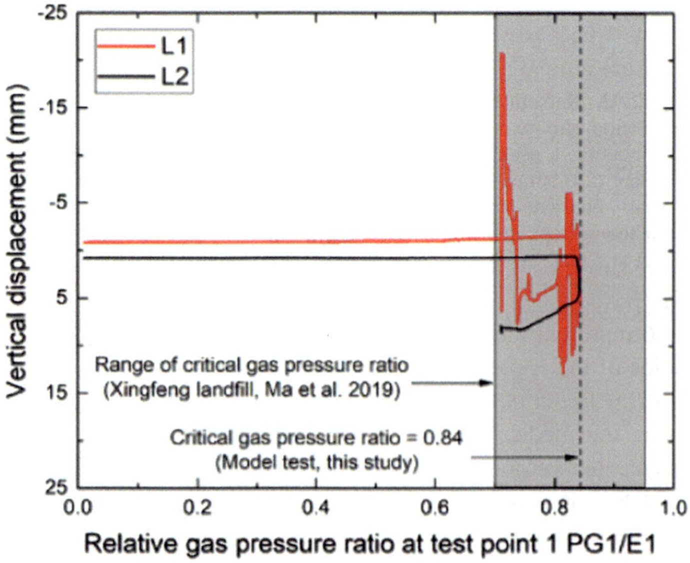

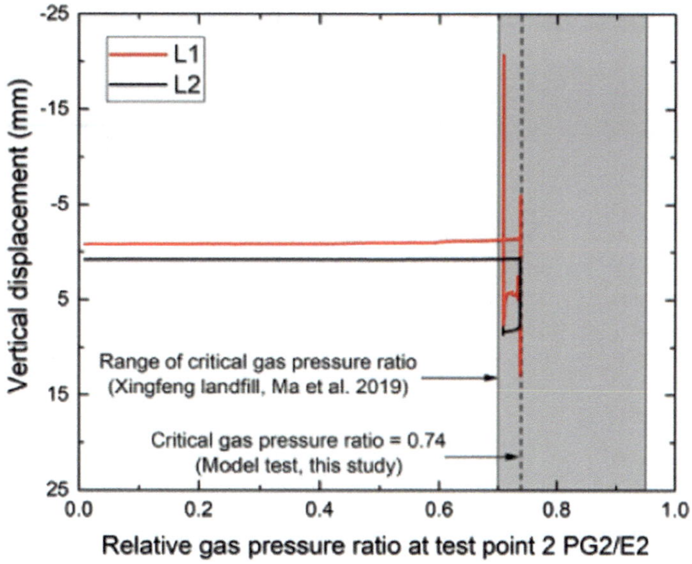

Fig. 7.46 Variation in vertical displacement with the relative gas pressure ratio at test point 2 (close to the slope side) (reprinted from Hu et al. 2024b, with permission from Canadian Science Publishing)

indicating sliding of the landfill slope. According to the curve of vertical displacement with respect to the relative gas pressure ratio, the critical gas pressure ratio at test point 1 (away from the slope side) was determined to be 0.84.

Figure 7.46 illustrates the variation in vertical displacement with the relative gas pressure ratio at test point 2 (close to the slope side). Notably, this trend at test point 2 resembled that at test point 1. In particular, when the relative gas pressure ratio at test point 2 reached 0.74, vertical displacements at both the slope crest and slope side abruptly increased. Hence, the critical gas pressure ratio at test point 2 was determined to be 0.74, indicating that the critical gas pressure ratio near the slope side was lower, rendering this area more susceptible to sliding under the influence of gas pressure. The gray areas in Figs. 7.45 and 7.46 represent the measured critical gas pressure ratios of the Xingfeng Landfill ranging from 0.7 to 0.95. The critical gas pressure ratio determined using the model test falls within this measured range, demonstrating that the test accurately reflects the critical triggering conditions of landfill instability under elevated gas pressures.

7.7 Expansion of Landfill Capacity Based on Stability Control

As illustrated in Fig. 7.47, during the landfilling process, each new layer of MSW imposes additional load on underlying layers, leading to primary and secondary compressions. As landfilling operations continue, the thickness and age of each MSW layer change continuously. Furthermore, variations in the main leachate level directly influence the effective stress acting on each MSW layer, causing varying degrees of settlement in each MSW layer and consequently increasing landfill capacity.

The expansion rate of landfill capacity (δ), defined as the ratio of the increased capacity to the designed geometric capacity, is calculated using the following formula:

$$\delta = \left(\frac{Q_m g}{\gamma_0} - 1 \right) \times 100\% \tag{7.18}$$

Based on the study conducted by Gao (2018), a case study was designed to examine the

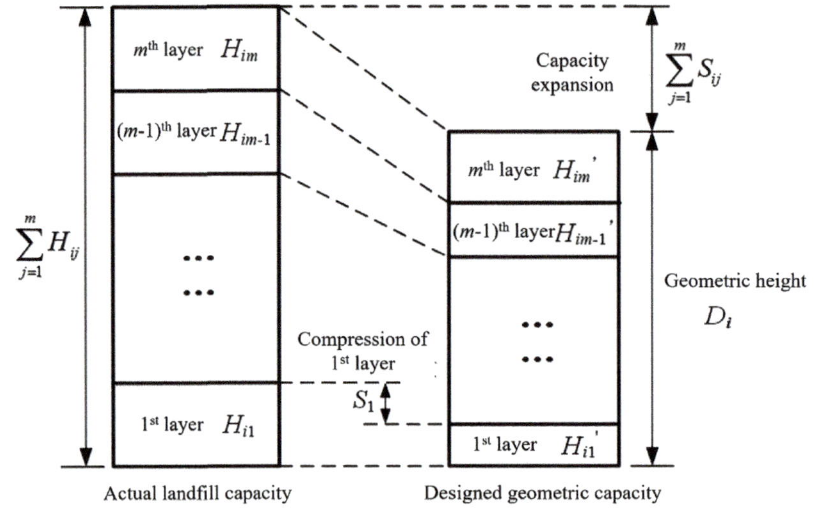

Fig. 7.47 Diagram illustrating landfill capacity expansion

relationships among landfill settlement, stability, and capacity, considering the differences in landfilling processes due to changes in leachate levels.

The case study involved a trapezoidal landfill, as depicted in Fig. 7.48, with a bottom length of 800 m, bottom elevation of 0 m, and slope gradient of 1:3. The landfill received 4,000 t of MSW daily and was filled incrementally from 0 m to a target elevation of 100 m at 10 m intervals. The leachate level ratio, defined as the ratio of the main leachate level to the target elevation, increased from 0% to 100% in 20% increments during the landfilling process to reach each target elevation. Using LANDFILL, the actual amount of MSW disposed, actual landfill capacity, and capacity expansion rate were computed under different conditions (various leachate level ratios and different target elevations). Furthermore, the waste compression model adopted the stress–degradation compression model proposed by Chen et al. (2010a).

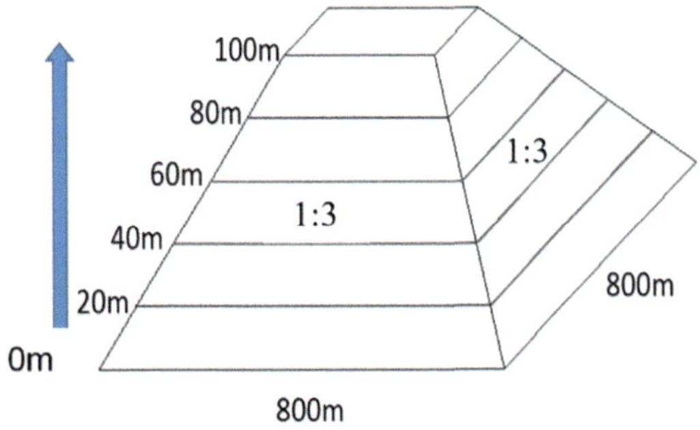

Fig. 7.48 Case study computing the landfill settlement, stability, and capacity

Figure 7.49a illustrates the correlation between the expansion rate and target elevation under different leachate level ratios. Notably, for a given leachate level ratio, the expansion rate increases as the target elevation of the landfill increases. Figure 7.49b illustrates the relationship between the expansion rate and leachate level ratio at different target elevations. In this case, for a given target elevation of the landfill, the expansion rate is higher at smaller leachate level ratios.

Figure 7.50 illustrates the relationship among the critical leachate level ratio, corresponding expansion rate, and target elevation. Evidently, as the target elevation increases, the corresponding leachate level ratio decreases significantly, whereas the corresponding expansion rate increases notably. At a safety factor threshold of 1.3, no critical leachate level ratio is observed at a target landfill elevation of 100 m. However, at a target elevation of 20 m, the corresponding critical leachate level ratio exceeds 90%, with an

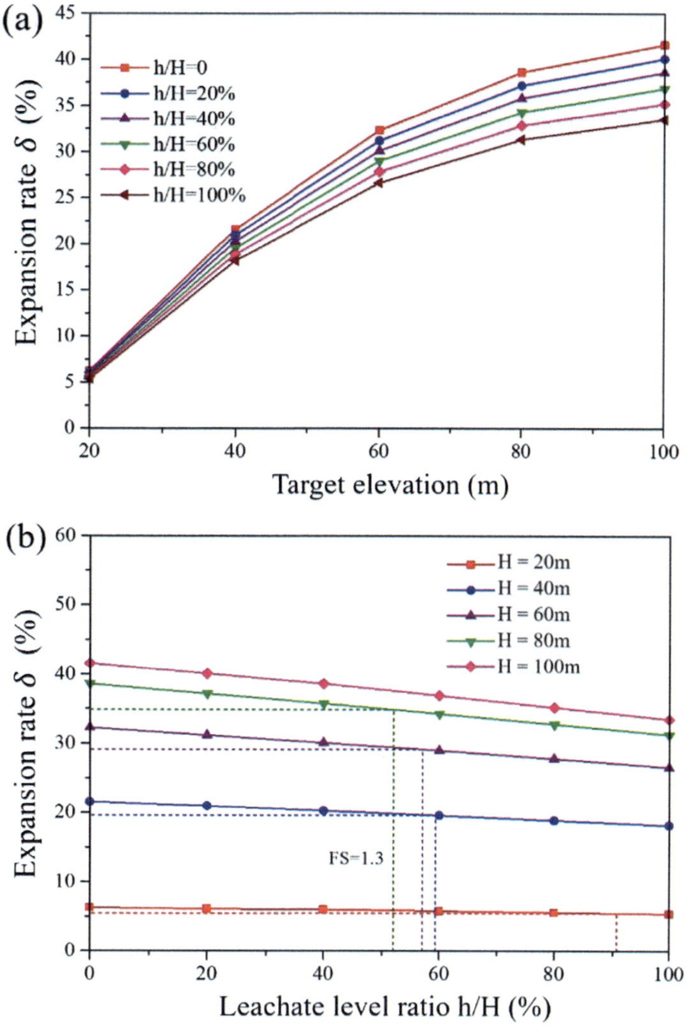

Fig. 7.49 Relationship between the **a** expansion rate and target elevation and **b** expansion rate and leachate level ratio

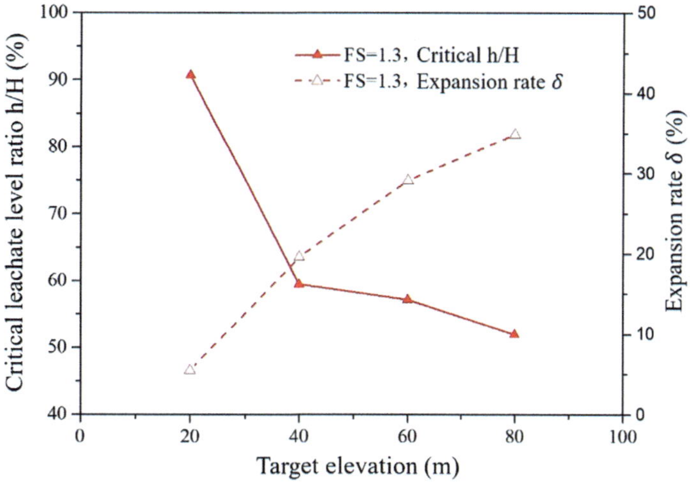

Fig. 7.50 Relationship among the critical leachate level ratio, target elevation, and corresponding expansion rate (δ)

expansion rate of only 5.5%. Conversely, for a target elevation of 80 m, the corresponding critical leachate level ratio drops to 51.9%, and the corresponding expansion rate increases to 34.9%. Notably, the expansion rate corresponding to the critical leachate level represents the minimum value necessary for landfill stability. Reducing the leachate level would result in even higher expansion rates.

This case study highlights that as modern landfilling advances toward large scales and intensive landfilling practices, increasing landfill heights lead to significant increments in the capacity expansion rate, optimizing land utilization. However, the stability of landfills diminishes with elevation. In particular, the level of leachate within the landfill significantly impacts its stability, as the critical leachate level ratio required for stability decreases with increasing elevation. This observation emphasizes the importance of stringent leachate level control measures. Maintaining the leachate level ratio below the critical threshold can yield further improvements in landfill capacity expansion rates.

In engineering practice, three-dimensional collection and drainage systems (see Sect. 6.4) are commonly employed in landfills to maintain leachate and gas pressures below their critical thresholds, ensuring the required stability safety factor. Combining small diameter dewatering wells, large diameter pumping wells, and horizontal drainage trenches has proven highly effective in lowering landfill leachate levels and enhancing slope stability (Zhan et al. 2015). Additionally, reinforcement techniques such as those relying on geosynthetics, pond slag backpressure, and raising of landfill dams are utilized to further enhance landfill stability (Ke et al. 2021).

For example, the Tianziling Landfill capacity expansion project in Hangzhou implemented measures such as improving liner system configuration to be lined on the sloping ground, establishing three-dimensional collection and drainage system to drawdown leachate level, raising the height of the downstream retaining dam, applying 10-m thick surcharge layer on the downstream slope, and setting up a safety monitoring system, including determination of the early crtical leachate level. The landfill stability analyses indicate these measures allowed the landfill to be filled to the design height of 130 m, ensuring its stability (Zhan et al. 2019).

References

Chen YM, Ke H, Fredlund DG, Zhan LT, Xie Y (2010a) Secondary compression of municipal solid wastes and a compression model for predicting settlement of municipal solid waste landfills. J Geotech Geoenvironmental Eng 136(5):706–717

Chen YM, Li JC, Yang CB, Zhu B, Zhan LT (2017) Centrifuge modeling of municipal solid waste landfill failures induced by rising water levels. Can Geotech J 54(12):1739–1751

Chen YM, Bian XC, Zhan LT, Ke H, Wang YS (2009) Landfill settlement and capacity calculation software. Copyright number: 2009SR08257 (in Chinese)

Chen YM, Zhan LT, Li YC (2010b) Development of leachate mounds and control of leachate-related failures at MSW landfills in humid regions. In: Proceedings of the 6th international congress on environmental geotechnics, New Delhi, pp 76–98

Chen YM (2014) A fundamental theory of environmental geotechnics and its application. Chin J Geotech Eng (in Chinese)

Fredlund DG, Morgenstern NR, Widger RA (1978) The shear strength of unsaturated soils. Can Geotech J 15 (3):313–321

Gao W, Chen YM, Zhan LT, Bian XC (2015) Engineering properties for high kitchen waste content municipal solid waste. J Rock Mech Geotech Eng 7(6):646–658

Gao W, Bian XC, Xu WJ, Chen YM (2018) Storage capacity and slope stability analysis of municipal solid waste landfills. J Perform Constr Facil 32(4):04018036

Gao W (2018) Research on time-dependent constitutive models for municipal solid waste and service performance of landfills. Ph.D. thesis, Zhejiang University, Hangzhou, China (in Chinese)

Hu J, Chen YM, Xu WJ, Ke H (2024a) A bio-hydro-chemical model for gas pressure development in municipal solid waste landfills. Géotechnique 74 (7):647–660

Hu J, Chen YM, Li JH, Ke H, Li JC, Zhu B (2024b) Centrifuge modeling of slope failure induced by elevated gas pressure in wet municipal solid waste landfill. Can Geotech J 61(6):1094–1108

Hu J (2020) Dual-porosity flow theory and foam-induced high gas pressure in MSW landfills. Ph.D. thesis, Zhejiang University, Hangzhou, China (in Chinese)

Ke H, Liu JL, Chen YM, Zhan LT, Xu XB (2010) Biodegradation-compression tests on municipal solid waste subjected to different vertical pressures. Chin J Geotech Eng 32(10):1610–1615 (in Chinese)

Ke H, Ma PC, Lan JW, Chen YM, He HJ (2021) Field behaviors of a geogrid reinforced MSW slope in a high-food-waste-content MSW landfill: a case study. Geotext Geomembr 49(2):430–441

Li JC, Zhu B, Lian BQ, Wang L, Ke H, Chen YM (2014) Strain-hardening mechanisms and methods for determining strength parameters of municipal solid waste. Chin J Rock Mech Eng 33(4):826–837

Li JC, Guan LH, Wang YB, Zhu B, Chen YM (2024) Centrifuge modelling on seismic failure of MSW landfills with high water level. Soil Dyn Earthq Eng 182:108753

Li JC (2018) Studies on static and seismic stability of landfills with high water level by centrifugal model tests. Ph.D. thesis, Zhejiang University, Hangzhou, China (in Chinese)

Lin WA, Zhu B, Chen YM, Zhan LT (2008) Tension analysis of geomembrane in landfill slope considering interface strain-softening. Rock Soil Mech 29 (8):2063–2069 (in Chinese)

Ma PC, Ke H, Lan JW, Chen YM, He HJ (2019) Field measurement of pore pressures and liquid-gas distribution using drilling and ERT in a high food waste content MSW landfill in Guangzhou, China. Eng Geol 250:21–33

Ministry of Housing and Urban-Rural Development of the People's Republic of China (2012) Technical code for geotechnical engineering of municipal solid waste sanitary landfill (CJJ 176–2012). China Architecture & Building Press, Beijing (in Chinese)

Reddy K, Hettiarachchi H, Gangathulasi J, Bogner J, Lagier T (2009) Geotechnical properties of synthetic municipal solid waste. Int J Geotech Eng 3(3):429–438

Singh MK, Sharma JS, Fleming IR (2009) Shear strength testing of intact and recompacted samples of municipal solid waste. Can Geotech J 46(10):1133–1145

Thusyanthan NI, Madabhushi SG, Singh S (2006) Centrifuge modeling of solid waste landfill systems—part 1: development of a model municipal solid waste. Geotech Test J 29(3):217–222

Vanapalli SK, Fredlund DG, Pufahl DE, Clifton AW (1996) Model for the prediction of shear strength with respect to soil suction. Can Geotech J 33(3):379–392

Xie Y, Chen YM, Tang XW, Ke H, Zhan LT (2005) Development and application of biodegradation-compression test apparatus for municipal solid waste. Chin J Geotech Eng 27(5):571–578 (in Chinese)

Zhan LT, Chen YM, Ling WA (2008) Shear strength characterization of municipal solid waste at the Suzhou landfill. China. Eng Geol 97(3–4):97–111

Zhan LT, Xu XB, Chen YM, Ma XF, Lan JW (2015) Dependence of gas collection efficiency on leachate level at wet municipal solid waste landfills and its improvement methods in China. J Geotech Geoenvironmental Eng 141(4):04015002

Zhan LT, Lan JW, Li W, Chen YM (2019) Slope stabilization and capacity expansion at Tianziling Landfill in Hangzhou, China. In: Proceedings of the 8th international congress on environmental geotechnics volume 2: towards a sustainable geoenvironment. Hangzhou, Springer, 2018, pp 26–34

8 Sustainable Landfilling of Municipal Solid Waste

Abstract

This chapter delves into approach to sustainable landfilling, leveraging insights derived from the degradation–consolidation theory to manage waste degradation and stabilization within MSW landfills. Beginning with an exploration of the sustainable landfilling concept, the chapter subsequently details several strategies, including large-scale intensive landfilling and landfill capacity recycling, to mitigate environmental impacts and maximize landfill capacity recycling. The chapter then addresses mass transformations and carbon emissions associated with landfilled MSW, presenting pathways for mass transformation throughout the lifecycle of MSW and assessment methods for carbon emissions. The discussion then focuses on the exploitation potential of MSW landfills, considering the degradation and stabilization effects of fine particles and pollutants, and presents assessment methodologies. Additionally, aerobic accelerated stabilization and in-situ dehydration treatment are explored as means of enhancing landfill exploitation potential. Finally, the chapter discusses the mining and screening of landfilled MSW based on the project of the Woqishan Landfill in Wenzhou, China, covering the physicochemical properties and resource utilization of fine fractions derived from these processes.

Keywords

Municipal solid waste · Landfill · Sustainable landfilling · Intensive landfilling · Mass transformation · Carbon emission · Exploitation potential · Fine fraction · Aerobic accelerated stabilization · Dehydration treatment · Screening technologies

8.1 General Introduction

Kavazanjian (2010) described that sustainable landfilling is an approach to minimizing the lifecycle impacts of landfills, including minimizing the associated soil and water pollution, greenhouse gas emissions, and dissipation of nonrenewable resources. To achieve sustainable landfilling is to operate landfills in a more sustainable manner (Kavazanjian 2010; Townsend et al. 2015).

The primary content of sustainable landfilling can be expanded from three aspects: (1) to prevent geoenvironmental issues at landfills, including landfill slope instability, LFG emission and leachate leakage; (2) to increase the amount of MSW filled on the unit land, and to get the storage capacity recycled through landfill mining; and (3) to utilize LFG and reuse the materials from landfill mining (Chen et al. 2019a, b).

This chapter first introduces approach to sustainable landfilling. It outlines strategies like

Y. Chen, *Soil Degradation-Consolidation Theory and Its Applications*,
https://doi.org/10.1007/978-981-97-7985-7_8

large-scale intensive landfilling and landfill capacity recycling to minimize environmental impact and maximize capacity. The chapter then covers mass transformations, carbon emissions, and assessment methods throughout the MSW lifecycle. It also examines the mining potential of MSW landfills, addressing fine particle and pollutant effects, and presents methodologies for aerobic accelerated stabilization and in-situ dehydration to improve the mining potential of landfills. Lastly, it discusses the mining and screening of landfill MSW based on the project of the Woqishan Landfill in Wenzhou, China.

8.2 Approach to Sustainable Landfilling

8.2.1 Large-Scale Intensive Landfilling

Large-scale intensive landfilling involves the optimal disposal of MSW within a confined landfill space while maintaining stable and secure landfill operation. To this end, sustainable geoenvironmental disaster prevention and control methods are employed, particularly focusing on regulating leachate levels. These methods predominantly encompass capacity-enhancing techniques based on stability control, measures for mitigating leachate pollution associated with groundwater and soil, and efficient techniques for minimizing and collecting LFG leakage.

(1) Capacity-enhancing techniques based on stability control

Maximizing the landfill capacity of the limited geometric space of a landfill entails optimizing waste disposal through efficient waste settlement. Such settlement primarily occurs in two phases: The first is instantaneous settlement resulting from the increase in load upon initial waste placement, while the second is degradation settlement originating from the weakening of the waste skeleton following solid waste degradation. To adhere to the requirements of stability and safety, the height of the landfill body is regulated by controlling leachate levels following landfill expansion. This approach, grounded within stability control, exploits the theoretical correlation among landfill settlement, capacity, and stability. Notably, this approach of expediting landfill stabilization and settlement to improve capacity expansion rates and controlling stability through leachate level reductions has been successfully implemented in numerous large-area landfill expansion projects across major cities in China, including Shanghai, Hangzhou, and Xi'an.

(2) Control measures for mitigating leachate pollution associated with groundwater and soil

Pollutant concentrations and hydraulic loads within the drainage layer of the landfill directly influence the risk of groundwater and soil contamination by leachates. Employing efficient cover layers and dewatering measures can help manage the hydraulic load within the drainage layer, thereby effectively mitigating the leachate pollution associated with groundwater and soil. Furthermore, horizontal and vertical composite barriers can prevent leachate penetration, thus minimizing groundwater and soil contamination.

(3) Efficient collection and minimizing emission of LFG

Employing improved temporary covers minimizes gas escape, while installing three-dimensional collection and drainage systems within the landfill body enhances gas transport performance. These measures considerably enhance the collection rate of LFG, thereby reducing its uncontrolled leakage into the atmosphere.

Thus, sustainable geoenvironmental disaster prevention and control methods can facilitate large-scale intensive landfilling activities. A comparative analysis between the key indicators of the Tianziling Landfill in Hangzhou, China, and the Puente Hill Landfill in the United States, reveals that Chinese landfills, even HFWC landfills, can successfully accomplish large-scale intensive landfilling (Table 8.1).

Table 8.1 Indicators of the Tianziling landfill in China and Puente hill landfill in the United States

Technical indicator	Hangzhou Tianziling landfill	Puente hill landfill
Area	The total area is 9.6×10^5 m^2, while the landfill area is 4.8×10^5 m^2	2.8×10^6 m^2
Cumulative landfill amount	2.926×10^6 t	130.0×10^6 t (1957–2013)
Height of the landfill body	130 m	150 m (500 feet)
MSW disposed per unit area	60.9 t/m^2	71.4 t/m^2
Power generation by LFG	The installed capacity is 16 MW. The LFG collection amount in 2019 was 4.6 × 104 m^3/h. The annual power generation is 125.39 was 10^6 kWh, while the revenue generated from electricity sales exceeded 83 million yuan	The installed capacity is 46 MW

8.2.2 Recycling of Landfill Capacity

At present, numerous Chinese landfills are nearing the end of their operational lifespan, leading to the potential shortage of efficient disposal options for a considerable amount of MSW in the near future. Fortunately, the recycling of landfill capacity offers a viable solution to this challenge.

The complete mineralization of MSW typically requires decades or even centuries. However, owing to the short construction periods of landfills in China, achieving complete mineralization and subsequent excavation is challenging. Notably, the biochemical degradation process of Chinese HFWC MSW can be categorized into rapid degradation, slow degradation, and post-stabilization stages. Once the MSW approaches the post-stabilization stage, its biodegradable components essentially deplete, thus reducing the potential release of pollutants including LFGs and leachates, in turn minimizing their impact on the surrounding environment. Hence, the exploitation, categorization, and resource utilization of the post-stabilization-stage MSW disposed within landfills can help minimize waste volumes and promote the recycling of landfill capacity. Figure 8.1 illustrates the conceptual process of landfill capacity recycling.

To achieve sustainable landfilling, appropriate measures must be implemented at varying degradation stages of the MSW disposed within landfills to expedite its stabilization, thus shortening the cycle of sustainable landfilling. Drawing insights from the degradation and stabilization processes of HFWC MSW in China, feasible specific measures have been introduced (Fig. 8.2).

During the rapid degradation stage of landfilled MSW, measures such as leachate recirculation can be implemented to alleviate acidification inhibition and accelerate hydrolysis and methanation. Meanwhile, during the slow degradation stage, oxygen ventilation can convert the anaerobic degradation environment of the landfill body into an aerobic one, thus accelerating the degradation of recalcitrant organic matter. Assessments and analyses of several degradation and stabilization indicators can aid in identifying whether the landfilled MSW has entered its post-stabilization stage. At this stage, excavation and sorting activities can be undertaken for MSW that no longer poses environmental pollution risks. The resulting fine-grained fractions can be repurposed as soil for agricultural and forestry applications, while lightweight combustible fractions can be utilized for electricity generation through incineration.

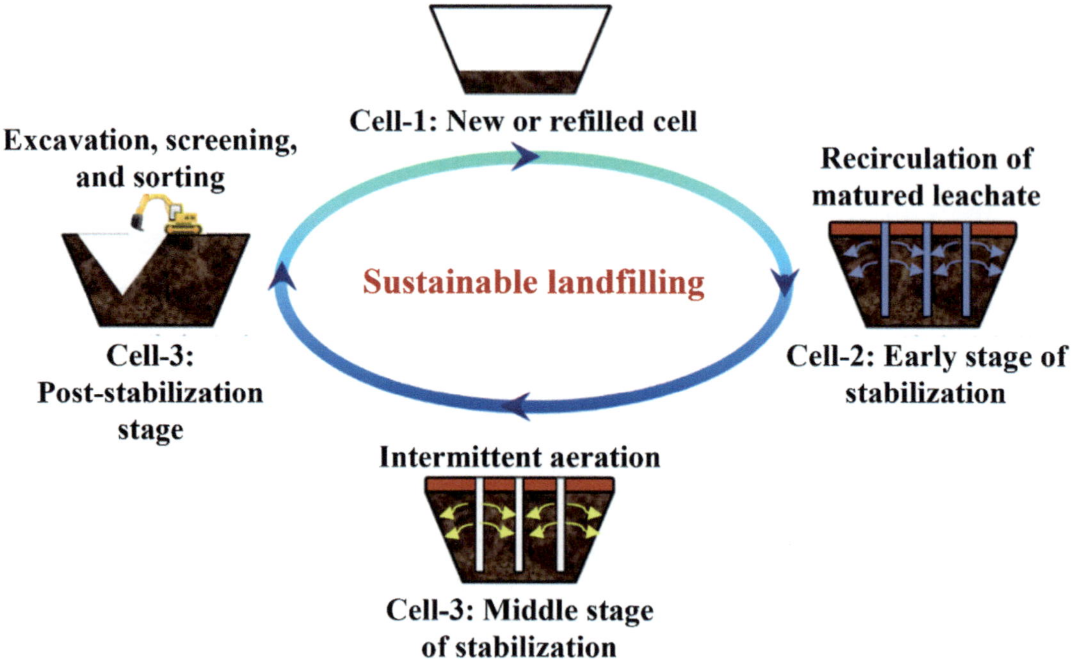

Fig. 8.1 Conceptual process of landfill capacity recycling (Guo 2017)

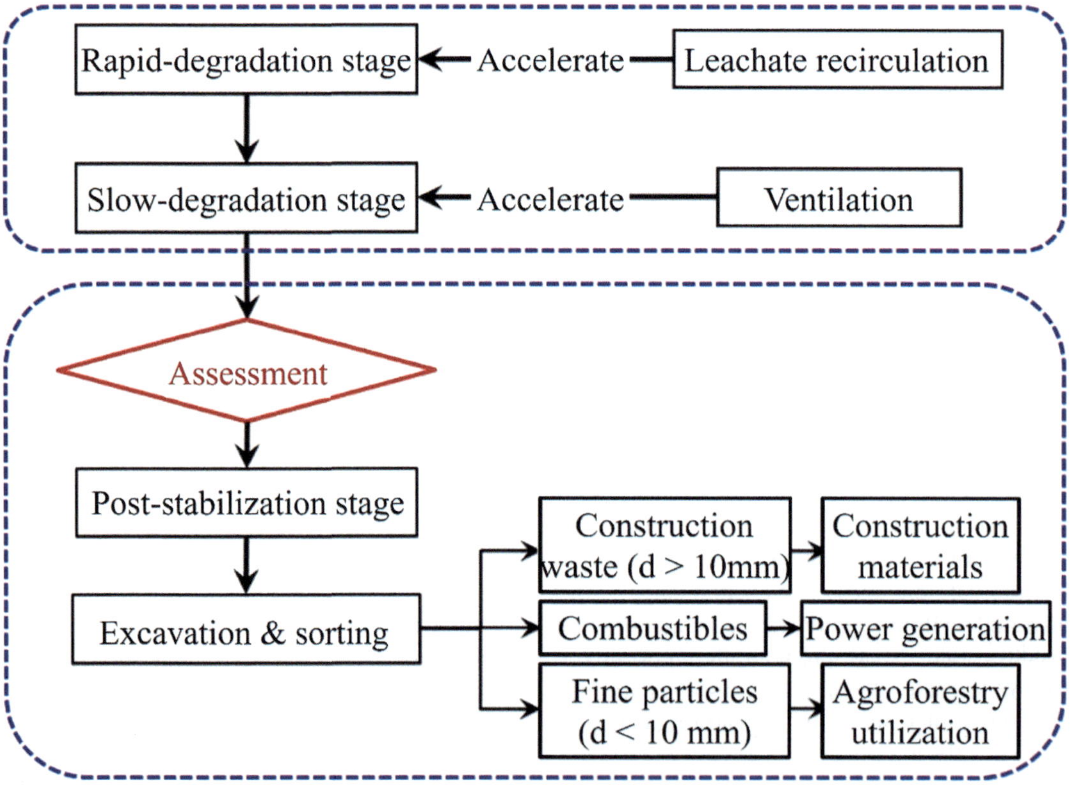

Fig. 8.2 Measures corresponding to each stage of sustainable landfilling

Meanwhile, coarse aggregate materials, after processing, can be employed as construction materials.

8.3 Mass Transformation and Carbon Emissions of MSW

8.3.1 Mass Transformation Pathways of Landfilled MSW Throughout Its Lifespan

The mass transformation pathways of landfilled MSW encompass various processes through which substances within the MSW can transform under the influence of several biological, chemical, and other factors. In the anaerobic degradation environments of landfills, biodegradable organic materials typically undergo a two-stage process of hydrolysis–acidogenesis and methanogenesis. Among these, hydrolysis–acidogenesis decomposes biodegradable components, such as sugars, cellulose, lipids, and proteins, into intermediate products such as VFAs under hydrolytic conditions. Methanogens then convert these VFAs into CH_4 gas.

Throughout its lifespan, the environmental conditions of landfilled MSW are not exclusively restricted to anaerobic degradation. For instance, implementing aeration to adjust oxygen levels within the landfill body can transform the anaerobic degradation conditions into aerobic degradation conditions. Through this transformation, aerobic microorganisms can facilitate the aerobic hydrolysis and synchronous nitrification and denitrification of biodegradable organics, ultimately producing CO_2 and water.

Once MSW reaches the post-stabilization stage, it can be excavated, screened, and reused. Subsequently, lightweight materials from the excavated MSW, which typically comprise nondegradable or slowly degradable plastic products, can be incinerated to generate electricity. During this process, organic materials undergo combustion, releasing CO_2 and fly ash.

Evidently, the mass transformation pathways of landfilled MSW throughout its lifecycle are not singular or static. Instead, environmental conditions and engineering interventions can lead to MSW mass transformation pathways extending beyond anaerobic degradation, including aerobic degradation and incineration.

8.3.2 Assessment Method for Carbon Emissions from Landfills

In landfills, MSW primarily undergoes anaerobic degradation, releasing LFG containing CH_4 and CO_2. Theoretically, these gases constitute equal proportions of the LFG. However, CH_4, in particular, is a primary non-CO_2 greenhouse gas, demonstrating a 28-fold stronger greenhouse gas effect compared to CO_2. Consequently, CH_4 emissions from landfills significantly impact global climate warming. In fact, the greenhouse gas carbon emissions of landfills worldwide account for approximately 6% of the total emissions.

Previous studies have analyzed the carbon emission potentials of all possible mass transformation pathways of MSWs within landfills, including anaerobic degradation, aerobic degradation, and incineration. Considering 1 t of Chinese MSW as a benchmark, the carbon emissions of different degradation pathways have been computed based on the involved chemical transformations. Given the electricity generation capability of both waste incineration and LFG combustion, these can offer additional carbon emission reduction effects. Notably, the carbon emission baselines for all degradation pathways summarized in Table 8.2 are computed through biochemical reaction processes based on the mass conservation principle (Wei et al. 2024).

During the carbon emission assessments of specific landfill sites, carbon emissions resulting from waste degradation and from daily landfill operations must be comprehensively considered. These assessments must also account for the emission reduction effects of LFG collection for power generation. The specific inventory is detailed in Fig. 8.3.

LFG emissions involve the direct release of CH_4 and CO_2 from the landfill surface into the atmosphere. To assess LFG emissions, the

Table 8.2 Carbon emissions of various mass transformation pathways

MSW disposal method	Carbon transformation pathway	Carbon emissions per unit mass/Net carbon emissions considering power generation reduction (kg/t)
Incineration	Combustible materials + $O_2 \rightarrow CO_2$	639.85/235.69
Aerobic degradation	Degradable carbon $\rightarrow CO_2$	382.96
Anaerobic degradation	Degradable carbon $\rightarrow CO_2 + CH_4$	2,115.2/206.12

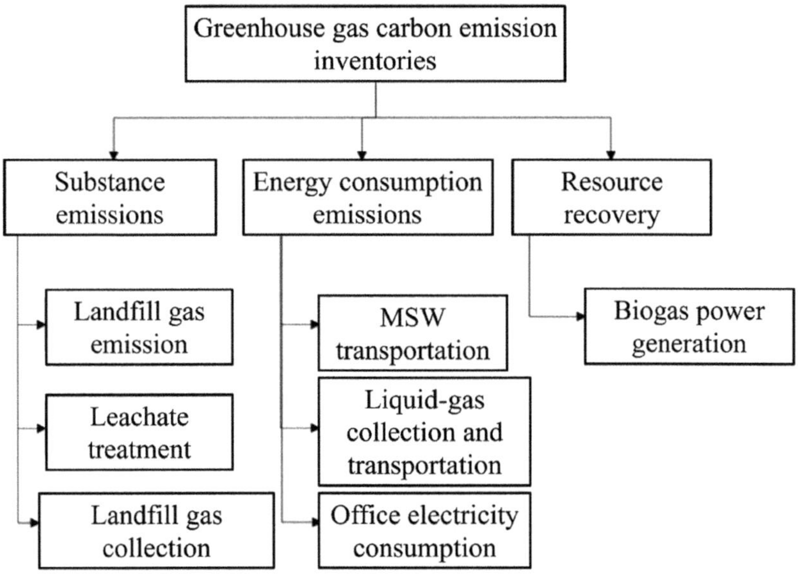

Fig. 8.3 Landfill carbon emission assessment inventory

landfill surface is typically segmented into three areas: closure, intermediate cover, and landfill operation areas. By monitoring the carbon emissions per unit area in each area type and considering changes in their respective areas, the total carbon emissions from gas leakage can be determined. Various methods, including the static chamber method, differential absorption lidar, gas tracer method, drone detection method, and satellite remote sensing, are commonly employed to monitor surface carbon emissions.

Landfill leachate contains biodegradable organic components. For HFWC MSW landfills in China, these organic components can constitute up to 10% of the total biodegradable organic matter within the MSW. These organic materials further react to produce CH_4 and CO_2 in leachate collection and treatment systems. Thus, carbon emissions from leachate treatments can be estimated based on the concentrations of biodegradable organic materials within the leachate.

In landfills equipped with LFG collection systems, sensors such as flow meters and gas composition analyzers installed within the LFG collection facilities can accurately monitor the collected LFG amount and its composition. Once collected, LFG is typically treated through either direct flaring or combustion for power generation. Both methods convert CH_4 from the LFG into CO_2 through combustion. Carbon emissions from LFG collection systems can be directly computed using gas volume and composition data from monitoring experiments. Furthermore, if LFG is utilized for power generation, carbon emission reductions can be directly assessed based on the amount of electricity generated, reflecting its carbon reduction effects. Activities such as waste collection and transportation, as well as on-site electricity usage, contribute toward the daily energy consumption of landfills. These energy consumption activities contribute to the overall carbon emissions of the landfill and should therefore be factored into assessments.

For instance, the carbon emissions of the Tianziling Landfill in Hangzhou have been calculated and assessed. Theoretical calculations using both the IPCC model and the two-stage gas generation model for HFWC MSW were conducted to estimate LFG production. The obtained LFG production amounts were subsequently compared with the actual collection amount of LFG recorded on-site, as depicted in Fig. 8.4.

The environmental conditions of Hangzhou, with an average annual temperature of 16.5 °C and a humid climate characteristic of the temperate zone, align with the default values specified by the IPCC (2006), featuring an LFG generation rate constant (k) of 0.09. For the two-stage model, the values of k_R (LFG generation rate constant of rapidly degradable components) and k_S (LFG generation rate constant of slowly degradable components) were selected within the recommended average ranges.

The results highlight that the two-stage gas generation model for HFWC MSW better captures the dual-phase gas generation pattern. This pattern features a peak at 1–2 years after landfilling initiation, followed by a rapid decline in gas generation rate over time. The actual collection amount of LFG recorded on-site closely correlates with operational management practices. Notably, in the initial stages of landfill construction, gas collection facilities were minimal. However, with the increasing installation of gas collection pipelines and gas generators over time, LFG collection peaked. However, following the termination of landfilling activities by the end of 2020, LFG collection rapidly declined, consistent with the expected gas generation pattern of HFWC MSW. Furthermore, the actual collection amount closely aligned with the theoretical calculation results.

Notably, the theoretical LFG production estimate can be considered as the maximum value of LFG production. Generally, of the total LFGs produced within a landfill, some LFGs can be released into the atmosphere through pathways such as surface fugitive emissions and losses via leachate drainage.

By considering the total energy consumed by landfill operations, waste collection facilities, and transportation activities, as well as the carbon emission reduction effects of LFG-based power

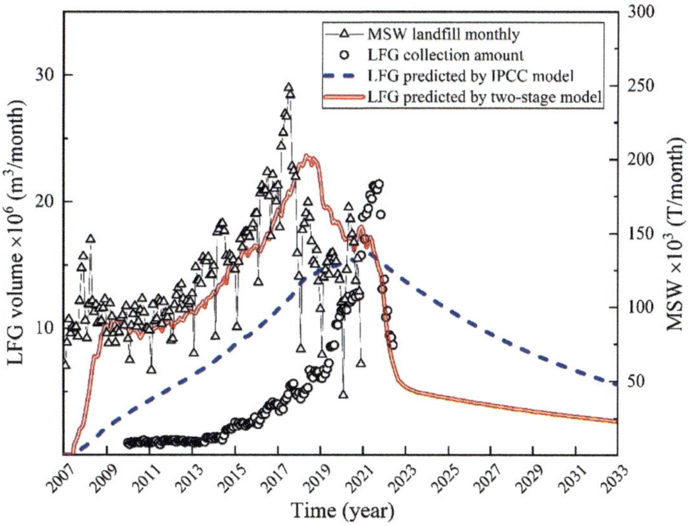

Fig. 8.4 Theoretical LFG production compared to actual LFG collection (Wei et al. 2024)

generation, the annual carbon emissions of a given landfill can be computed, as depicted in Fig. 8.5. As indicated, the total carbon emissions of the landfill amounted to 25.52 Mt, with significant contributions from LFG leakage. The LFG collection system effectively reduced the fugitive LFG, resulting in a carbon emission reduction of 9.47 Mt from 2018 to 2022. Furthermore, carbon emission reductions of 0.57 Mt and 0.48 Mt were achieved through LFG-based power generation and leachate treatment, respectively, yielding a total carbon emission reduction of 10.52 Mt.

In conclusion, when assessing the total carbon emissions of a given landfill, carbon emissions from the generated LFG, carbon emissions from the daily activities of the landfill, and carbon emission reductions through LFG-based power generation must be comprehensively considered. A modern landfill facility with well-established operational management can efficiently collect and utilize data from routine monitoring to reasonably assess the current and future carbon emissions of the landfill.

8.4 Assessing and Improving Mining Potential of Landfills

When landfilled MSW enters its slow degradation of post-stabilization stages, it can be excavated, sieved, and reused. Following sieving, fine particles from the waste can be utilized for agricultural or forestry greening or for landfill capping applications. Meanwhile, lightweight fractions from the waste can be used for waste-to-energy incineration, and coarse aggregate materials can be utilized to produce construction materials. However, the successful excavation and utilization of landfilled MSW necessitates fulfillment of certain prerequisites. For instance, the water content of the waste must not be excessive to facilitate efficient sieving. Furthermore, the content of heavy-metal ions and pollutants in the fine particles of waste must meet environmental standards for reuse. Moreover, the degree of degradation and stabilization of the waste must meet reference values, implying that the waste must undergo adequate decomposition before its excavation and utilization.

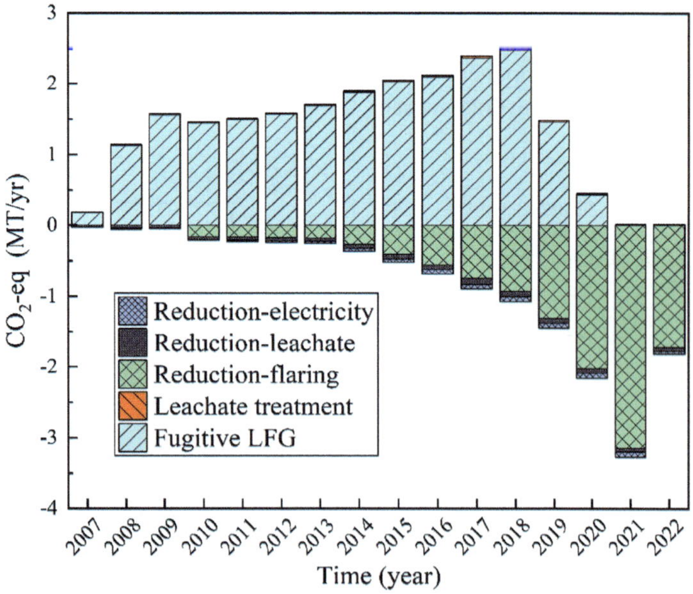

Fig. 8.5 Contributions of various factors to the total carbon emissions of a landfill (Wei et al. 2024)

8.4.1 Effects of Stabilization on Fine Particles and Pollutant Content

During the biochemical degradation of landfilled MSW, its solid phase components undergo gradual decomposition, resulting in an overall reduction in the size of solid particles. Ultimately, as degradation progresses, the content of fine particles increases. Furthermore, leachate migrating through the landfill body carries these fine particles toward the interior of the waste body. Conversely, pollutants within waste, such as heavy-metal ions, do not decompose during degradation; however, they migrate through the landfill body owing to leachate percolation. Furthermore, owing to the varying adsorption capacities of different heavy-metal ions by the solid phase of waste, the contents of heavy-metal ions within waste differ with the degree of waste degradation.

Chen et al. (2019a) conducted sampling tests and analyses on landfilled MSW at different burial depths and ages. They utilized the normalized ratio between cellulose and lignin (β) content as a parameter assessing the degree of degradation and analyzed the content of fine particles and heavy-metal ions. As depicted in Fig. 8.6, when the β value increased from 0.76 to 0.86, the content of fine particles also increased from 40% to around 48–60% (i.e., a higher β value corresponds to a greater content of fine particles). This fine particle content demonstrates a significant increase compared to the fine particle content (of 10.91%) of the initial fresh waste. Evidently, as the stabilization level of a landfill improves, decomposable components such as food waste and paper continuously degrade into fine particles, leading to a corresponding increase in the content of fine particles.

Figure 8.7 illustrates the variation in the arsenic (As) and lead (Pb) concentrations within fine particles with changing β values. As depicted in Fig. 8.7a, the concentration of As increases from 16 mg/kg to 28 mg/kg with increasing β values. Notably, As primarily originates from recalcitrant components, and during the slow degradation stage, the decomposition of degradable components leads to the gradual accumulation of As in fine particle fractions. Figure 8.7b indicates that the concentration of Pb decreases from 200–250 mg/kg to 100–150 mg/kg as β increases, indicating that Pb exhibits strong leaching characteristics. During the slow degradation stage, Pb gradually precipitates owing to

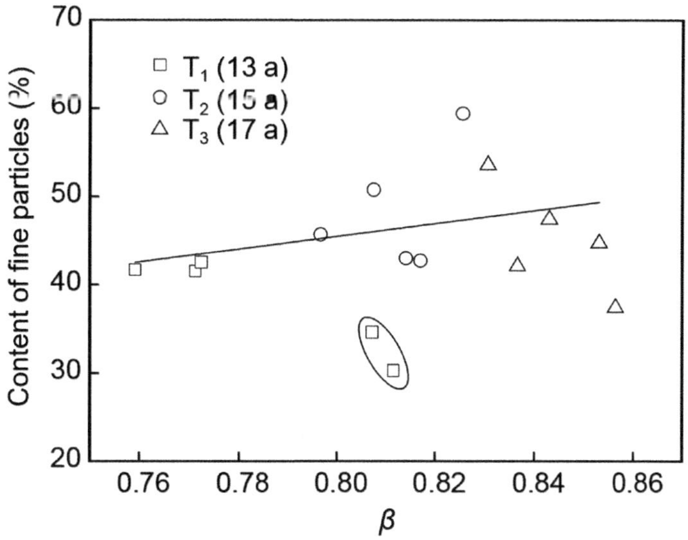

Fig. 8.6 Relationship between the normalized degradation–stabilization index and fine particle content (on a dry basis) (reprinted from Chen et al. 2019a, with permission from Scientia Sinica Technologica)

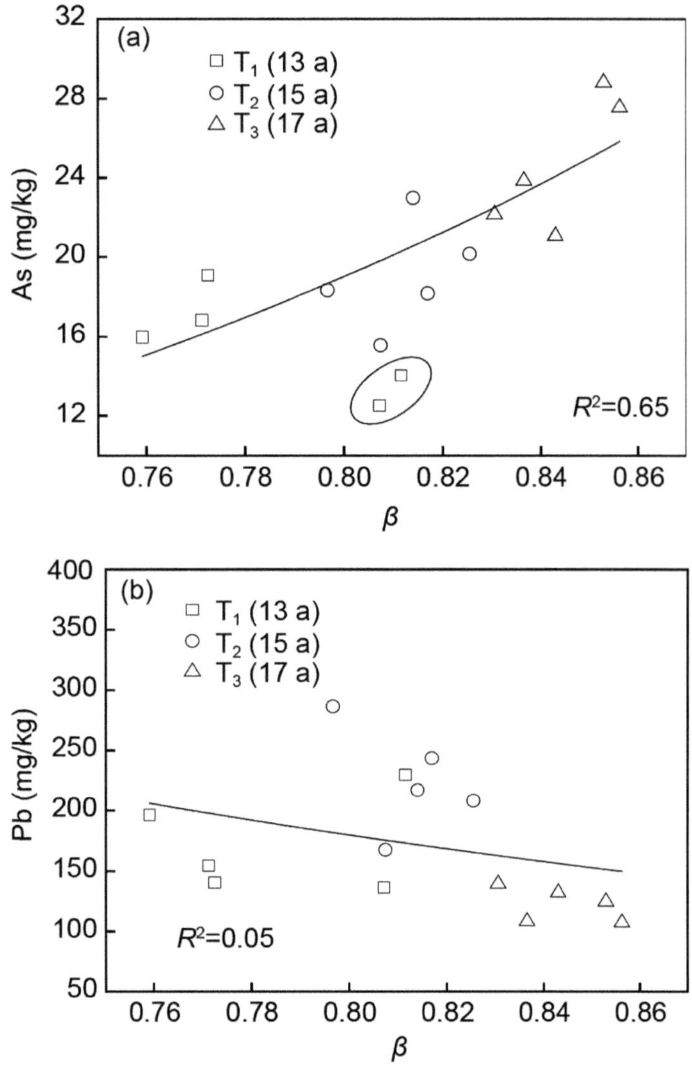

Fig. 8.7 Heavy-metal ion concentrations of fine particle fractions at different degradation levels: **a** As and **b** Pb (reprinted from Chen et al. 2019a, with permission from Scientia Sinica Technologica)

the flushing and soaking of leachate, leading to a gradual decrease in its concentration. Meanwhile, the concentrations of other heavy-metal ions, such as zinc (Zn), mercury (Hg), copper (Cu), and cadmium (Cd), remain relatively unchanged with increasing β. This is primarily because these heavy-metal ions are transferred from degradable substances to fine particle fractions during the rapid degradation stage and are adsorbed onto fine particles, exhibiting relatively poor leaching characteristics. Consequently, the measured concentrations of these heavy-metal ions demonstrate insignificant changes with variations in β.

8.4.2 Assessment Method for Mining Potential

To assess the feasibility of excavation at a landfill site, systematic survey, testing, and analysis procedures are required. First, a survey must be conducted to gather basic information regarding the spatial distribution of landfilled MSW, including its age, composition, degradation level, and other factors. Second, systematic testing must be performed on waste samples collected during the survey. Figure 8.8 illustrates the recommended procedure for this testing, including tests on the water content, sieving, assessments of the chemical properties of leachate from fine particles, examination of the degree of solid phase degradation, analysis of the pollutant concentration within fine particles, and sorting of coarse particles.

Based on the systematic testing of landfilled MSW samples, further analysis can assess the feasibility of excavation at the landfill site. For instance, waste disposed of in the Tianziling Landfill in Hangzhou contained 43.83% fine particles, 42.97% combustible fractions, 13.20% construction materials, and had a water content of 40.76%. Waste excavation and screening activities would require the drainage of approximately 2.46 Mt of leachate, yielding 1.56 Mt of fine particles, 1.53 Mt of combustible fractions, and 0.47 Mt of construction materials. The resulting combustible fractions can be directly incinerated for power generation, while construction materials can be directly utilized as building materials or recycled. However, the resulting fine particles cannot be directly used as soil for greening owing to deviations from the requirements of the "Soil Environmental Quality—Risk control standard for soil contamination of agricultural land" (GB 15618-2018) and the "Standard for Pollution Control on the Landfill Sites of Municipal Solid Waste" (GB 16889-2008). Nevertheless, at the Tianziling Landfill, containing approximately 9 Mt of initial fresh MSW, even if all fine particles are not utilized and instead backfilled into the landfill site, the reduction rate following excavation and screening can still exceed 80%, effectively achieving landfill capacity reuse.

High leachate levels are common in HFWC MSW landfills in China. While such high leachate levels reduce the strength and compromise the stability of the landfill body, thus complicating waste excavation. In addition, high water contents within landfilled MSW reduce the efficiency of mechanical screening. Therefore, effective drainage of leachate is a key aspect of landfill excavation feasibility and forms the focus of the assessment. Generally, if landfilled MSW exhibits low permeability or if on-site precipitation conditions prevent the efficient drainage of leachate, the given landfill site is not feasible for excavation. Conversely, if the landfill body is capable of achieving leachate drainage through engineering measures, such as the combined use of small-diameter wells (short construction periods and cost-effective) and large-diameter wells (long service lives and clogging prevention), then the landfill site can be deemed feasible for excavation.

8.4.3 Aerobic Accelerated Stabilization and In-Situ Dehydration Treatment

For anaerobic landfills failing to meet stabilization standards, aerobic accelerated stabilization treatments can be implemented to satisfy the requirements of excavation feasibility. The fundamental principle of aerobic accelerated stabilization involves treating the landfill as a large-scale bioreactor, housing buried aeration, injection, extraction, and ventilation wells. In this case, high-pressure blowers are used to inject fresh air into the landfill body, while extraction blowers are used to remove anaerobic LFG such as CH_4 and CO_2 from the waste. These processes are accompanied by continuous monitoring of the waste temperature and gas composition. Furthermore, leachate is either extracted or reinjected into the landfill body to achieve optimal water content, creating an ideal aerobic degradation environment. This process activates aerobic microorganisms within the waste, thus accelerating the degradation of organic matter and reducing leachate and gas pollution loads. In particular, aerobic degradation can shorten the degradation duration of organic matter within the landfill by 5–30 times. In practice, following aerobic stabilization treatment, landfills often stabilize within 2–4 years (Li et al. 2021). Compared to anaerobic degradation, which produces strong greenhouse gases such as CH_4, NH_3, and H_2S, the products of aerobic degradation (CO_2 and H_2O) are less harmful; hence, aerobic degradation contributes toward greenhouse gas reduction. Under aerobic degradation conditions, the concentrations of pollutants within leachate, indicated by measures such as

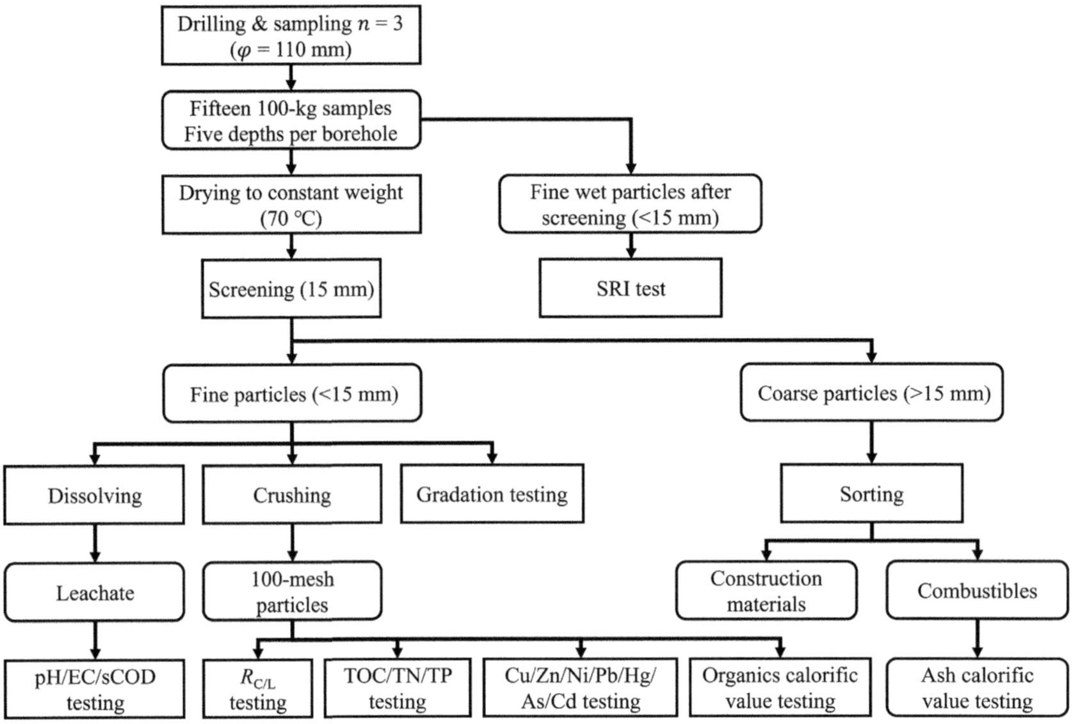

Fig. 8.8 Systematic testing process of landfilled MSW samples (reprinted from Chen et al. 2019a, with permission from Scientia Sinica Technologica)

the BOD, COD, and NH_4-N concentration, decrease rapidly, increasing the solid phase settlement rate and thus resulting in reduced deformation after landfill cover.

Table 8.3 lists typical landfill projects subjected to aerobic accelerated stabilization. In 1997, the United States pioneered a large-scale aerobic accelerated stabilization project at the Columbia County Landfill in Georgia. Subsequently, with support from the United States Environmental Protection Agency, over 20 aerobic treatment projects were implemented at landfills such as the New River Landfill in Florida, Williamson County Landfill in Tennessee, and Yolo County Landfill in California (Read et al. 2001; Powell et al. 2006; Ko et al. 2013). In the following years, other countries such as Germany, Italy, and Austria gradually initiated large-scale aerobic treatment projects in the early twenty-first century, conducting detailed tests on the biochemical characteristics of waste before and after aerobic treatment (Raga and Cossu 2013; Hrad and Huber-Humer 2017; Stegmann 2019). However, aerobic engineering projects in China began relatively late. Here, the first aerobic treatment project was launched at the Heishitou Landfill in Beijing in 2007. Remarkably, after aerobic treatment, the primary indicators satisfied the requirements of the "Technical Requirements for Site Utilization after Stabilization in Municipal Solid Waste Landfill" (GB/T 25179-2010) for moderate utilization conditions of stabilized landfill sites, rendering it suitable for the construction of public facilities such as parks, sports fields, golf courses, and wildlife parks (Tang 2012). In 2014, the largest in-situ aerobic remediation project worldwide was initiated at the Jinkou Landfill in Wuhan, treating a waste volume of 5.02 Mm^3. Following remediation, the site later hosted the 10th China International Garden Expo (Liu et al. 2018).

Initially, the environment of a landfill is anaerobic, with high CH_4 contents in the produced LFG. Aerobic stabilization remediation

Table 8.3 Typical aerobic accelerated stabilization landfill projects

Country	Landfill (aerobic remediation time)	Amount ($\times 10^6$ m^3)	Remediation time and effectiveness
U.S.	Columbia County Landfill (1997)	19.5	Aerobic remediation spanned 18 months. The maximum settlement rate of the landfill was 9%. Its leachate volume decreased by 86%, BOD reduced by 70%, VOC concentration decreased by 75–99%, and CH_4 content decreased by 50–90%
	Live Oak Landfill (2000)	5.7	Aerobic remediation spanned 9 months. The maximum settlement rate of the landfill was 10%. Its biodegradation rate increased by 110%, leachate volume decreased by 50%, and leachate VOC concentration decreased by 50%, and CH_4 content decreased by 50–90%
	New River Landfill (2003)	88.0	Aerobic remediation lasted for 36 months. Air injection into the deep, high-moisture waste areas of the landfill was difficult, and frequent shutdowns occurred owing to rapid temperature increments to 75 °C. The resulting remediation effect was significantly inferior compared to expectations
China	Heishitou Landfill, Beijing (2007)	140.0	Aerobic remediation lasted for 24 months. The average content of biodegradable organic matter within waste decreased to 3.11%. CH_4 concentration dropped below 1%, and the settlement rate of the landfill reached approximately 6%
	Jinkou Landfill, Wuhan (2014)	502.0	Aerobic remediation lasted for 12 months. The content of organic matter within waste decreased to 9–10%, while the concentration of CH_4 decreased to less than 3.6%
	Yiling Landfill, Yichang (2013)	17.8	Aerobic remediation lasted for 11 months. After treatment, CH_4 concentration was less than 5%, the settlement rate was less than 0.83 mm/d, the BOD/COD ratio was 0.1, and the cellulose/lignin (C/L) ratio was less than 0.4
Germany	Kuhstedt Landfill (2003)	22.0	Aerobic remediation lasted for 75 months. After 2 months of aeration, the BOD, total organic carbon content, and halogenated hydrocarbon content within waste decreased by 85%, 60%, and 50%, respectively. After 22 months of aeration, the aerobic respiration index (RI_4) decreased from 6.2 mgO_2/gDM to 2.2 mgO_2/gDM. Stable conditions were achieved after long-term passive aeration
Austria	Landfill in Vienna (2013)	20.0	Aerobic remediation lasted for 64 months. The aerobic respiration index (RI_4) and anaerobic gas potential (GP_{21}) decreased from initial values of 1.7 mgO_2/gDM and 0.5 NL/kgDM to 1.5 mgO_2/gDM and 0.3 NL/kgDM, respectively. The concentration of ammonia nitrogen was below the target value of 400 mg/kgDM
Italy	Landfill in Northern Italy (2014)	6.0	Aerobic remediation lasted for 12 months. During this aeration, the temperatures at all monitoring points increased by an average of 6–7 °C, while the water content decreased from 27% to 50% before treatment to 23%. After aerobic treatment, the values of RI_4 and GP_{21} were 2.8 mgO_2/gDM and 7.1 NL/kgDM, respectively

entails the injection of a significant amount of air into the landfill. However, when injected O_2 mixes with the CH_4 already present within the landfill, fire or explosion incidents can occur. As depicted in Fig. 8.9, Ko et al. (2013) proposed several criteria for gas combustion and explosion risk assessments during aerobic remediation, emphasizing the importance of monitoring and evaluating CH_4 and O_2 concentrations within the LFG. According to these criteria, if the composition of the gas mixture falls within the explosive range, air injection must be immediately terminated, and the area must be monitored. Additionally, aerobic degradation of waste generates considerable heat, which can elevate landfill temperatures, potentially causing smoldering or even fires. Based on practical experiences from aerobic projects worldwide, if areas with temperatures exceeding 70 °C are observed, the rate of air injection must be adjusted. Furthermore, when temperatures approach the high-temperature threshold, terminating air injection while simultaneously injecting leachate into surrounding wells may be essential to lower temperatures.

In addition to low degradation and stabilization levels, China's stockpiled waste also features high water contents, causing fine particles to adhere to lightweight components such as plastics. This adherence leads to low separation efficiency during screening. To tackle these challenges, highly efficient, in-situ dewatering equipment for stockpiled waste has been developed (Patent number: CN202210571641.7) (Hu et al. 2022), as illustrated in Fig. 8.10.

This equipment comprises multiple extraction and injection wells that are strategically placed at regular intervals beneath the surface of the landfill body. At the bottoms of these extraction wells, submersible pumps are provided for leachate extraction, while air vents are positioned on the sidewalls at the top of the wells. Conversely, injection wells are equipped with segmented injection heads designed for pressurized liquid or gas injection. Above the surface of the waste body, various other systems are installed, including vacuum pumps, injection pumps, blowers, and extraction and injection power systems, along with liquid and gas storage tanks, pipeline systems, and exhaust gas treatment systems. Depending on operational requirements, operations such as liquid injection, air injection, leachate pumping, and gas extraction can be performed independently or simultaneously. The use of segmented injection heads for local liquid and gas injection significantly minimizes pressure attenuation within and around the well casing, thereby extending the range and efficacy of injection efforts.

Furthermore, the in-situ dewatering equipment initially reduces the water content of waste from a saturated state to a field capacity state by

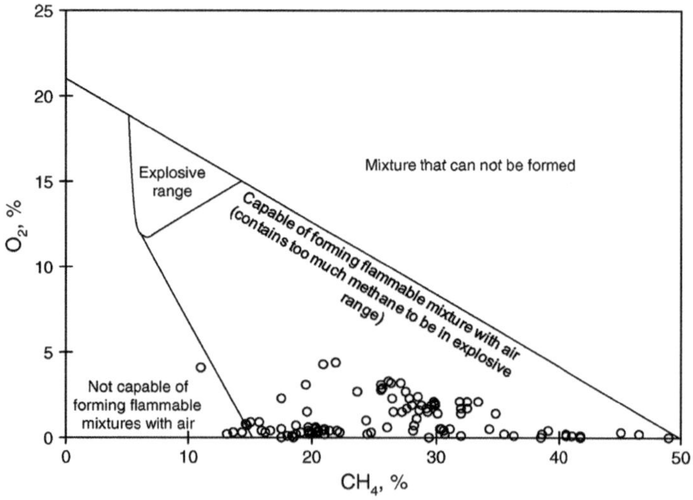

Fig. 8.9 Distribution of LFG composition on an explosivity chart (reprinted from Ko et al. 2013, with permission from the American Society of Civil Engineers)

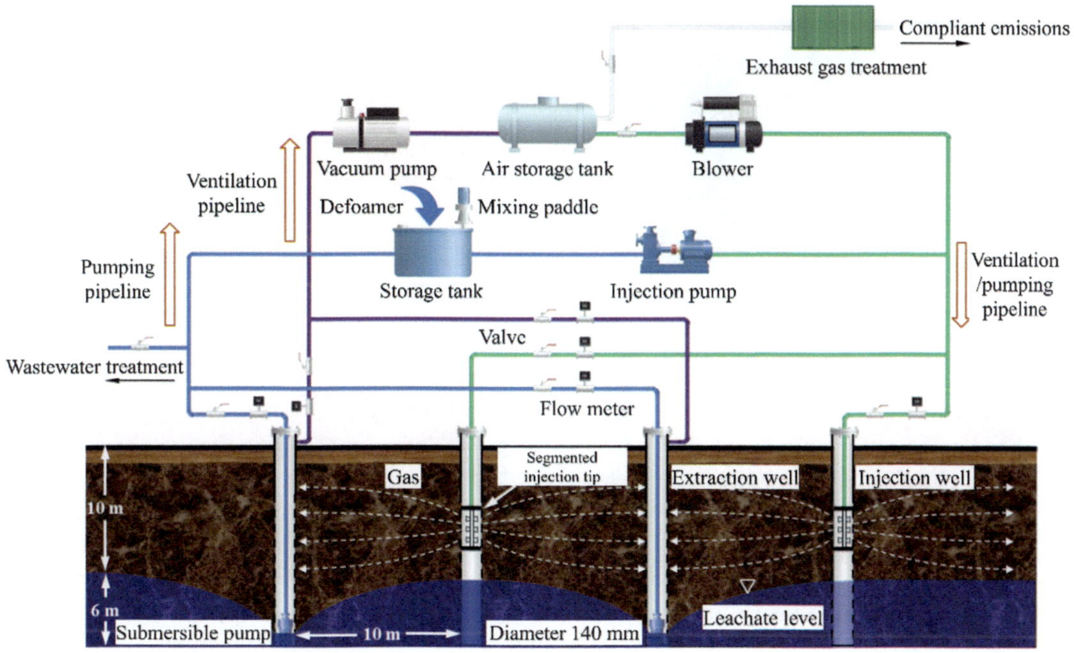

Fig. 8.10 Efficient in-situ dewatering equipment for stockpiled waste

Fig. 8.11 Dual-porosity water retention curve of MSW

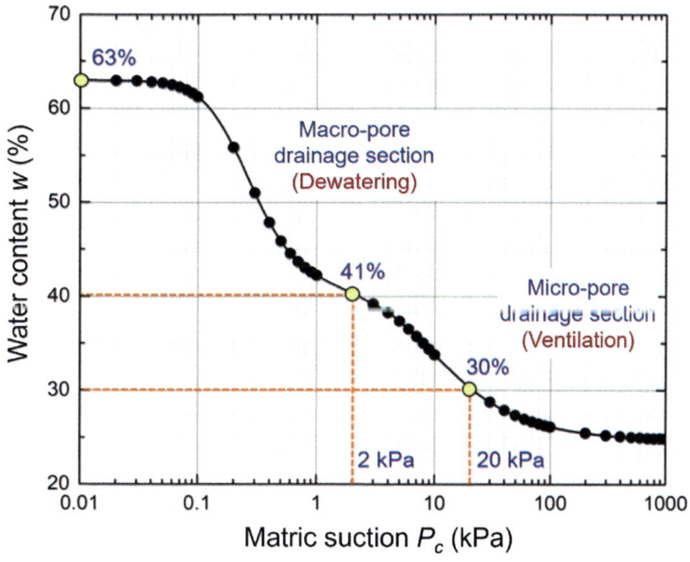

pumping, following which it further displaces residual pore water through air injection. The change in the water content of waste is closely linked to the water retention curve of waste (see Sect. 4.5.1). Figure 8.11 depicts the measured dual-porosity water retention curve of MSW.

During the pumping phase, the equipment effectively drains water from large pores, reducing the initial saturated water content of the waste from 63% to a field capacity of 41%. The subsequent air injection phase further displaces moisture from small pores. When the air pressure

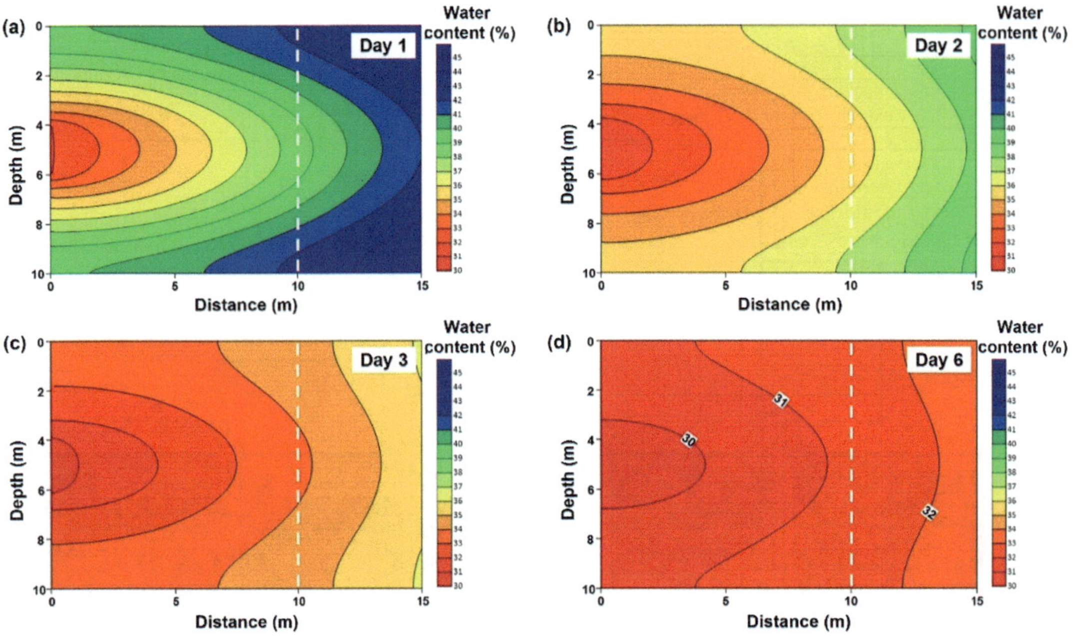

Fig. 8.12 Contour maps of the distribution of MSW water content at different times during the air extraction and injection processes: **a** after 1 day of air extraction and injection, **b** after 2 days of air extraction and injection, **c** after 3 days of air extraction and injection, and **d** after 4 days of air extraction and injection

within waste pores reaches 20 kPa, the water content of the waste is reduced to 30%, achieving rapid in-situ dewatering.

Figure 8.12 presents a distribution map depicting the variation in the water content of waste at different stages of the air extraction and injection processes. The left side of the model represents an air injection well with an injection pressure of 30 kPa, while the right side represents an air extraction well with an extraction pressure of −20 kPa. After 6 days of air injection, the water content within a 10 m range from the injection well is reduced to around 30%.

8.5 Mining and Screening of Landfilled MSW

Figure 8.13 illustrates the screening process of MSW. Specifically, MSW with low water contents is first segregated into materials exceeding 160 mm using a preseparation unit comprising a feeding machine and bar screen. Meanwhile, materials smaller than 160 mm proceed to the first separation unit, where a trommel screen segregates them based on particle size. Here, materials exceeding 60 mm are further sorted by the first air classifier, which separates light materials such as plastics and fabrics, suitable for incineration or processing into plastic granules, from heavier materials, which are used as sand, gravel aggregate, and road subgrade filling materials. Materials with dimensions smaller than 60 mm are directed toward a 20 mm drum screen for further screening, and materials above the screen are segregated by the second air classifier into light and heavy materials. Here, materials smaller than 20 mm primarily comprise fine soil, which can be further sorted into particles with dimensions smaller than 5 mm and between 5–20 mm using a flip-flow or vibrating screen. Coarser particles with dimensions between 5–20 mm are treated using matrix utilization technology, while finer particles smaller than 5 mm are subjected to heavy-metal stabilization treatment. Both components can be

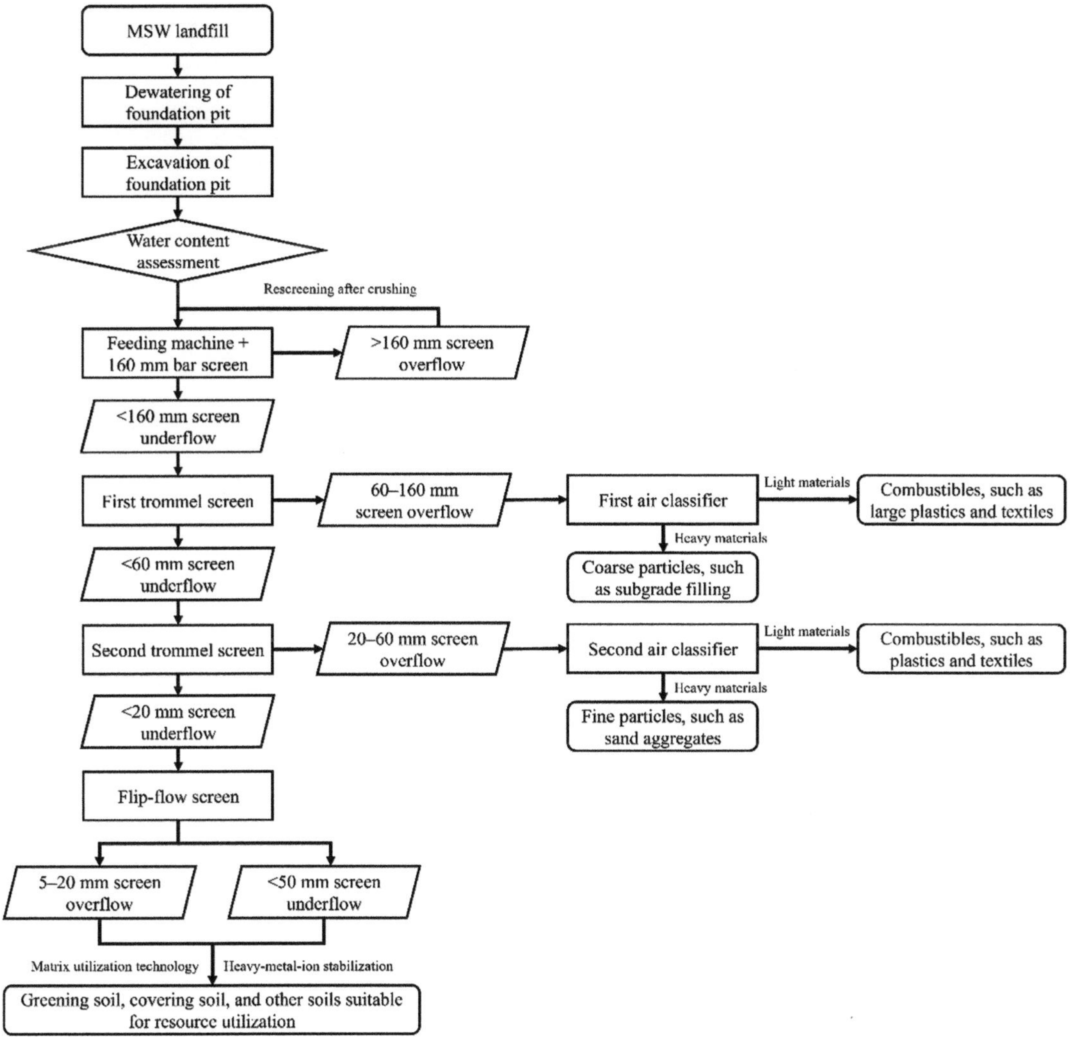

Fig. 8.13 Waste screening process

utilized as soil for greening or as cover materials for landfill sites, contributing toward resource utilization.

Figure 8.14 shows the screening equipment corresponding to the screening process in Fig. 8.13. This equipment consists of key equipment such as a three-dimensional screen, two-stage trommel screens, two-stage air classifier, and flip-flow screen. By using vibrating rods and cutting-blowing components, it solves the problem of fine-grained soil clumping and fabric blocking the screen holes, achieving a 30% reduction in fine-grained humus soil content and a reduction of plastic content to less than 1.5%.

The equipment in Fig. 8.14 was used to screen 1.44 million m^3 of MSW from the Woqishan Landfill in Wenzhou, China (as shown in Fig. 8.15). The mass proportions of coarse aggregate, lightweight material, and fine humus soil were approximately 28%, 32%, and 40%, respectively. Fine grained humus soil is the most difficult to treat due to its high degree of pollution, and its physicochemical properties and potential applications need to be evaluated.

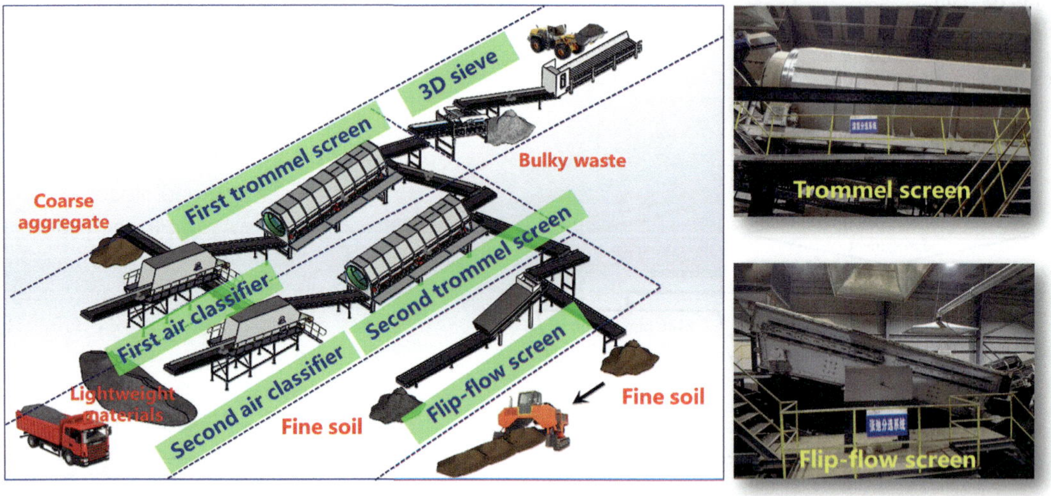

Fig. 8.14 Waste screening equipment

Fig. 8.15 Mining and screening project of the Woqishan Landfill

Figure 8.16 illustrates the particle size distribution curves of the fine fractions of different-aged MSW samples sourced from the Woqishan Landfill (Zheng et al. 2022). Evidently, the contents of clay and silt particles (<0.075 mm) range from 5.1% to 13.0%, of sand particles (0.075–2 mm) range from 40.9% to 44.1%, and of gravel particles (2–15 mm) range from 42.9% to 53.9%, indicating fine sandy soil.

A comparison between the fine fractions of different-aged MSW samples reveals a gradual decrease in overall particle size with increasing landfill age. For instance, the content of fine fraction particles larger than 2 mm decreases

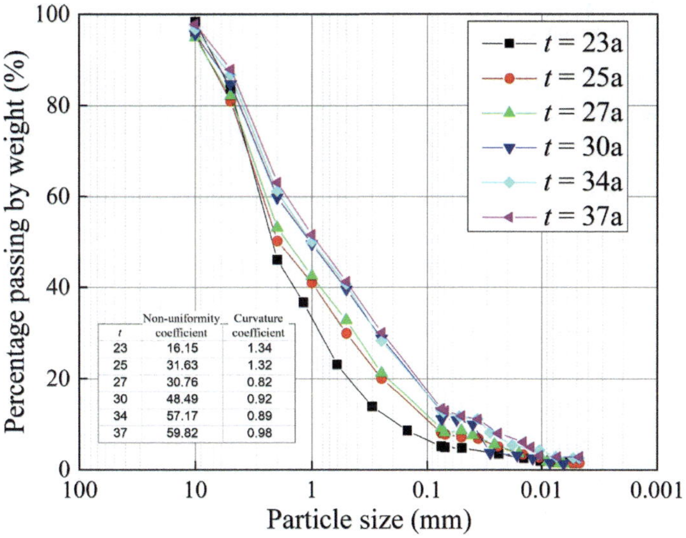

Fig. 8.16 Particle size distribution curves of MSW fine fractions with different ages (reprinted from Zheng et al. 2022, with permission from China Environmental Science)

from 53.94% to 36.95%, while that of particles smaller than 0.075 mm increases from 5.12% to 12.36%. This change is attributed to biochemical reactions, such as hydrolysis, fermentation, and residue degradation, occurring in landfilled waste, leading to mass loss and altering the particle size distribution. Generally, as the landfill age increases, a substantial amount of organic matter within MSW continuously degrades, resulting in the formation of fine fractions composed of various substances such as mixed slag and glass.

The water content of fine fractions is influenced by various factors including composition, pore structure characteristics, groundwater level, water immersion degree, air temperature, and humidity. Test results reveal that the water content of fine fractions ranges from 37.04% to 47.49%, with an average value of 42.02%. Relatively fresh fine fractions, typically located at the top of landfills, are influenced by rainfall and air humidity, thus containing higher water contents. Conversely, older fine fractions, located within deeper layers, are affected by groundwater, also exhibiting higher water contents.

The specific gravity of fine fractions ranges from 1.845 to 1.883, with an average of 1.867. Notably, the specific gravity gradually increases with increasing landfill age, demonstrating a noticeably faster increase for fine fractions aged more than 30 years compared to those aged less than 30 years. This trend aligns with the increase in fine particle content with increasing landfill age.

The results of the chemical composition testing of fine fractions are summarized in Table 8.4. Evidently, fine fractions predominantly comprise S_iO_2, along with certain amounts of CaO, Al_2O_3, Na_2O, and Fe_2O_3 and trace quantities of MgO and K_2O. This composition closely resembles that of soil minerals, reflecting that fine fractions are primarily composed of sand, gravel, and silicate clay minerals.

Table 8.5 details the contents of N, P, K, and organic materials in fine fractions of different ages. Notably, the total nitrogen (TN) content ranges from 0.50% to 0.82%, total phosphorus (TP) content ranges from 0.21% to 0.25%, total potassium (TK) content ranges from 1.14% to 1.32%, and the total nutrient (NPK) content ranges from 1.98% to 2.31%. Furthermore, the organic material content ranges from 18.17% to 19.10%. These results indicate that the N, P, and K contents within fine fractions are independent of the landfill age.

However, with increasing landfill age, the organic material content of the fine fractions gradually decreases. This suggests that organic materials such as fabrics and wood continue to degrade into fine fractions aged over 20 years.

Table 8.4 Components of MSW fine fractions (reprinted from Zheng et al. 2022, with permission from China Environmental Science)

Chemical composition	SiO_2	CaO	Al_2O_3	Fe_2O_3	Na_2O	MgO	K_2O	Others
Content (%)	47.73	14.52	8.07	1.78	3.73	1.51	0.87	21.79

Table 8.5 Concentrations of N, P, K, and organic materials within the fine fractions

Indicator	Organic matter (%)	TN (g/kg)	TP (g/kg)	TK (g/kg)	NPK (g/kg)	Hydrolysable nitrogen (mg/kg)	Effective phosphorus (mg/kg)	Available potassium (mg/kg)
S1 (23 years)	19.10	5.64	2.52	11.6	1.98	1,120	11.8	581
S2 (25 years)	18.91	7.92	2.47	12.2	2.14	1,540	18.1	725
S3 (27 years)	18.78	8.18	3.50	11.4	2.31	2,960	29.3	1,060
S4 (30 years)	18.52	6.11	3.28	11.9	2.28	2,180	17.5	833
S5 (34 years)	18.29	4.98	2.11	13.2	2.03	1910	32.0	769
S6 (37 years)	18.17	5.23	2.85	12.4	2.09	2,520	25.7	804
Organic medium for greening (GB/T 33891-2017)	>25	/	/	/	≥ 1.5	$\leq 3{,}000$	$\leq 1{,}200$	$\leq 4{,}000$
Planting soil for greening (CJ/T 340–2016)	1.2–8.0	/	/	/	/	40–200	5–60	60–300

Nevertheless, the degradation rate stabilizes after 30 years, exhibiting a certain consistency with the trends of increasing fine particle content and changes in specific gravity.

Furthermore, the total nutrient content of the fine fractions meets the requirements of the "Organic Medium for Greening" (GB/T 33891-2017), with the contents of hydrolysable nitrogen and available potassium far exceeding the requirements for the "Planting Soil for Greening" (CJ/T 340–2016). The organic material content exceeds the requirement of the "Planting Soil for Greening" (CJ/T 340–2016) but fails to meet the requirements for the "Organic Medium for Greening" (GB/T 33891-2017). Hence, the fine fractions extracted from landfills can be processed and sorted to create materials satisfying the quality standards for use as planting soil or organic substrates for greening.

Table 8.6 details the pH and electrical conductivity values of leachate extracted from fine fractions of different ages. Notably, the pH values of leachate extracted from fine fractions range from 7.26 to 8.30, indicating weak alkalinity. Meanwhile, their electrical conductivity values range from 1.08 mS/cm to 2.51 mS/cm, exhibiting values significantly lower than the limiting electrical conductivity (4.00 mS/cm) that can inhibit crop growth. Overall, the pH values of the fine fractions satisfy the requirements of both the "Planting Soil for Greening" (CJ/T 340–2016) and "Organic Medium for Greening" (GB/T 33891-2017), while their electrical conductivity values surpass the requirements of the

Table 8.6 pH and electrical conductivity data of leachate extracted from fine fractions

Indicator	S1 (23 years)	S2 (25 years)	S3 (27 years)	S4 (30 years)	S5 (34 years)	S6 (37 years)	Planting soil for greening (CJ/T 340–2016)	Organic medium for greening (GB/T 33891-2017)
pH	7.33	7.26	7.39	8.30	7.70	7.85	5.0–8.3	4.0–9.5
Electrical conductivity (mS/cm)	1.08	2.39	2.51	2.23	2.06	2.19	0.15–0.9	0.3–3.5

"Planting Soil for Greening" (CJ/T 340–2016) but complies with the requirements of the "Organic Medium for Greening" (GB/T 33891-2017). Hence, fine fractions are deemed suitable for use as growth substrates for plants.

Table 8.7 summarizes the concentrations of heavy-metal ions of leachate extracted from fine fractions. Notably, the leaching concentrations of Cr, Pb, and Cu exceed the Class V limits stipulated in the "Environmental Quality Standards for Surface Water" (GB 3838-2002) by 1–5 times. Meanwhile, the leaching concentration of Cd far exceeds the Class V limit by 30 times, with the leaching concentration of Zn slightly exceeding the standard.

Evidently, the direct utilization of fine fractions will contribute to environmental pollution, thus posing health risks to humans. Considering particle size distributions, fine particles smaller than 2 mm constitute 46.06–61.15% of the total fine fraction content. Furthermore, the leaching of heavy-metal ions including Cd, Cr, Pb, Cu, Zn, and Ni collectively represents approximately 94.24%, 87.45%, 99.30%, 78.46%, 78.80%, and 98.48% of the total leaching, respectively. This highlights that a significant proportion of heavy-metal ions exists within fine particles smaller than 2 mm. This is because the increased surface areas of smaller soil particles enhance the adsorption capacity for heavy-metal ions, leading to the predominant accumulation of metals within fine particles. Hence, when classifying fine fractions into finer categories, particular attention must be focused on the leaching concentrations of heavy-metal ions from fine particles smaller than 2 mm.

In summary, the fine fractions examined by Zheng et al. (2022) resemble fine sandy soil, with an organic matter content similar to that of fertile soil, meeting the requirements of the "Control Standard for Urban Wastes for Agricultural Use" (GB 8172-87). Furthermore, the TN and TP concentrations of these fine fractions exceed those of conventional soil, aligning with national standards for organic fertilizer nutrient contents. Leachate extracted from these fractions exhibits weak alkalinity, and the results of electrical conductivity tests indicate the suitability of the fine fractions for plant growth. However, the heavy-metal ion contents of the fine fractions exceed the limits set by the "Environmental Quality Standards for Surface Water" (GB 3838-2002) in China. Hence, after being subjected to appropriate treatments such as enzyme induced carbonate precipitation method (Chen et al. 2024; Bian et al. 2024), these fine fractions can be effectively utilized for greening and planting purposes, serving as covering layers for vegetation restoration on urban hilly slopes, in abandoned mines, and in landfills.

Table 8.7 Leaching potential and pollution index values of heavy-metal ions within the fine fractions (reprinted from Zheng et al. 2022, with permission from China Environmental Science)

Concentrations of heavy-metal ions in leachate (mg/L)	Cd	Cr	Pb	Cu	Zn	Ni
<15 mm	0.310 ± 0.192	0.308 ± 0.174	0.330 ± 0.047	2.140 ± 1.127	2.078 ± 1.616	0.310 ± 0.087
<2 mm	0.324 ± 0.127	0.342 ± 0.153	0.357 ± 0.084	2.648 ± 0.931	2.521 ± 1.284	0.431 ± 0.084
2–15 mm	0.0196 ± 0.051	0.049 ± 0.022	0.003 ± 0.002	0.720 ± 0.263	0.671 ± 0.332	0.007 ± 0.002
Environmental quality standards for surface water (GB 3838-2002) Class V	0.0100	0.1000	0.1000	1.0000	2.0000	/

References

Bian Y, Chen YB, Zhan LT, Guo HW, Ke H, Wang YZ, Wang QY, Gao YF, Gao YQ (2024) Effects of enzyme-induced carbonate precipitation technique on multiple heavy metals immobilization and unconfined compressive strength improvement of contaminated sand. Sci Total Environ 947:174409

Chen YB, Liu B, Bian Y, Gao YF, Liu Y, Cheng MY, Wang LY, Hang L (2024) Effects of soybean urease induced carbonate precipitation on the seed emergence and seedling growth of Caragana korshinskii Kom and its application in wind erosion control. Plant Soil 1–25

Chen YM, Liu XC, Xu WJ, Li YC, Lan JW, Zhan LT, Li H, Li CM (2019a) Analysis on stabilization characteristics and exploitability of landfilled municipal solid waste: case of a typical landfill in China. Scientia Sinica Technologica 49(2):199–211 (in Chinese)

Chen YM, Zhan LT, Gao W (2019b) Waste mechanics and sustainable landfilling technology: comparison between HFWC and LFWC MSWs. In: Proceedings of the 8th international congress on environmental geotechnics, Volume 1: Towards a sustainable geoenvironment. Springer, Hangzhou, pp 3–37

Guo RY (2017) Research on the biochemical and mechanical behaviors of high food content MSW. Ph.D. thesis, Zhejiang University, Hangzhou, China (in Chinese)

Hrad M, Huber-Humer M (2017) Performance and completion assessment of an in-situ aerated municipal solid waste landfill—final scientific documentation of an Austrian case study. Waste Manag 63:397–409

Hu J, Zhang CS, Ke H, Chen YM, Zhan LT, Lan JW (2022) A device and method for in-situ pretreatment of stockpile municipal solid waste. Patent number: CN202210571641.7, China (in Chinese)

Inter-governmental Panel on Climate Change (IPCC) (2006) IPCC guidelines for national greenhouse gas inventories. Institute for Global Environmental Strategies

Kavazanjian E Jr (2010) Sustainable landfilling. In: The 6th international congress on environmental geotechnics, New Delhi, pp 113–124

Ko JH, Powell J, Jain P, Kim H, Townsend T, Reinhart D (2013) Case study of controlled air addition into landfilled municipal solid waste: design, operation, and control. J Hazard Toxic Radioact Waste 17(4):351–359

Li L, Peng Y, Tan HY, Yang PJ, Ru LY, Wang XM, Peng XY (2021) Application status and research process of in-situ landfill aeration. China Environ Sci 41(6):2725–2736 (in Chinese)

Liu L, Ma J, Xue Q, Shao J, Chen Y, Zeng G (2018) The in situ aeration in an old landfill in China: multi-wells optimization method and application. Waste Manag 76:614–620

Ministry of Ecology and Environment of the People's Republic of China (1987) Control standard for urban wastes for agricultural use (GB 8172-87) (in Chinese)

Ministry of Ecology and Environment of the People's Republic of China (2002) Environmental quality standards for surface water (GB 3838-2002) (in Chinese)

Ministry of Ecology and Environment of the People's Republic of China (2008) Standard for pollution control on the landfill sites of municipal solid waste (GB 16889-2008). China Environment Publishing Group, Beijing (in Chinese)

Ministry of Ecology and Environment of the People's Republic of China (2018) Soil environmental quality—risk control standard for soil contamination of agricultural land (GB 15618-2018). China Environment Publishing Group, Beijing (in Chinese)

Ministry of Housing and Urban-Rural Development of the Republic of China (2010) Technical requirements for site utilization after stabilization in municipal solid waste landfill (GB/T 25179-2010) (in Chinese)

Ministry of Housing and Urban-Rural Development of the Republic of China (2016) Planting soil for greening (CJ/T 340-2016) (in Chinese)

National Forestry and Grassland Administration (2017) Organic medium for greening (GB/T 33891-2017) (in Chinese)

Powell J, Jain P, Kim H, Townsend T, Reinhart D (2006) Changes in landfill gas quality as a result of controlled air injection. Environ Sci Technol 40(3):1029–1034

Raga R, Cossu R (2013) Bioreactor tests preliminary to landfill in situ aeration: a case study. Waste Manag 33 (4):871–880

Read AD, Hudgins M, Harper S, Phillips P, Morris J (2001) The successful demonstration of aerobic landfilling: the potential for a more sustainable solid waste management approach? Resour Conserv Recycl 32 (2):115–146

Stegmann R (2019) Development of waste management in the last 30 years. In: Proceedings of the 8th international congress on environmental geotechnics, Volume 1: Towards a sustainable geoenvironment, pp 172–185

Tang R (2012) Research of technology and application of aerobic degradation on the accelerating stabilization process of closed non-regular landfill. Ph.D. thesis, China University of Geoscience (Beijing), Beijing, China (in Chinese)

Townsend TG, Powell J, Jain P, Xu QY, Tolaymat T, Reinhart D (2015) Sustainable practices for landfill design and operation. Springer, New York

Wei YF, Xu WJ, Chen YM, Peng Y, Ke H, Zhan LT, Lan JW, Li H, Zhang YX (2024) Evaluation of greenhouse gas emission and reduction potential of high-food-waste-content municipal solid waste landfills: a case study of a landfill in the east of China. Waste Manag 189:290–299

Zheng KQ, Chen P, Qiu YF, Guo QM, Chen YB, Yuan MX, Zhan LT (2022) Physicochemical properties and reuse of municipal solid waste fine fraction: case of an aged landfill site in Zhejiang Province. China Environ Sci 42(7):3254–3264 (in Chinese)

GPSR Compliance

The European Union's (EU) General Product Safety Regulation (GPSR) is a set of rules that requires consumer products to be safe and our obligations to ensure this.

If you have any concerns about our products, you can contact us on ProductSafety@springernature.com

In case Publisher is established outside the EU, the EU authorized representative is:

Springer Nature Customer Service Center GmbH
Europaplatz 3
69115 Heidelberg, Germany

Batch number: 10705782

Printed by Printforce, the Netherlands